The GMO Deception

Also edited by Sheldon Krimsky and Jeremy Gruber:

Biotechnology in Our Lives
Genetic Explanations: Sense and Nonsense

The GMO Deception

What You Need to Know about the Food, Corporations, and Government Agencies Putting Our Families and Our Environment at Risk

Sheldon Krimsky and Jeremy Gruber, editors

Foreword by Ralph Nader

Skyhorse Publishing

This book is dedicated to members of the Council for Responsible Genetics Board of Directors and Advisory Board who are no longer with us, but whose work and vision continue to inspire new generations of social and environmental justice activists: David Brower, Barry Commoner, Marc Lappe, Anthony Mazzocchi, Albert Meyerhoff, Bernard Rapoport, George Wald, and William Winpisinger.

Skyhorse Publishing books may be purchased in bulk at special discounts for sales promotion, corporate gifts, fund-raising, or educational purposes. Special editions can also be created to specifications. For details, contact the Special Sales Department, Skyhorse Publishing, 307 West 36th Street, 11th Floor, New York, NY 10018 or info@skyhorsepublishing.com.

Visit our website at www.skyhorsepublishing.com.

10 9 8 7 6 5 4 3 2 1

Library of Congress Cataloging-in-Publication Data is available on file.

Cover design: Victoria Bellavia and Rain Saukas
Front cover illustration: Rain Saukas

ISBN: 978-1-51070-266-0
Ebook ISBN: 978-1-62914-020-9

Printed in the United States of America

CONTENTS

ACKNOWLEDGMENTS

We are grateful for the public interest commitment and superb work of *GeneWatch* editors and editorial boards past and present: Sam Anderson, Philip Bereano, Kostia Bergman, Phil Brown, Sujatha Bvravan, Nancy Connell, Donna Cremans, Christopher Edwards, Leslie Fraser, Phyllis Freeman, Judith Glaubman, Terri Goldberg, Barbara Goldoftas, Colin Gracey, Susan Gracey, Daniel Grossman, Jeremy Gruber, Ruth Hubbard, Kit Johnson, Brandon Keim, Sophia Kolehmainen, Sheldon Krimsky, Herbert Lass, Evan Lerner, Gary Marchant, Wendy McGoodwin, Shelley Minden, Eva Ng, Rayna Rapp, Barbara Rosenberg, Peter Shorett, Seth Shulman, Christine Skwiot, Martin Teitel, Suzanne C. Theberge, Jonathan B. Tucker, Shawna Vogel, Nachama Wilker, Kimberly Wilson, and SusanWright.

Special thanks to Hector Carosso, for his consultation on the title.

We would like to thank Holly Rubino, senior editor for Skyhorse Publishing, who provided editorial guidance that helped the editors clarify the book's goals and audience.

We also wish to acknowledge the substantial support of both the Cornerstone Campaign and the Safety Systems Foundation.

FOREWORD

"Science is science," declared my college biology professor, alluding to its own rigorous standards, openness, and integrity. Today, my response would be "not quite." For in the autocratic, commercially driven hands of multinational corporations, "science" becomes the instrument of an overall business plan that results in serious corruptions of scientific attitude, method, and peer-reviewed accountability. This confidential, proprietary "corporate science" closes off Alfred North Whitehead's definition of science as "keeping options open for revision." It becomes, in this book's context, the central chattel in a comprehensive business strategy to corporatize global agriculture. This is accomplished through a remarkable matrix of controls and public subsidies that takes monopolizing corporate behavior and its wildcat offshoots to historically unforeseen depths of danger to people and planet.

To better absorb the significance of this wide-ranging anthology of articles by independent scientists and science writers,* a brief summary of the biotechnology industry's emergence over time is in order.

Two events from the public realm were crucial initiators. In 1972, came the publication of scientific papers by Cohen & Boyer and associates of transgenic DNA splicing across phylogenetic species, including bacteria, viruses, and insects.[1] The second event in 1980 was a US Supreme Court 5–4 decision allowing patents on life forms. Those two "assets" facilitated the start of a broadly conceived strategic planning process for genetically modified (GM) food by the fledging biotechnology industry led by Monsanto.

Corporations engage in strategic planning as the essence of their quest for more revenue and profits. More globalized than ever, they are continually expanding such planning to include, in their favor, our elections, government, environment, education, media, research and development, energy, tax and credit systems, trade agreements, transport, land use, food, and our genetic

* brought together by Professor Sheldon Krimsky and Jeremy Gruber, president and executive director of the Council for Responsible Genetics (CRG) founded more than thirty years ago by university scientists, public health professionals, union organizers, and environmental activists.

inheritance. They equate their planning with exercising the control and predictability necessary for their definition of stability. They do not always get their way, but no other institutions have been their match for more than half a century. As artificial entities, from the late nineteenth century onwards, courts have given them the status of "persons" for the purposes of constitutional rights. The world has never seen such an ingenious, power-concentrating machine as the modern, global corporation. Ideally, many of them would like to view themselves as both amoral and "anational" (meaning not domiciled in any particular country).

Their corporate lawyers have constructed a proliferating system of privileges and immunities which must be viewed as a remarkable intellectual achievement, apart from the purposes served. Because these corporate giants plan the subordination of civic values to the supremacy of commercialism, any resistance is to be countered, preempted, undermined, or destroyed.

Let us envision, for example, what Monsanto decided to do, once armed with the twin tools of transgenic capability and patentable life forms, to move toward domination of agricultural sub-economies in the US and around the world.

Monsanto had to create a narrative as to why its GMO patented seeds were needed and superior to traditional seeds in the first place. In a massive, relentless marketing campaign, genetically modified foods were touted as safe, cheaper, higher yielding, more nutritious, requiring lower chemical inputs, and resistant to drought and blight. Because these objectives were shared by federal agencies that funded the basic research, the Monsanto campaign, abetted by an uncritical mass media, made their hope spring eternal.

The prospect of alleviating world hunger through these measures of productivity was irresistible to the media, which tend to report scientific discoveries by establishment promoters who tout potential benefits without mentioning potential drawbacks. Outlets like the *New York Times* and *Science Magazine* have been prone to falling for this propaganda year after year. One has only to recall reports on how the starving masses would allegedly be saved by GMO rice ("Golden Rice"), or GMO cassava, or a GM virus-resistant sweet potato or edible vaccines, followed for years by no such realities in the fields, to see the successes of the biotech industry's deception.

Jonathan Latham, crop geneticist and founder of the Bioscience Resource Project, calls "Golden Rice" the "Emperor of GMOs." He cites food writer Michael Pollan, who called Golden Rice a "purely rhetorical technology."[2] Another way of describing unproven claims and benefits of genetically engineered foods is that the "engineering" on the ground has rushed far ahead, both in public relations and in misapplication, of the "science" that must be its ultimate discipline.

In his article "Imaginary Organisms: Media Tout Benefits of GMOs That Never Were" published by *Extra!* in March 2014, Mr. Latham concludes, "These misreports of biotechnology are endlessly useful to the industry. . . . The great value of 'fakethroughs' is to confirm, in the eyes of the world, the industry's claim to be ethical, innovative and essential to a sustainable future."

This manufactured credibility is connected with a Washington lobbying force that sways Congress, greased with campaign cash, which destroys through Congressional budget-making what should be balanced research priorities of the National Institutes of Health. The results are non-regulation, unenforceable guidelines, and a supine Department of Agriculture. In this corporate government, the Food and Drug Administration cedes Monsanto its demand *not* to label genetically-engineered food sold in the markets that the company is presumably proud to sell.

Thus fortified by its political engineering, Monsanto mutes or compromises academic science through joint ventures with university departments, lucrative consultancies, and a piece of the stock action. Independent academic scientists, who wish to replicate or test the industry's claims, find a paucity of available grants, obstructed access to the products, and a litigiously backed refusal to disclose the basis of Monsanto's claims that are cloaked in the alleged cover of trade secrecy.

In July 2009, *Scientific American* described the intolerable situation concisely:

> Unfortunately, it is impossible to verify that genetically modified crops perform as advertised. That is because agritech companies have given themselves veto power over the work of independent researchers. . . . Research on genetically modified seeds is still published, of course. But only studies that the seed companies have approved ever see the light of a peer-reviewed journal. In a number of cases, experiments that had the implicit go-ahead from the seed company were later blocked from publication because the results were not flattering. . . . It would be chilling enough if any other type of company were able to prevent independent researchers from

> testing its wares and reporting what they find. . . . But when scientists are prevented from examining the raw ingredients in our nation's food supply or from testing the plant material that covers a large portion of the country's agricultural land, the restrictions on free inquiry become dangerous.

Research on the migration of GM pollen from farms to non-GMO-farms; the level of developing bacterial, viral, and insect resistance to GMO-linked herbicides; and longer-run studies of the consequences of GMO seeds and crops on the environment, or even the multiple effects at the cellular level from the newly inserted gene, are hampered and grossly underfunded, whether by government or foundations. But, as the articles in this book demonstrate, enough is known to require the Monsantos to bear their burdens of proof behind their many claims—*before* marketing their products.

In the concluding statement to this anthology, Sheldon Krimsky cites disturbing findings that counter the health and safety assertions of the industry and writes, "In the absence of evidence that genetically modified foods are cheaper, produce greater yields, or even work particularly well, lies one widely recognized conclusion: GMO foods provide no added nutritional or cost benefit to the consumer."

Why then have farmers accepted much higher priced GM seeds, rooted in radically one-sided censorious contracts with Monsanto called "Technology/ Stewardship Agreements"? These "agreements" shift responsibility to the growers, along with a government that does not advise farmers with independent extension research, nor moves under the antitrust laws to break up the ever tighter, more expensive seed oligopoly. The lure started with convenience and an innocent belief in the vendors' other claims. More than a decade ago, an Iowa corn farmer told me he liked Bt corn primarily because it allowed him to spend more time with his wife—meaning less time needed for weeding. Now that weed resistance to Round Up Ready is emerging, these bad superweeds need more Round Up Ready or other herbicides.

The impact of GMO seed invasions in developing countries, such as India, is described here in brutal detail. There, an emerging large industrial monoculture disrupting traditional seed saving, sharing, and selling by seed monopolies and royalties, and leading to spiraling debt and displacement of small farmers, is assisted by the promotional support of US government agencies. Co-optation of local regulatory officials, along with campaign-like *ad hominem* attacks on the few independent scientists and agronomists who raise warnings, are further signs of corporate power abuses.

After a few years, the traditional seeds are not available or are rendered polluted with cross-pollination contamination by nearby GMO fields. Protesting Indian farmers have marched in immense numbers against what they see as the GMO chokehold as it spreads to other crops such as the proposed Bt eggplant.

In his essay "Breathing Sanity into the GM Food Debate" published by *Issues in Science and Technology* in 2004, Jerry Cayford provided dispassionate treatment of the critics of biotech food as being about "control of the food supply" and the "concentration of industry power." He wrote, "For better or worse, then, the biotech debate is a political debate, not just a scientific one." There are social, political, and economic issues, he noted, recognized by leading plant scientists but "few go far into them." Cayford continued by saying, "Beyond the dangers of farm concentration and cultural dislocation, of unreliable seed supplies, and the threat of famine, and of increasing poverty and dependence in our poor countries [on imported licensed seeds], critics also worry that patented and centralized seed production endangers biodiversity." He related that according to critics, massively subsidized and government-protected industrial agriculture threatens non-GMO agriculture and expanding organic farms, and then added, "It is because industrialized countries have elected to consider GM plants patentable that biotechnology threatens to take control of the food supply out of the public domain and hand it to multinational corporations," with "the power of the state behind" them.

That was almost a decade ago, and subsequent events have only proved these critics to be prophetic. Millions of acres have been planted with GM corn, soy, canola, and sugar beets with unproven claims of consumer and environmental benefits.

Although there are clearly differences, the history of other corporate technologies reveal common patterns of power, control, immunity, and deception. These ways of harmful domination—in the pursuit of growing sales, profits and bonuses—are instructive for purposes of long-range risk assessments and concerns over corporate secrecy.

For decades, mass-produced cigarettes were marketed as having flavors that tasted good and kept smokers alert. This powerful industry, of course, did not fund research into possible adverse health effects. Instead it paid scientists to dubiously dismiss any claims of causing cancer and made sure that Washington did nothing other than continue the tobacco subsidy and look the other way. According to the US Surgeon Generals, more than 400,000 Americans died every year from tobacco-related diseases.

For decades, the gasoline industry put tetraethyl lead into gasoline to allegedly reduce "knocking" and asserted its safety, without funding research on the traceable effects of lead poisoning, especially on children. Asbestos was widely used as an obstacle to spreading fires before research discovered its connection to widespread, deadly cancers. There was nearly a six decade gap between the first mass production of motor vehicles, with their internal combustion engines, until Cal Tech's Arie Jan Haagen-Smit discovered their contribution to photochemical smog.

The nature of corporate marketing is to use secret "corporate science" to promote the benefits, while the corporations politically block peer-reviewed academic or independent science from discovering the costs and risks. In all these cases, publically funded research conducted by independent, critical scientists and citizen action finally enabled truthful science to catch up with these technologies and lay bare the immense, deadly casualties and costs of their reckless deployment.

The same gaps now prevail with the GMOs. Similar power plays to promote marketing without researching or recognizing risks are evident and detailed by the contributions to this volume. Professor Krimsky summarizes the major points in his conclusion.

The silent violence of civilian, chemical, and biological technologies allows their promoters lengthy periods of license before the reckonings start registering. Unfortunately, the more time contaminating GMO crops remain in the genome, the growing areas, and the ecosphere, the more difficult it will be to contain them. Impunity is further exacerbated because the law does not compel the biotech industry's executives to have accountable "skin in the game" when matters are revealed to have gone terribly wrong—such as knowing about migratory pollution of non-GMO seeded lands. Tort law and the law of trespass may someday catch up with these corporate executives, whose one-sided contracts with farmers already immunize themselves and their companies.

Absent a rigorous ethical and legal framework for the tumultuous biotechnology industry, the scientific and legal professions that produce and empower the knowledge required must assert their codes of ethics to press for answers to the following questions:

- Who decides?
- Who benefits?
- Who controls the public domain?
- Who has the burden of proof?
- Who will be held accountable for what?
- Who has the right to know and when?
- How can precautionary principles be put into practice?
- What are the yardsticks and mandatory corporate disclosures by which various industry actions are to be judged?
- Why should taxpayers directly or indirectly subsidize GMO technology without a net cost/benefit advantage over continual progress in traditional agronomy, breeding, and ecological agriculture practices?
- What are the standards of failure for the GMO industry?
- As GMOs move up the animal food chain, what are the limits?

Presently, these questions, which befit a democratic society, are not being asked in public arenas (save for a few symposia) nor by public authorities. The centralization of control over the critical agricultural sub-economy, here and abroad, is relentless. Two or three dominant corporations are the norm in the beef, pork, and chicken processing industries. Vertical and horizontal integration of the food supply into ever fewer conglomerates are routinely taken for granted in agribusiness trade publications. Monsanto and the rest of the biotechnology industry use more than genetics to exercise their deepening control; they deploy economic, political, and social interventions backed by what can be considered abusive litigation, even against free speech by their challengers. Together, they represent a commercial autocracy that has declared war on democratic competition, regulation, free and open science, biological diversity, small farmers, and evolving standards of accountability.

Farmers who enter into contractual arrangements with companies such as Monsanto find that they are unable to extricate themselves from an even tighter web of external control. Always strategically planning the next move, Monsanto and DuPont, which dominate the seed market, are pressing farmers for data harvesting from tractors and combines in order to collect allegedly favorable information for greater yields, additional machinery, and chemical inputs. On February 26, 2014, the *Wall Street Journal* report on this "prescriptive farming" recounted skepticism from some farmers and the American Farm Bureau itself about the suppliers steering farmers to buy more of their products that benefit them as vendors, but not necessarily the farmers. The struggle over the loss of another measure of control—that over data and who controls and uses the data—is underway. In the long run, farmers who may wish to return to their former ways, having experienced the onset of superweeds and other internalizing costs not mentioned in the rosy introductory promotions, will discover the stringency of the biotech webs that have enveloped and trapped them into dependency.

The history of man's disruption of the natural world—what used to be exclaimed as "the conquest of nature"—is replete with touting its benefits and ignoring its costs. Nature, long-abused, turns on its abusers. Deforestation, land erosion, the poisoning of water and air, extinctions, and climate change are some of the manifestations of this. The biotechnological disruption of nature—driven by contrived goals, and without any, in Professor Krimsky's words, "scientific consensus on the safety and agricultural value of GM crops"—is laying the basis for major, uncontrollable blowbacks. Animate and inanimate nature is far too complex for the transgenic penetration of corporate hubris to manage.

This book sheds light on how, why, and when this industry has kept the public in the dark. At a meeting in December 8, 1999 Robert Shapiro, the former CEO of Monsanto, told a small gathering of consumer and environmental advocates that Monsanto was not opposed to labelling GM food products; it just disagreed on how to do it. I understood that what Mr. Shapiro really meant was that his company did not want the start of a public conversation that labelling might generate. This impression has been strengthened in subsequent years by the absence of any Monsanto proposal for labelling. After all, the biotechnology industry's growing control of what goes into the world's food supply and just how it is going about doing this may jolt people's sensibilities and strip away the falsified curtain of deception.

Together with resisting farmers, challenging scientists, and liberated civil servants, an aroused public will recognize that its own interests and those of posterity must be preeminent over these corporate monopolists and their short-range, narrow commercial pursuits. Widely read, this fertile book can be a major step toward public mobilization.

—Ralph Nader
March 21, 2014

INTRODUCTION: The Science and Regulation behind the GMO Deception

Agriculture had its origins about ten thousand years ago. Throughout most of that period farmers shared seeds, selected desired phenotypes of plants, and with keen observation and experience sought to understand the environmental factors affecting crop productivity. Through selective breeding, farmers chose plants that were best adapted to their region. By saving seeds of the more desired varieties they were able to achieve shortened growing seasons, larger fruits or vegetables, enhanced disease resistance and varieties with higher nutritional value. The birth of botany as a discipline can be traced to ancient Greece. Theophrastus of Eresos (371–287 BCE), a student of Aristotle, wrote two botanical treatises summarizing the results of a millennium of experience, observation, and science from Egypt and Mesopotamia.

By the Enlightenment of the seventeenth century, experimental science had emerged. The fields of agronomy and plant breeding developed out of that milieu. In 1865 Gregor Mendel published *Experiments on Plant Hybridization*. Plant hybridization (or cross breeding) involves crosses between populations, breeds, or cultivars within a single species, incorporating the qualities of two different varieties into a single variety. Thus, the pollen of plants with one desired trait was transferred to a plant variety with other desired traits. This could only be achieved with plant varieties that were closely related. Hybridization created plants more suitable for the palette of modern humans.

With the discovery in the first half of the twentieth century that radiation and chemicals could create mutations (or changes in the DNA code) in plant cells and germ plasm, plant breeders deliberately induced mutations that they hoped would produce more desirable plant varieties. This was a long, tedious, and unpredictable process. Nevertheless, it is estimated that more than 2,500 new plant varieties were produced using radiation mutagenesis.[1]

Tissue culture engineering was another technique used in plant breeding. Once a desirable plant variety was found, the plant could be cloned by extracting a small piece of the plant tissue and inducing it to grow in cell culture with the appropriate media. Plant cloning from somatic cells was a forerunner to cloning of animals like Dolly the Sheep.

After the discovery of recombinant DNA molecule technology (aka gene transplantation) in the early 1970s, the new field of plant biotechnology was launched less than a decade later. Scientists were now capable of cutting and splicing genes and transferring them from one biological entity to another, thereby crossing broad species barriers. Plant biotechnology made its debut at an international symposium in Miami, Florida, in January 1983.[2] Three independent groups of plant geneticists described experiments in which foreign genes were inserted into plants, leading to the creation of normal, fertile, transgenic plants, which means that they contained artificially inserted genes. The first plant used in these experiments was tobacco. And the vehicle for introducing the foreign genes into the plants was a bacterium called *Agrobacterium tumefacients (A. tumefaciens)*.

The recombinant DNA debates had precipitated one of the greatest public science education periods in modern history, occurring at a time when environmental issues were of paramount social importance in the United States. The American consumer had begun to think about the quality, safety, and purity of food, and the organic food movement was underway. National environmental groups petitioned the government to remove dangerous pesticides from farming. In the 1980s when agricultural biotechnology was born, there was already a skepticism building among consumers and food activists that genetically modified organisms (or GMOs) would not contribute to these priorities.

While there have been longstanding controversies between vegetarians and omnivores or organic versus conventional farming, rarely has there been a time when food has divided society into two major warring camps. But that is the situation that people now find themselves throughout the world in response to genetically modified food. One camp proclaims that GMOs represent the future of food. They echo the words of Francis Bacon, the seventeenth-century philosopher and scientist, who more than four hundred years ago in the *New Atlantis*, prophesied a future of biotechnology:

> And we make by art, in the same orchards and gardens, trees and flowers to come earlier or later in their seasons, and to come up and bear more speedily, than by their natural course they do. We

> make them also by art greater much than their nature; and their fruit greater and sweeter, and of differing tastes, smell, colour, and figure from their nature.[3]

Bacon saw the biotic world around him as providing the feedstock, or starting materials, for recreating plant life on the planet according to human design and utility. In more contemporary terms, the plant germ plasm holds the building blocks for new food crops just as the chemical elements of the Periodic Table were the starting material for synthetic chemistry that has brought us plastics, pesticides, and nanotechnology.

The Debate

In the view of the modern agricultural Baconians, farms are like factories. Food production must be as efficient as an assembly line. This means that the producers of food must reduce the uncertainty of inputs and speed up food production to cultivate more crops per given acre, per unit of time, per unit of labor, and per unit of resource input. They proclaim the need for higher food productivity to provide for a growing population of more than seven billion people on the planet. The American Council on Science and Health, a GMO-philic organization, believes that skeptics or non-believers are irrational luddites: "It's truly mind-boggling that this technology, which has already provided so many benefits and will continue to do so, is being demonized to such a great extent. It's a sad commentary on how susceptible a population deficient in scientific understanding can be to fear mongering activists with a scary agenda."[4]

The opposing camp is comprised largely of food purists, skeptics of industrial, high chemical input farming, critics of agribusiness, and scientists who are not convinced that genetically modified food is as safe and as ecologically sustainable as its proponents claim it to be. They point out that the altruistic promises of GMO proponents have had no relationship with the actual use of genetic engineering techniques in modern agricultural production. In fact, there have been only two commonly applied major innovations in GMO agriculture: 1) crops resistant to herbicide, and 2) crops that contain their own insecticide. Both methods were designed to find synergies with their corporate sponsor's existing pesticide, herbicide, and fertilizer businesses in order to maximize profits. For example, a farmer who buys Monsanto's Roundup

Ready soybeans would also need to buy Monsanto's Roundup Ready herbicide. The GMO skeptics further claim that neither method has any direct benefit to the consumer and that both are failing to achieve productivity expectations as both weeds and insects develop resistance to these toxins.

GMO skeptics harken back to the transformation of small-scale agriculture, where crop rotation, agro-ecological diversity, family farming, animal husbandry, taste, freshness, and purity were core values. Their perspective on GMOs can best be characterized by use of the acronym GAUF: Genetically Adulterated Unlabeled Food. Consumers, especially those not on the edge of poverty and famine, are asking more from their food than its price, its plentitude, its perfect geometry, its homogenous color, and its shelf life. They are demanding that their food be grown without the use of poisons, that animal protein not be harvested at the expense of the humane treatment of sentient beings, that agricultural practices not destroy the substrate of the natural ecology (the soil), and that modern agriculture not put an end to agrarian life by turning land-based food production into industrially based cell culture and hydroponics. Finally, they believe the entire agricultural system should be sustainable. In the end, they find that the GMO technology has so far only benefited a handful of corporations, that it is expensive, polluting, it may be unsafe for humans, and it is beginning to fail to meet its own instrumental objectives.

The two camps represent opposing world views about the role and structure of agriculture in modern civilization in the post industrial age. When you ask people from the pro-GMO camp what are their core values, they will likely say productivity, profit and safety. They have no intention of producing food that will make people sick as they declare, "We cannot stay in business if our product harms people."

The GMO-skeptics have a more nuanced view of adverse consequences. Their concerns include whether GMOs will induce subtle changes in long-term health and nutritional quality, increase food allergies, incentivize non-sustainable farming practices, create dependency on chemical inputs, justify a lack of transparency in evaluating food quality and safety, or transform farming practices into a political economy resembling serfdom where the seed is intellectual property leased by the farmer. They also include advocates for a return to conventional methods of food production, which have been marginalized because they don't offer corporations a higher profit margin. They note, for example, that conventional methods for creating drought resistance in crops actually create higher yields than GE methods.[5]

A 2013 report in the *Village Voice* summed up the new political economy of GMO agriculture: "Monsanto's thick contracts dropped like shackles on the kitchen table of every farmer who used the company's seed, allowing Monsanto access to the farmers' records and fields and prohibiting them from replanting leftover seed, essentially forcing farmers to buy new seed every year or face up to $3 million in damages."[6]

As agricultural biotechnology has developed, three methods have been widely used for the introduction of foreign genes into plants: 1) using a bacterium or virus to carry genes into a plant, 2) using electrical shock to get pure DNA into the plant cell's nucleus, or 3) using microprojectiles coated with DNA.

According to a report by the International Service for the Acquisition of Agri-Biotech Applications (ISAAA), a trade organization of the biotech industry, by 2012 genetically modified crops were planted on nearly a quarter of the world's farm land, on 170 million hectares, by some 17.3 million farmers in 28 countries.[7]

Why has a new food technology that splices DNA sequences into plant germ plasm with what is widely believed to be more precision than hybridization or mutagenesis created such controversy? The answer can be found in the essays contained in this volume which cover three decades of commentary and precautions regarding GMOs. The controversy over GMOs has divided scientists and food activists, some of whom consider GMOs benign while others believe they can reduce chemical inputs into agriculture and increase yield. The public and scientific criticism and skepticism over GMOs come from many directions. But they are connected by some form of risk evaluation. The following list of questions illustrate the range of concerns:

- Are GMOs a health danger to consumers?
- Do GMO crops offer farmers higher productivity or pest resistance at the expense of other crop benefits such as nutrition or taste?
- Do GMOs reinforce and expand monoculture and destroy biodiversity?
- Do GMOs create greater control by seed manufacturers over farmers?
- Do GMOs result in one seed variety for all agricultural regions?
- Do GMOs contribute to a more sustainable, locally empowered, and farmer-directed agriculture?

- Do GMOs result in more or less dependency on toxic chemical inputs in agricultural production?
- Do GMOs play a role in reducing world hunger?
- Are the scientific studies of these questions carried out by independent scientists who have no financial interests in the outcome and who publish their findings in refereed journals?

The conventional view about GMOs held by the pro-GMO camp is expressed in this statement by Ania Wieczorek and Mark Wright:

> All types of agriculture modify the genes of plants so that they will have desirable traits. The difference is that traditional forms of breeding change the plants genetics indirectly by selecting plants with specific traits, while genetic engineering changes the traits by making changes to the DNA. In traditional breeding, crosses are made in a relatively uncontrolled manner. The breeder chooses the parents to cross, but at the genetic level, the results are unpredictable. DNA from the parents recombine randomly. In contrast, genetic engineering permits highly targeted transfer of genes, quick and efficient tracking of genes in the varieties, and ultimately increased efficiency in developing new crop varieties with new and desirable traits.[8]

The deception in this statement is that it mistakenly assumes that genetic engineering of plants is a precise technology for transplanting genes. The fact is that the insertion of foreign DNA is an imprecise and uncontrolled process. One of the common mistakes made by the pro-GMO advocates is that they treat the plant genome like a Lego construction where the insertion or deletion of a gene does not affect the other genes. They argue that adding new genes just adds new properties to the organism. This understanding of genetics was long ago proven obsolete in human biology where scientists have come to understand that most characteristics are influenced by complex interactions among multiple genes and the environment acting together. Yet proponents of GMOS continue to assert their safety based on such antiquated science. In fact the plant genome and that of any other complex living thing is like an ecosystem. This means that introducing or deleting new genes can affect other genes in the plant. This is called "pleiotropy" by biologists. There is another effect called "insertional mutagenesis" in which the added foreign genes causes a mutation in the DNA sequence proximate to it. For example, small changes in the plant genome can affect the expression of genes for nutrition or mycotoxins. There is only one way to tell and that is to test the plants that have been gene adulterated.

The fact that Americans have been consuming large amounts of GMO corn and soybeans does not mean that GMO crops are highly desirable nutritionally unless we know that other changes in the crop have not taken place. One report concludes, "Using the latest molecular analytical methods, GM crops have been shown to have different composition to their non-GM counterparts . . . even when the two crops are grown under the same conditions, at the same time and in the same location."[9] Studies have found that GMO soy contains lower amounts of isoflavones, GMO canola contains lower amounts of vitamin E, and GMO insecticidal rice has higher levels of sucrose, mannitol, and glutamic acid than its non-GMO counterparts. These are all results consumers should know about.

A 2009 paper in the *International Journal of Biological Sciences* analyzed blood and system data from trials where rats were fed three varieties of commercial varieties of genetically modified maize. They reported new side effects associated with the kidney and liver, which are the body's primary detoxification organs. The authors of this paper strongly recommend additional long-term studies of the health risks.

Government and industry risk assessments have focused mainly on the foreign DNA introduced into the plant rather than the pleotropic effects on the plant's genome. According to the European Commission, scientists have not found a deleterious protein introduced by genetic modification into a plant. "It can be concluded that transgenic DNA does not differ intrinsically or physically from any other DNA already present in foods and that the ingestion of transgenic DNA does not imply higher risks than ingestion of any other type of DNA."[10] And the ISAAA proclaims the commercialization of GMOs, which began in 1996, "confirmed the early promise of biotech crops to deliver substantial agronomic, environmental, economic, health, and social benefits to large and small scale farmers worldwide."[11] Yet in 2013 more than two hundred scientists were signatories to an open letter titled "No Scientific Consensus on GMO Safety."[12]

This takes us to another deception, and that is that GMOs are highly regulated.

The US regulatory framework for GMOs was issued in 1986 under the Coordinated Framework for Regulation of Biotechnology (US OSTP, 1986). The regulatory authority for plant biotechnology was divided among three federal agencies: the US Department of Agriculture (USDA), the Environmental Protection Agency (EPA), and the Food and Drug Administration (FDA). Under the authority of the Plant Pest Act and the Plant Quarantine Act, the USDA's role is to insure that the GMO crop is not a plant pest—that it does not harm other crops. The EPA's authority under the Federal Insecticide, Fungicide and Rodenticide Act (FIFRA) is over those GMO crops that possess pesticidal properties called PIPs or plants with insecticidal properties." Thus, the EPA oversees crops bioengineered with genes that code for toxins in *Bacillus thuringensis* (Bt), a natural pesticide used by organic farmers. The FDA's authority over GMO crops is to insure human and animal safety on consumption. In 1992 the FDA issued voluntary guidelines for GMO crop manufacturers and affirmed that foreign genes introduced into crops (independent of the source) do not constitute a food additive. For the purpose of regulation, GMO crops were not to be treated any differently than hybrid crop varieties. GMO crop producers were advised to consult with the FDA according to a guidance flow chart. The FDA subsequently promulgated a mandatory pre-market notification that included a requirement for GMO plant manufacturers who planned to release GMO crops, non-binding recommendations for early food safety evaluations[13] and they also issued a draft guidance for labeling in January 2001.[14] As of November 30, 2012 the EPA had forty-one PIPs registered including varieties of Bt corn, Bt soybeans and Bt cotton.[15]

The FDA's approach to risk assessment of GMO crops is based on the concept of "substantial equivalence." While the concept has never been well-defined, it is based on the idea that when a few components of the GMO crop such as certain nutrients and amino acids are similar in content to its non-GMO counterpart, the GMO crop is declared "substantially equivalent." The vagueness of the concept and its application to risk assessment has been widely criticized.[16] The late Marc Lappé wrote, "At the core of the debate between anti-biotechnology activists and its proponents is the assertion that no meaningful differences exist between conventional and genetically engineered food. Establishing the truth of this assertion was critical to deregulating various commodity crops."[17] As previously mentioned, there is growing evidence that GMO crops can have different protein composition than non-GMO varieties.

Under its nonbinding recommendations for industry of "Non-Pesticidal Proteins Produced by New Plant Varieties Intended for Food Use introduced on June 2006," the FDA requires the manufacturer to provide the name, identity, and function of any new protein produced in a new plant variety. It also requires information as to whether the new protein has been safely consumed in foods, the sources, purpose, or intended technical effect of the introduced genetic material; the amino acid similarity of the new protein with and known allergens and toxins; and the stability and resistance to enzymatic degradation of the protein. Within 120 days after submission for a GMO crop, the FDA will alert the manufacturer whether it has been accepted or whether there are questions about the submission (see note 6). The FDA's data requirements do not include changes to the other gene expressions in the plant, with mutational mutagenesis and pleiotropy. As of April 2013, the FDA completed ninety-five consultations including thirty for corn, fifteen for cotton, twelve for canola, twelve for soybean, and twenty-four for all other crops including alfalfa, cantaloupe, flax, papaya, plum, potato, squash, sugar beet, tomato, and wheat. These consultations have accepted the assurances from the biotechnology industry that their safety assessment is reliable. The information the FDA receives is shrouded in confidential business information and is not transparent to the independent scientific community on the methods used and the application of the "weight of evidence."

In a report titled "Potential Health Effects of Foods Derived from Genetically Engineered Plants: What Are the Issues?" published by the Third World Network (an international NGO), two scientists proposed a *de minimus* list of tests that should serve as the basis for GMO crop health assessment. The report rejects the *a priori* claim that GMO and hybrid crops should be treated the same. The requirement cited in the report includes a full biochemical, nutrition, and toxicological comparison between the transgene product implanted in the germ plasm of the recipient plant and the original source organism of the transgene. The report also includes a molecular examination of the possible secondary DNA inserts into the plant genome; an assessment of the variation of known toxins of GMO plants grown under different agronomic conditions; and an investigation of the nutritional, immunological, hormonal properties and the allergenicity of GMO products. Some of these tests should involve laboratory animals. The authors state, "Compositional studies and animal tests are but the first in GM risk assessment. Next, long-term, preferably lifetime-long metabolism, immunological and reproduction studies with male and female laboratory and other animal species should also be conducted under controlled conditions."[18] The gap between what is being proposed by independent scientists and what is actually being done in hazard assessment of GMO crops is gargantuan. Nowhere is it greater than in the United States where the food safety requirements for GMO crops are similar to the chemical food additives designated as "generally regarded as safe" (or GRAS). In both cases the evaluation of health effects largely has been left to the manufacturers.[19]

About This Book

This volume is comprised largely of articles originally published in *GeneWatch*, the magazine of the Council for Responsible Genetics, which was launched in 1983. The editors also added some new essays that were written exclusively for this volume to give the reader both a historical perspective and a current view of the development of genetically modified food crops. The contributors to this volume consist of biologists, social scientists, and public health and environmental policy experts who have been committed to studying the new technologies and communicating their questions and findings to the general public. They have left their comfort zones in academia or in government-funded science to take on the role of public science.

Part 1 is devoted to essays on human health and ecology. In Part 2, the labeling of GMO crops and consumer activism are discussed. Part 3 focuses on GMOs in the developing world and their relevance to the problems of world hunger. Part 4 examines the relationship between GMO technology and the growth of corporate seed oligopolies. In Part 5 the authors examine GMO policy, regulation and law. Part 6 explores the effects GMOs are having on ecology and sustainability. The ethics of GMOs is addressed in Part 7. In Part 8, the authors discuss the role that the genetic modification of food animals is having on agriculture. The final chapter explores different scenarios on the role GMOs are likely to have on the future of food.

There are two trends headed in a collision course. The first is the rise in an organic food movement in which consumers are demanding greater quantities and varieties of organic produce and processed food. They also want to know more about how and where their food is produced and what it contains. The second trend is the expansion of GMO crops, which are currently unlabeled and classified as non-organic under US regulations. Thus far, US citizen movements to alleviate these tensions and expand GMO labeling have not succeeded. The one exception is the community of Honolulu, Hawaii, which enacted Bill 113 into law, thereby prohibiting biotech companies from introducing any new genetically altered crops on the island beyond the GMO papayas that have been grown there. Thus far, the only meaningful food label that excludes GMOs is the organic label. The essays in this volume will afford readers the opportunity to understand why there remain two warring camps in the struggle over GMO crops.

—Sheldon Krimsky and Jeremy Gruber
February 2014

What Is Genetic Engineering? An Introduction to the Science

By John Fagan, Michael Antoniou, and Claire Robinson

John Fagan, PhD, *is founder and, until November 2013, was Chief Scientific Officer of Global ID Group, through which he pioneered the development of innovative tools to verify food purity, quality, and sustainability. These tools include DNA tests for genetically engineered foods, the first certification program for non-GMO foods, and a leading program dedicated to certifying corporate social and environmental responsibility in the food and agricultural sectors. Dr. Fagan currently works primarily through Earth Open Source, a non-profit that focuses on science-policy issues related to GMOs, pesticides, and sustainability in the food and agricultural system and on rural development. Dr. Fagan holds a PhD in molecular biology, biochemistry, and cell biology, from Cornell University.* **Michael Antoniou, PhD,** *is reader in molecular genetics and head of the Gene Expression and Therapy Group at the King's College London School of Medicine. He has twenty-eight years' experience in the use of genetic engineering technology investigating gene organization and control, with more than forty peer-reviewed publications of original work, and holds inventor status on a number of gene expression biotechnology patents. Dr. Antoniou has a large network of collaborators in industry and academia who are making use of his discoveries in gene control mechanisms for the production of research, diagnostic and therapeutic products, and human somatic gene therapies for inherited and acquired genetic disorders.* **Claire Robinson, MPhil,** *is research director at Earth Open Source. She has a background in investigative reporting and the communication of topics relating to public health, science and policy, and the environment. She is an editor at GMWatch (www.gmwatch.org), a public information service on issues relating to genetic modification, and was formerly managing editor at SpinProfiles (now Powerbase).*

The use of genetic engineering, or recombinant DNA technology, to genetically modify crops is based on the understanding of genetics and gene regulation that was current twenty to thirty years ago. The ensuing decades have provided a much deeper, more comprehensive and nuanced understanding that explains many of the limitations and problems with the use of genetic engineering or genetic modification. However, the simplicity of this model makes it a convenient starting point for explaining what genetic engineering is and how it is done.

If you look inside any living organism, you find that it's made of cells. As illustrated in Figure 1, whether the organism is a human being or a soybean, if you look inside the cells of the organism, you find a nucleus and within the nucleus you will find strands of DNA. According to this simplified model, essentially all the information needed to specify the structure and function of every part of any organism is understood to be stored in the DNA of that organism in the form of units called genes. Genes are understood to function as the blueprints for proteins.

How do these blueprints function? DNA is made of four chemical units, and genetic information is stored in the DNA of the organism in the sequence of these four chemical units. Just as information is stored on this page in the sequence of the twenty-six letters of the English alphabet, genetic information is stored in the DNA in the sequence of these four chemical units or letters of the genetic alphabet. According to this model, information for the structure and function of every component of the physiology of each living organism is encoded in the sequence of the letters of the genetic alphabet carried in that organism's DNA.

Based on this model, the rationale for genetic engineering is this: Given that genes are the blueprints for every structure and function of every part of every living thing, by modifying these genes one should be able to modify the structure or function of any part of any organism.

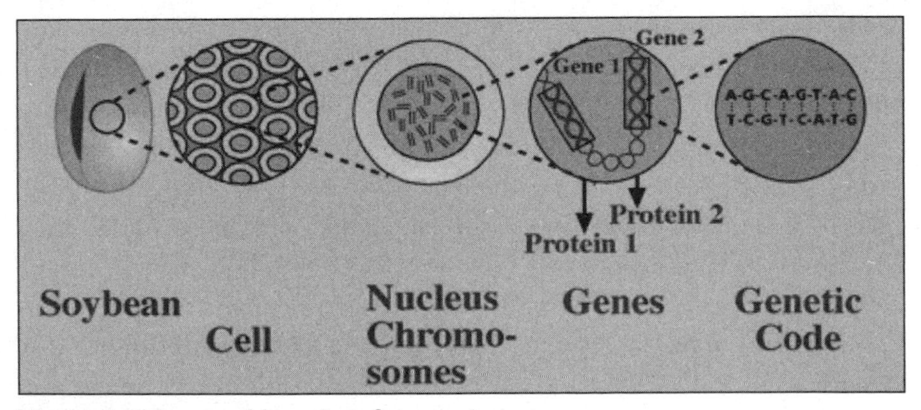

Figure 1. Genes are blueprints for proteins.

How does one actually carry out the genetic engineering of an organism? The first step of the process is to isolate genes. They can be isolated from any organism. Once isolated, they can be modified in the laboratory, cut and spliced to the genetic engineer's specifications. Once the genetic engineer has modified a gene in the laboratory, the second step in the process of genetically engineering an organism is to insert the modified gene into the DNA of the target organism. The result is to re-program the genetic functioning of that organism so that it produces a new protein based on the DNA blueprint inserted.

For example, as shown in Figure 2, one might isolate genes from a bacteria, a virus, a plant, and an animal like a pig. Pieces of those genes can then be spliced together and recombined to create a new gene. The result is termed a genetically modified gene, or GMO gene. It is also called a "recombinant" gene because this process re-combines elements from multiple different genes.

This recombining process can be done very precisely; you can cut and splice DNA with the same letter-by-letter precision with which you can cut and splice sentences with your word processor. Proponents point to the precision of this step in the genetic engineering process as being the proof that

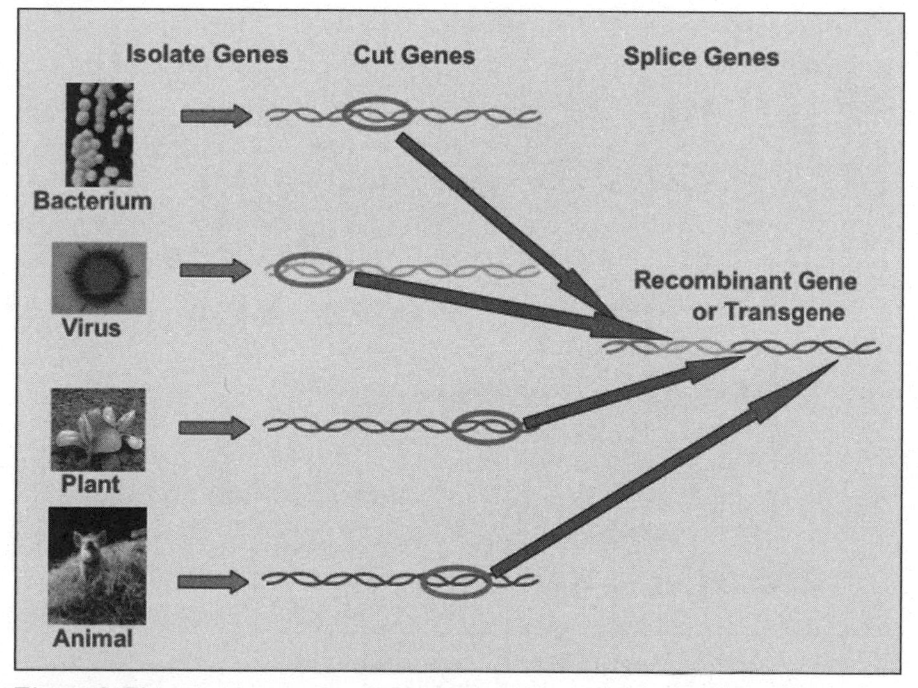

Figure 2. First step in creating a GMO.

genetic engineering is precise, reliable, predictable in outcome, and, therefore, safe. But, somehow, they forget to mention that the subsequent steps in the genetic engineering process are far from precise, reliable, and predictable in their outcomes.

The Imprecise, Uncontrollable, Unpredictable Steps of Creating a GMO

Once the recombinant gene or GMO gene has been created in the laboratory, it must be inserted in a functional form into the cells of the target organism. This is illustrated in Figure 3 where we see that the GMO gene must first enter the nucleus of the cell. There are multiple methods for achieving this, but we will not go into these details here. Then, once the DNA enters the nucleus of the cell it becomes inserted into the cell's own DNA at some low frequency. Scientists do not deeply understand the mechanism by which the DNA insertion process occurs, and they have no control over it. As described in more detail in *GMO Myths and Truths*[1], this process always causes mutations to the DNA of

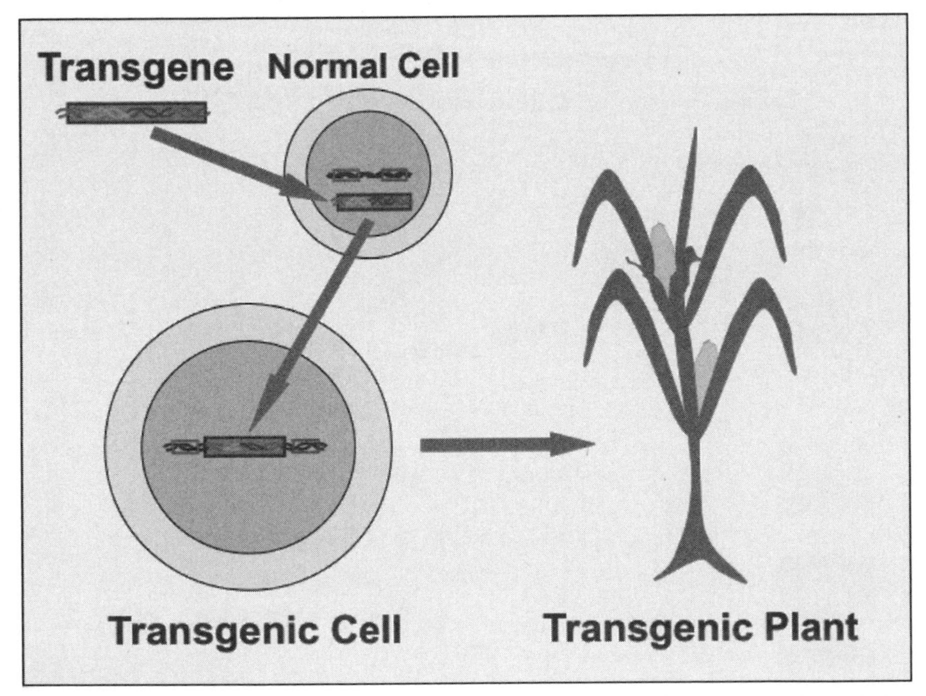

Figure 3. Inserting the engineered gene or transgene into the DNA of the recipient organism.

the organism that is being engineered. This is problematic, because these mutations can give rise to unintended, unexpected damage to the functioning of the organism.

Subsequent to the insertion process, the next step is to treat the cultured plant cells with a compound that kills any cells that have not incorporated the GMO gene sequences into their DNA in a functional form. This selects from the millions of cells that were originally in the culture only that tiny population of cells that has incorporated the GMO gene, in functional form, into their DNA. Finally, this small population of genetically engineered cells is treated with plant hormones that stimulate each of these cells to develop into a small sprout or plantlet which can be transplanted and grown into a genetically modified plant.

All of these steps, up to transplantation, take place in tissue culture or cell culture. It is well known that this process, itself, causes mutations. This adds further unpredictable effects to the genetic engineering process.

In addition to mutagenesis caused by gene insertion and by cell/tissue culture, there is another source of unintended, potentially harmful effects. This

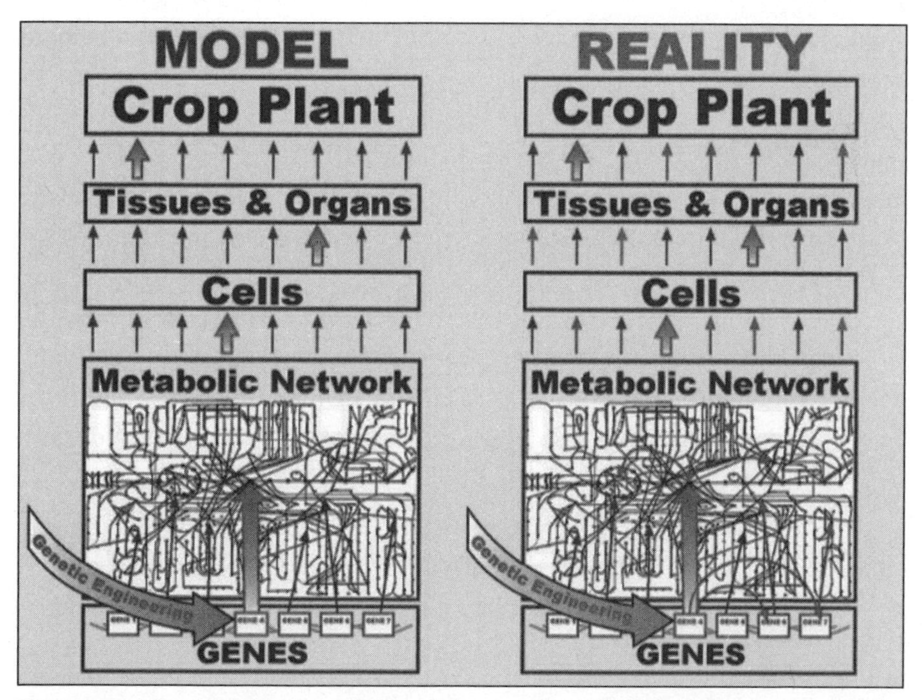

Figure 4. The idealized model and the reality of inserting a gene into the genome of a crop plant

is the fact that living organisms are highly complex and their components are highly interactive. When you change one component of an organism, it can give rise to a multiplicity of changes in the functioning of the organism. This is illustrated in Figure 4.

The left-hand panel of Figure 4 illustrates the model of the genetic engineering process that is put forward by proponents of GMOs. This model assumes that when you insert a single gene into the DNA of another organism, it will cause a single change in the biochemical makeup of that organism—a change in one protein—and they assume that this results in a single change at the cellular level of the organism, a single change at the tissue and organ levels, and finally a single change at the level of the plant as a whole.

This model is not accurate because it fails to recognize the complexity and interconnectedness of the many components of living organisms. It also ignores many new discoveries that have come to light since the early years of genetic engineering twenty to thirty years ago. For instance, it is now known that most genes encode not one protein but two, three, four, or more. Another example is that it is now known that regulatory sequences associated with a gene can also influence the expression of neighboring genes. Another example is the recent observation that sequences which have been known for many years to encode proteins can also carry information relevant to the regulation of expression of a gene. This has given rise to the concept of duons, DNA elements that have dual functions; they carry both structural information (the blueprint for a protein) and regulatory information (information that controls how and when a gene will be actively expressed).

As a result of these recently discovered features of the genetic system, and as a result of the interconnectedness of the many components of living systems, the effects of inserting a gene into the DNA of an organism correspond to the model illustrated in the right-hand panel of Figure 4, not the simplistic model described in the left-hand panel of Figure 4.

In the right-hand panel, we see that when a new gene is inserted into the DNA of an organism, that gene is likely to influence the expression, not of one gene, but multiple genes. Then, when the newly inserted gene is expressed as a protein, that protein will not have just a single effect at the cellular level. It will influence multiple cellular processes and, subsequently, multiple processes at the organ, tissue, and whole plant levels. Therefore, instead of having a single effect, insertion of a single gene results in multiple effects. Because of the complexity of living systems, it is not possible to predict what all of these effects will result from insertion of even a single gene into the DNA of a crop plant.

We conclude that the process of genetic engineering is not precise and controlled, but is imprecise, mutagenic, and uncontrolled and because of the complexity of living systems, these effects are highly unpredictable.

The lack of precision, control, and predictability means that the genetic engineering process can, and almost always does, result in unintended effects. These unintended effects translate into potential harm to health and the environment. Because of the complexity of living systems and the unpredictable nature of the genetic engineering process, the technician who carries out the genetic engineering process has limited control over the outcome and is unable to prevent unintended harmful effects from occurring.

Thus, if one is going to genetically engineer crops, these manipulations should be carried out with caution and before the resulting genetically engineered organism is commercialized, it should be studied very carefully to make sure that it will not cause harm to health or the environment. Every GMO is different, and therefore each must be tested independently and in depth to assure safety.

PART 1

Safety Studies: Human and Environmental Health

One of the most troubling concerns about genetically engineered crops and food is the fact that so much is unknown and, at this time, unknowable. Though scientists have the skill to remove and insert gene sequences in living organisms, they are not able to control the many variables in the process. Scientists with a genetic map of a plant cannot yet predict what each gene does. In addition, genes interact with other genes and with their environment in complex ways, making it impossible for a scientist to be able to predict completely the overall changes in an organism resulting from the transference of even just one foreign gene.

Proponents of GMOs frequently state that they have been fully studied and proven safe. It's an often repeated claim, not just by industry but by many otherwise independent and free-thinking people. But repeating a claim does not make it true.

More often than not, the risks of GMOs have been understated, and the scientific justification has been selective and commercially influenced. The truth is we do not know conclusively what the long-term effects of growing and consuming GM crops will be. There have been very few systematic and independent animal studies that have tested the safety of GM crops. Since 1992 the FDA policy has considered the insertion of foreign genes into the plant genomes of crops as the equivalent of hybrid crops—in other words, crosses within the same species—and therefore exempt from the regulations on food additives.

Yet we know enough to be cautious. You simply cannot predict the safety of foreign gene inserts unless you do the testing. Most GM food studies have been generated by industry and it is the industry itself with sole access to so much of the data. There is little funding of independent studies on the effects of GM foods, and those few scientists who have engaged in such studies and reported concerns are discounted.

The essays in this section raise many safety concerns with GMOS that have yet to be resolved; and will not be conclusively resolved without serious and independent scientific study.

—Jeremy Gruber

1

The State of the Science

By Stuart Newman

Stuart Newman, PhD, *is Professor of Cell Biology and Anatomy at New York Medical College and a founding member of the Council for Responsible Genetics. This article originally appeared in* GeneWatch, *volume 26, number 1, January–March 2013.*

When scientists first learned in the late 1970s how to sequence DNA and transfer it from one kind of organism to another, improving foods and other crop plants by introducing foreign genes was among the first applications proposed. Given contemporaneous findings in molecular genetics, such as the recognition that a mutation in a single gene could promote a cell's transformation to cancerous state,[1] it was unsurprising that concerns were raised about the capability of the transgenic methods to dramatically change the biochemistry or ecological stability of plants. Some critics suggested that the quality and safety of fruits and vegetables could be impaired, making them allergenic or toxic to humans and nonhumans who consume them, or that "superweeds" might be created which could disrupt wild or farmed ecosystems.

By 2005, however, when more than 90 percent of the annual soybean crop and 50 percent of the corn crop in the United States had come to be genetically engineered—a transformation in agricultural production that took less than a decade[2]—efforts at testing and regulation of genetically modified (GM) foods were increasingly portrayed as irrational. A perusal of the summaries of recent policy articles on the PubMed database turns up dozens in which reservations about the massive introduction of GM food into the food chain are represented as scientifically ignorant, economically suicidal, and cruel to the world's hungry. One abstract in the journal *Nature* reads: "Unjustified and impractical legal requirements are stopping genetically engineered crops from saving millions from starvation and malnutrition."[3]

These papers—many by European commentators decrying the successful efforts to keep GM foods out of the markets there, and some by US commentators bemoaning the necessity to test these products at all—mainly support their cases by referencing short-term feeding studies of animals. But this type of study is not adequate to allay valid concerns. One group, reviewing the relevant areas, has written, "It appears that there are no adverse effects of GM crops on many species of animals in acute and short-term feeding studies, but serious debates of effects of long-term and multigenerational feeding studies remain."[4]

According to another group that has looked into these issues:

> The most detailed regulatory tests on the GMOs are three-month long feeding trials of laboratory rats, which are biochemically assessed. The test data and the corresponding results are kept in secret by the companies. Our previous analyses . . . of three GM maize [varieties] led us to conclude that [liver and kidney] toxicities were possible, and that longer testing was necessary.[5]

Another team actually performed such long-term studies, with the findings that mice that were fed for five consecutive generations with transgenic grain resistant to a herbicide showed enlarged lymph nodes and increased white blood cells, a significant decrease in the percentage of T lymphocytes in the spleen and lymph nodes and of B lymphocytes in lymph nodes and blood in comparison to control fed for the same number of generations with conventional grain.[6]

A central issue for crop foods, of course, is their effects on humans. The most comprehensive review of this subject as of 2007 stated:

> . . . the genetically modified (GM) products that are currently on the international market have all passed risk assessments conducted by national authorities. These assessments have not indicated any risk to human health. In spite of this clear statement, it is quite amazing to note that the review articles published in international scientific journals during the current decade did not find, or the number was particularly small, references concerning human and animal toxicological/health risks studies on GM foods.[7]

The same group revisited the literature four years later, reporting that whereas the number of citations found in databases had dramatically increased in the intervening period, new information on products such as potatoes, cucumber, peas or tomatoes, among others was not available. Regarding corn, rice, and soybeans, there was a balance in the number of studies suggesting that GM corn and soybeans are as safe and nutritious as the respective conventional non-GM plant, and those raising still serious concerns. They also note that "most of these studies have been conducted by biotechnology companies responsible [for] commercializing these GM plants."[8]

Given the uncertainties of the long-term health impact of GM foods, it is significant that so far, virtually all genetic modification of food and fiber crops has focused on the economic aspects of production (i.e. making crops resistant to herbicides and insect damage, increasing transportability and shelf-life) rather than the more elusive goals of improving nutrition or flavor. Introducing biological qualities that enhance production, transportability and shelf life can compromise palatability, as seen with the Flavr Savr tomato, the first GM crop to be approved by the FDA for human consumption, two decades ago.[9]

To protect its investment against a skeptical public, the biotech food industry has depended on compliant regulators,[10] on its proponents' ridicule of biotech industry critics' supposed scientific ignorance,[11,12] and on expensive campaigns against labeling of prepared foods that would draw undue attention to the presence of GM components, which they claim to be natural and ordinary.[13] (These are the same components that when presented to the Patent Office and potential investors are portrayed as novel and unique.) A food crop that actually benefited the people who eat it rather than only those who sell it would likely open the floodgates of greatly weakened regulation. Golden Rice, designed to provide Vitamin A to malnourished children, has failed to overcome the hurdles for approval for dietary use since it was first described in 2000. Though very limited in its ability to alleviate malnutrition, it has some merit in the prevention of blindness, and seems poised for approval in the next year or so.[14] If so, it will almost certainly help agribusiness tighten its grip on the world food supply and increase its capacity to foist products that are much more questionable on their captive clientele—that is, everyone.

2

Antibiotics in Your Corn

By Sheldon Krimsky and Timo Assmuth

Sheldon Krimsky, PhD, *is the board chair of the Council for Responsible Genetics and professor of Urban and Environmental Policy and Planning at Tufts University. Dr. Krimsky served on the National Institutes of Health Recombinant DNA Advisory Committee from 1978–1981, chaired the Committee on Scientific Freedom and Responsibility of the American Association for the Advancement of Science, and has been a consultant to the US Congress Office of Technology Assessment. He is the author of numerous articles and books on regulation and the social and ethical aspects of science and technology.* **Timo Assmuth** *studied environmental sciences at the Universities of Turku and Helsinki, Finland, where he is affiliated as adjunct professor, and on a Fulbright grant at Tufts University Department of Urban and Environmental Policy and Planning in 2007. He has worked for twenty-five years in the Finnish Environment Institute, a government R&D organization, on wastes, soil protection, chemicals, GMOs, and environmental and health risks, with increasing interest in the human aspects of risk assessment and management. This article originally appeared in* GeneWatch, *volume 22, number 1, January–February 2009.*

Biotechnology, including technologies based on genetic engineering or genetic modification, is becoming increasingly important in the global economy, ecology and politics. Agricultural biotechnology for food production has been the subject of much interest and debate in international politics, but in terms of market value, health care represents the largest sector of biotechnology, with pharmaceutical substances playing the major role. Most biotechnological pharmaceuticals are produced in microbes, but the use of genetically modified plants (often called "pharmacrops") to this end has gained increasing attention.[1] The first field trial permit for GM plants based on an application using the term "pharmaceutical" was issued in January 1991 in the US By the year 2006 there had been 237 applications for field trials in the US alone; however, no commercial products have resulted to date.[2] Among the drugs being produced in plants are vaccines, antibodies, antigens, hormones, growth factors and structural proteins.

The possible advantages of plants over other systems in producing drugs include the production of larger volumes of drugs, more flexibility and cost-effectiveness in manufacture, better suitability of plant cells for production, and the potential of using plants and seeds for drug storage and delivery.[3] Plants have also some safety advantages over other pharmaceutical production systems, such as safety from contamination with human pathogens, endotoxins and tumorigenic DNA sequences.[4]

On the other hand, pharmacrops present important new risks and safety issues. By definition, they are used to produce substances that have potent biological effects on humans and other higher animals. Pharmacrops contain higher concentrations of active substances than these animals are ordinarily exposed to in GM plants. Several genetic modifications are often carried out simultaneously, increasing risks.[5] Risks arise not only from biological but also socio-economic factors.

Pharmacrops consequently introduce special challenges to regulation. This is inherent in their position between agricultural, medical and general industrial biotechnology, and the special ecological-physical and socio-technological characteristics of these technologies. Some of these regulatory challenges include:

- The extension of new-generation GM crops to novel processes wherein the plants are not intended to be utilized as food crops but rather as plant-based 'drug factories.'
- Emergence of new forms of biopollution in possible gene transfers of GM pharmacrops to conventional crops.
- New methods required to evaluate drugs derived from plants that are grown in open fields.
- The need to align environmental, food and agricultural, as well as pharmaceutical and medicinal policies and regulatory procedures.
- The subsequent introduction of new actors, new interests and new contested issues regarding the development and application of the technology.

Risk issues are particularly urgent when pharmaceuticals are produced in plants that are potential food crops.[6] The need to control these risks has been stressed, by both consumers and food processors.[7] Currently pharmacrop risks are still addressed mainly within the conceptual frameworks of other GM food plants, and it is unclear how to accomplish the protection and management of food supplies as the distinction between food and pharmaceuticals becomes blurred.

The main direct risks associated with pharmacrops can be categorized in terms of causative agents (for instance, the drug being produced); dispersal processes (especially gene flow) and environmental fate of the produce; exposed organisms or systems (such as animals which feed on pharmacrops in field trials); and biological (toxic, allergenic, ecological), agricultural and social effects. All of these need to be accounted for in the life-cycle of the technology (see Figure 1).

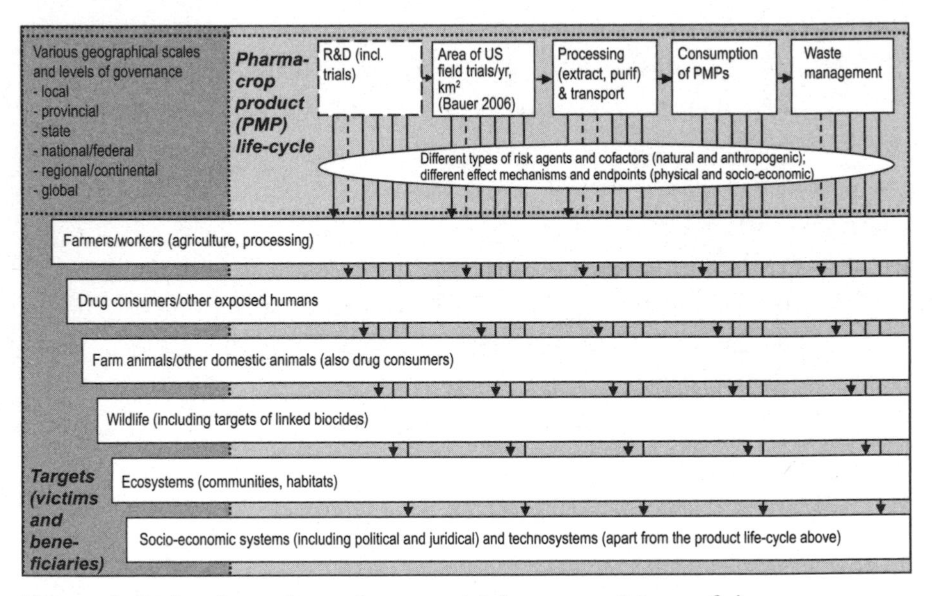

Figure 1. Risks along the cycle, to multiple targets. Many of these targets are routinely excluded from assessment.

Indirect risks arise in complex socio-ecological processes, also from attempts to control risks which inadvertently create new risks, for instance when using unproven 'terminator' technology to render GM plants sterile, combating vandalism by non-disclosure of information on field trial sites, or causing losses of relative benefits from pharmacrops as compared with conventional drug manufacture.[8] Some risks may be irreversible, especially in regard to gene flow into the environment. Accidental outbreaks from field trials and associated food chain contamination scandals indicate that the transgenes cannot be totally contained.[9] The crucial questions become how the different kinds of risks are judged and weighed against each other, what risks are deemed acceptable and on whose criteria, and what are feasible and justified risk abatement or prevention options.

The risks of pharmacrops are unevenly distributed geographically. The map of field trials in the United States (Figure 2) shows that transgenic corn with pharmaceutical proteins has been tested mainly in the Corn Belt. Threats to food production systems, biodiversity, worker safety and rural development also vary according to location. If a pharmacrop is grown near fields of the same species, the risk of transferring the "drug gene" to a conventional crop is increased. The benefits are unevenly distributed as well; those who stand to

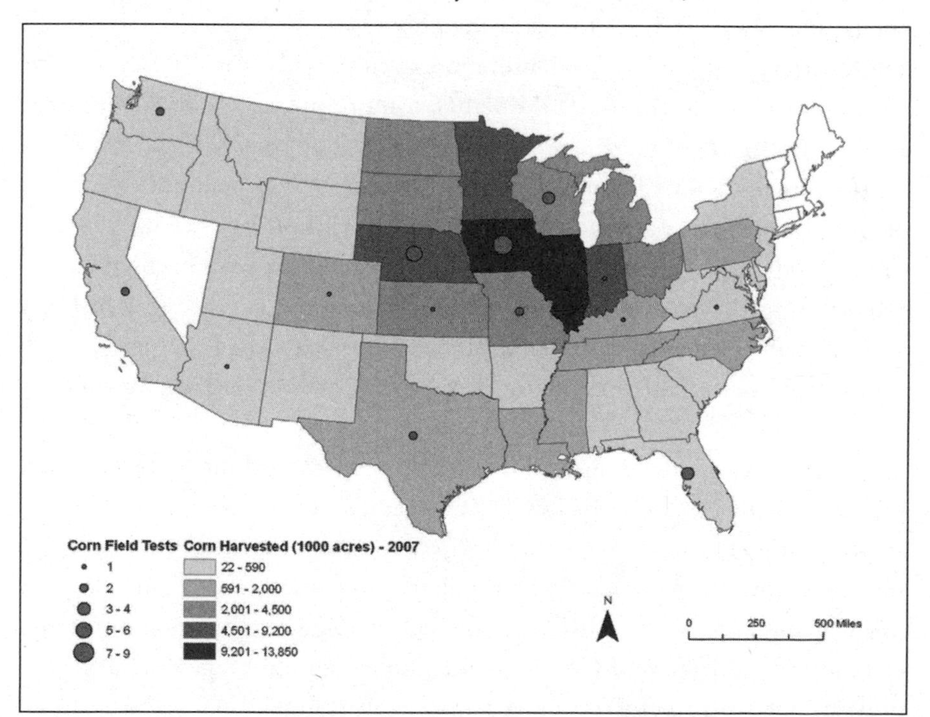

Figure 2. Pharmacrop corn field trials by state through November 2007 (APHIS data; map courtesy of Barbara Parmenter, Tufts University)

directly benefit from a field test (such as pharmaceutical companies or landowners paid to allow test plots on their land) do not necessarily share the risks, and thus the presence of potential risks does not necessarily inform decisions such as location of field plots. For example, Iowa and Nebraska—two of the top corn producing states in the United States—have some of the highest numbers of corn pharmacrop test plots, despite the heightened risk of contamination or cross-pollination.

Castle (2008) singled out informed consent, risks to agricultural policy and intellectual property rights as the key global challenges for ethical production of vaccines in plants. The awareness, willingness (political) and capacity to respond to these issues varies between and within societies, and as a result there can be a mismatch between risks and responses. The geographical heterogeneity of risks and regulation increases from small nations to the US, the EU and the global systems.[10]

In the US, the policy toward pharmacrops has been relatively lax, but the regulatory procedures for assessing and managing their risks have been upgraded in response to contamination accidents. In the EU, even after passage of a moratorium on GM plants, a more precautionary stance contributes to the lag of pharmacrop applications. The specific regulatory risk management options for pharmacrops are focused on technical measures at production sites, particularly containment, while options in other stages of the product life-cycle and risks of other dimensions have been given less attention (see Table 1).[11] Additionally, critics point out that the seemingly self-evident options of restricting pharmacrops to closed systems and to inherently safer self-pollinating or non-food species have been a secondary consideration.[12]

In Europe, pharmacrop field trials have been carried out since 1995, but the number of trials declined after 1996; the cultivation acreage was nearly zero in 2002–2004. The onset of a more cautious approach to GM plants in general influenced these fluctuations. The regulatory approach in EU countries was pro-GM until the 1990s, only later to be replaced by a de facto moratorium on commercial cultivation of GM crops for human consumption, due largely to growing concerns among consumers and Member States.[13]

Table 1. Alternative framings of pharmacrops and plant-made pharmaceuticals (PMPs) along key dimensions of risks and governance.

Extent of frame	***Dimension of framing***				
	Unit object of regulation	*Scope of risks considered*	*Geographical scale*	*Scope of regulation*	*Type and agency of intervention (extent of normativity)*
Minimal or small	Pharmacrop species or PMP	Human health, acute	Local (field cultivation area)	Product	Voluntary information-based
Intermediate	GM plants or biologics with pharmaceutical use	- Human health, long-term - Ecological, acute/rudimentary - Socio-economics, rudimentary	- Regional (state) - National	- Class of products - Process	- Statute-based economic or information steering - Traditional regulation ('big government')
Maximal or large	GMOs or generic pharmaceuticals	Ecosystems (incl. humans), long-term also indirect including social	Global (incl. export and import areas)	Products and processes	Governance (broad; variable extent, agency and stringency)
Notes on choice of frame	*Depends on context*	*Large frame preferable for safe and efficiently coordinated regulation*	*Depends on context; often a combination of frames is needed*	*Large frame is preferable*	*Often a combination of interventions at different levels is advisable*

However, even if the European stance toward GMOs has been precautionary overall for over a decade, pharmacrop risks have not been officially singled out. Increased R&D activities suggest that the EU may seek to switch back to a more pro-pharmacrop policy despite official caution, due in part to reasons of global trade policy and competition. Biopollution and other risk issues have been debated in connection with the proximity of GM crops to organic or conventional farms and with the buffer distance required to ensure safe coexistence of GM and non-GM plants.[14] These issues are potentially even more pronounced with pharmacrops, because crops can be contaminated with the pollen and residues of pharmacrops and because pharmacologically modified plants (PMPs) carry particular potency; yet such distances have not been specified in the EU for pharmacrops (see Table 1).

The politics and practices of pharmacrop development and application involve the interplay and also tensions and clashes between different concepts of and approaches to risks, technology and regulation, and between interests and actors in various sectors and geographical regimes. Some of the polarization and conflicts in GMO politics influences pharmacrop policies, even if differently and as yet more subtly, due in part to the promises of producing wonder cures. Framing and evaluations of the risks from new-generation pharmacrops and other GMO "industrials" are only emerging, and the confidence in their safety vary greatly between and even within regulatory cultures.[15] Because of the complexity of the processes and influential factors, the trajectories of the technology and of regulation remain uncertain.

Although pharmacrops have been pursued actively, especially in the United States, some caution seems to hold commercialization back. It remains to be seen whether a fertile hybrid of pharmaceutical, agricultural and industrial technology will arise, and how the particular risks of pharmacrops will be dealt with. The development is likely to be uneven and turbulent. It will introduce the need to integrate activities on pharmacrops in partly new forms of communication, cooperation, negotiation and conflict resolution. These take time and effort to develop, due to differing concepts and traditions among actors and different views of the value-laden issues. Whatever action is taken needs to allow the legitimate involvement of a broader range of stakeholders. Even so, regulatory practices are as yet poorly equipped to deal with pharmacrops and their multi-dimensional largely unknown risks on a commercial scale. Meanwhile, certain concrete steps—such as restricting pharmacrops to closed systems and self-pollinating or non-food species—could provide a more immediate buffer against the risks, but even such solutions require the active engagement and interaction of concerned citizens including scientists and experts as well as regulators, consumer representatives and others.

3

A Conversation with Dr. Árpád Pusztai

By Samuel W. Anderson

Dr. Árpád Pusztai *has published nearly three hundred papers and several books on plant lectins [a group of proteins on the cell membrane that bind to particular carbohydrates]. Since the "Pusztai affair" described below, he has given nearly two hundred lectures around the world and received the Federation of German Scientists' whistleblower award. He was commissioned by the German government in 2004 to evaluate safety studies of Monsanto's Mon 863 corn. This article originally appeared in* GeneWatch, *volume 22, number 1, January–February 2009.*

Dr. Árpád Pusztai became the center of a political firestorm in Britain in the late 1990s when, on a television program, he expressed his concern with the results of a study he and a colleague had conducted on genetically modified potatoes. In the study, rats were fed either a) potatoes that had been genetically engineered (by a biotech company now called Axis Genetics) to express a protein called snowdrop lectin, b) conventional potatoes, or c) conventional potatoes mixed with snowdrop lectin. To Dr. Pusztai's surprise, the group of rats that had been fed GM potatoes showed damage to their intestines and immune systems, while the other groups did not.

With the permission of his employer, the Rowett Institute in Aberdeen, Scotland, Dr. Pusztai raised his concerns in a TV interview. The day after the program aired, the Rowett Institute suspended him and dismantled his research team, and he was ordered by the government not to speak on his research. According to one Rowett Institute colleague, the Institute had received phone calls from the British government, and the line of communication could be traced to Monsanto via the US government. The incident, referred to in the press as "the Pusztai affair," sparked fierce debate in the scientific community, with many criticizing the study even though it had not yet been published. Many people credit (or blame) Dr. Pusztai for tipping public opinion in Britain against GM foods.

Today Dr. Pusztai continues to work and remains one of the world's foremost experts on lectins. He spoke with *GeneWatch* by phone from Hungary, where he teaches.

Dr. Árpád Pusztai: So you're interested in whether there have been any attempts to repeat our experiment?

GeneWatch: Yes, you said that nobody has had the courage to do it.

Pusztai: I don't think that there has been any attempt. It would need a very . . . how shall I put it . . . a very brave person. I don't think that anybody will have the, I can say, the audacity to try to repeat our experiments—because they know perfectly well that they will get something very similar, if not identical results.

GeneWatch: You said that the methodologies you've established are not necessarily specific to GM materials—so what did you find was different in those studies, if anything?

Pusztai: Any new source of protein has to be tested. And you can just regard GM material as a new source of protein. And most of the time, what you do is you try to assess the nutritional value of this new protein source. There are quite a number of protocols for this, and the essence of all of them is the comparison. So for that reason, you can only compare things that are what you think is, protein-wise, nitrogen-wise, energy-wise, identical or very similar in these various tests. . . . But the first essential part of any evaluation is to feed the animals with that diet in comparison with the appropriate non-GM material.

And then you do all sorts of more sophisticated tests—you do immunity studies, you do allergenicity studies, you see how much of that nitrogen you're putting in is retained . . . see what is happening metabolically. If you are, for example, exposing animals early on to chemical carcinogens, then you can compare the effect of the GM diet versus non-GM diet, how long it takes for the tumors to develop. These are all using models that are already accepted and are already being used for this testing.

Now, with the GM, it's very seldom done.

GeneWatch: Why is that?

Pusztai: Because there is a problem of finances. Most of these studies are either financed by the biotechnology companies, or at least you need their agreement to carry out such studies. And not just the agreement, but to get the material from them, bona fide GM material—and very importantly, the appropriate parent line for the comparison. You are at the mercy of these companies, there is no other way to describe it. If they don't give you that material, you are going to have real difficulties. And that was the reason why we did use GM potatoes, because they were developed by a Cambridge team together with our friend in Durham to transfer the transgene into potatoes to make them resistant to aphid attacks. Because we could get this material and the parent-line potatoes in sufficient quantities—they were grown side-by-side in the UK under controlled conditions—so we had the material to carry out these studies.

Most of the people who incidentally tried to get to the bottom of this, whether they think the GM is as good as the non-GM, or what are the advantages or disadvantages—they are at the mercy of the biotech companies, such as, for example, Monsanto. Monsanto would never give you any material to do independent studies—or if they agreed to it, you have to sign a contract with them to say that all the results belong to them. Not just that they belong to them, but you would not be allowed to publish it without their consent. This is something that you have to always take into account when they are talking about safety studies and all that. The companies' interpretation of GM safety is not necessarily the last word in this matter.

GeneWatch: In other words, you can't just go and actually buy the product from, say, Monsanto, if you're going to conduct studies on it?

Pusztai: You go and try to do it! Particularly if—the seed, it has to be a direct comparison, so you need the isogenic line. You may be able to buy the GM material somewhere, by hook or crook, but you will never get the isogenic line. And I speak from experience. This is the reason we used the GM potatoes.

GeneWatch: So potatoes weren't your ideal crop?

Pusztai: I knew perfectly well that potatoes, on their own, are not one of the best materials, because they contain little protein, less than ten percent. Most of the animals that you are testing for nutritional value of the crops you feed them on require at least ten percent protein input. So with the potatoes, alone as the protein source in a full balanced diet, we had some difficulties. Nevertheless, that was the thing that was available.

So I was told by the Ministry that it was 1.6 million pounds, 3 million dollars—you have to use potatoes. And that's it. You never say no to such a proposition.

GeneWatch: So your funding depended on using the potatoes?

Pusztai: Yes. I mean, it had an economic importance for Britain, particularly for Scotland, so it was in their interest that we do the study on potatoes. Remember, I said it many times, I really did believe that the idea [of GM crops] was great, and it was only during the testing process that we found too many snags and started to think about what could be the reason for the snags. So I'm a late convert to skepticism.

Because when we started, I thought that it was great, a great idea. I was still at the university when that guy got a Nobel Prize for the genetic determinism, that you take one gene and that gene is expressing a particular phenotype and whatever. It sounded all right to me, it's just that as we were going ahead with our studies, we started to get results that did not fit into this pattern. And now I know—and anybody in the business, whether they are admitting it or not—they know perfectly well that you cannot splice a gene construct into another crop without making major changes in the genome of the crop that you spliced into.

So now we know what would explain our results. The genome of the potato is in any case quite an unstable genome. I remind you that every bit of the potato plant is poisonous except the tuber, and even the tuber can become poisonous under some conditions. So what we did is by splicing in a gene from the snowdrop plant, we disturbed the potato genome, and it became just as poisonous as any other part of the potato plant. When you look at the stupid idea of "substantially equivalent" . . . you can't say that it is substantially equivalent because you change it. This is published in a respectable American journal, the *Journal of Agricultural and Food Chemistry*—you splice the bean alpha-amylase gene in peas, and as a result of it you change all the genome and you are producing something that is chemically, immunogenically, allergenically different from what you intended.

What we found in 1998 is, in my opinion, commonplace. It has recently been explained by what we now call insertional mutagenesis.

GeneWatch: So you think you might find the same results among commercialized GM crops?

Pusztai: Well, yes—I mean, Monsanto 810 [a type of Bt corn] is commercialized, accepted even in Europe. Admittedly, the Hungarians, the Austrians, the Greeks, and some other parts of Europe have a moratorium on the growing of the stuff. But, I mean, this is now the accepted wisdom. You will never hear this from a biotechnology company, but why should they say it? They have invested a lot of money into this, so they are obviously going to defend their position and deny the existence of these things.

GeneWatch: Why do you suppose, considering that so many people know this but a lot of people will just never tell you—what do you suppose is keeping these ideas out of consumer consciousness? Maybe not in Europe, but certainly in the US.

Pusztai: Because US consumers don't know what they eat! It's not just with GM. For a very long time now, the American production system tried to put a huge gap between production and consumption, so that you would not be able to get this farm-to-fork idea. You would nicely package something—I used to live in the States, so I know it exactly—beautiful packaging, but what's actually in the package is not very well known. And there's no great interest in it either.

GeneWatch: What's the difference in Europe?

Pusztai: In Europe, there are some traditional values that don't seem to agree with this uncertainty. We don't know—I mean, many times people ask me, 'What do you think is the main danger of GM?' And the main danger is that we do not know what the main danger is. We need reasonable hypotheses that could be tested by experiments.

GeneWatch: And sometimes the argument you might hear in favor of accepting GM foods is the opposite—that you can't prove anything is wrong with those foods.

Pusztai: I think one of your great thinkers said, "'This is an irreversible technology." Therefore, when you are approaching it, either conceptually or in practice, you have to take into account that this is in essence irreversible—and unpredictable. You don't know what the consequences will be. If you put a plant into the ground, that plant is a living thing. Through the roots, you are communicating with the soil; through the leaves you are communicating with the air, with other organisms. You cannot look at it in isolation. This is a living thing, and that living thing is going to produce new DNA which gets into the ground, gets into the gut of animals and everything.

So it's something that you can't predict. You can't even predict how to test for it! It's common knowledge—probably a conservative estimate—that we don't know 98 percent of the living organisms in the soil. So how can you do an experiment? I mean, you can pick out some organisms, but what about all the others? So I think this is an extremely dangerous experiment—with our globe, with our Gaia, with our people—and if you ask me what are the likely consequences, I can only say that I haven't the faintest idea.

Once you've got this out, you can't turn around and say, "'Oh, I've made a mistake." This is an abrupt change. We have had no time for the system to adapt to the changes.

GeneWatch: Have we already passed the point of trying to warn people that it's irreversible, now that GM crops are so widely commercialized?

Pusztai: Well, it depends on the country. In your country, in the USA, a very large portion of the lands have been converted to it—but when you look all over the world, it's still only about two to three percent of the soil that has been exposed to it. So I don't know—I don't know much about population genetics—the only thing I know is that this is an experiment that has unpredictable, unmeasurable consequences, and I don't know what will happen. I am 78, so it shouldn't really concern me, but I am concerned because we are leaving something to our descendants to deal with, and they will be put in a situation where they have no choice, they just have to deal with it. And I suspect even then they may not know much about it. We know now that DNA constructs can survive for thousands of years in the ground.

I'm not saying that there is going to be a cataclysmic consequence of this. What I'm saying is that the cataclysmic thing about it is that we don't know what is going to happen.

GeneWatch: It seems so difficult for anyone to even obtain these materials for independent studies—is there any way for government to step in and make sure these studies happen. Do you have any sense of whether this has happened?

Pusztai: You're quite right—the trouble is that the US government is not doing it. Most of the stuff comes from the US, so it's very, very unlikely. Even your president-elect Obama has people on his advisory team who are coming from Monsanto. So what can you expect?

Humanity is mostly stupid. They only take something seriously once a disaster occurs.

GeneWatch: So they have to see the results?

Pusztai: This is history. We'll have to just wait for some sort of disaster to happen.

GeneWatch: Or even if it's not a disaster, it seems people need to just be able to see the problems with their own eyes.

Pusztai: We have to consider this, if it gives any advantage to the consumer. The consumers are carrying all of the risks but they aren't getting any of the benefits. That's one of the reasons why the biotech companies are now touting this idea, "the world is short of food, we're going to provide it."

GeneWatch: Bill Freese writes about those promises, and about how in reality the crops actually being commercialized carry traits that are not beneficial for consumers . . . it seems that those promises really rest on technologies which haven't even been developed yet.

Pusztai: We always say that everything is in the future. Promises, promises. You cannot exclude the possibility that they are right, but you have to take in the present situation. I can only say something about what is available, what has been looked at. When it comes, we'll have a look at it, but at the moment, there are no crops that can tolerate abiotic stress. That is a fact. Now Monsanto says that they will have this in twenty years' time. We'll have a look at it when it comes.

The future is, maybe, very rosy. You know, I am a Hungarian refugee who lived in Britain for 52 years. I remember back before I took refuge in Britain in 1956. Our "great leader," our communist leader here in Hungary, used to say, "We must not eat the chicken or the hen that's going to produce the golden egg of the future." We've been waiting for this golden egg for hundreds of years, and I believe that GM is not going to be that golden egg.

Scientifically, these are the key words: insertional mutagenesis. When you are inserting the transgene construct, you are changing the whole genome. Anything can happen. It could be that the disaster is just around the corner.

GeneWatch: You talked about being concerned about people who are overly certain about these technologies. These people may mostly be those in favor of the genetically modified foods, those who are overly certain of the benefits and the risks—but do you think there is also a problem with those who oppose the technologies, but not based on science?

Pusztai: I know that in 1998 Monsanto spent over one million pounds advertising GM crops, and they were always saying that people weren't accepting it because they don't know anything about it, and it was their job to illuminate the subject. But the fact is that at the end of the million pounds spent, there were more people disbelieving Monsanto's story.

Now everyone agrees—even the biotech companies agree—that they were shooting themselves in the foot. If they had gone about it quietly, they might have been fine, but instead they made a big deal about it. And people are not stupid. The issue is made out of all this, and people began to ask the question: what is Monsanto getting out of this?

GeneWatch: So in advertising they just drew attention to themselves.

Pusztai: Yes. And I have great respect for the British general public. They can ask these very uncomfortable questions of the biotech industry: who is going to benefit from it? They know perfectly well that I didn't benefit from it! But they made a huge hullaballoo about this, and the companies know now that the right way for them would have been to put a lid on it, to keep quiet about it. But the public knows that something is happening, no matter how much they try to explain it. I know that people may not be nutrition-wise or science-wise very clever, but they have a common sense. They do understand that . . . look, we are doing something that is fundamentally different from what we've done before. Therefore, just like the FDA's scientists whose sentiment was that we are doing something different, therefore the risks will be different—it is our responsibility to determine what these risks are. And if we can't come up with an acceptable answer, the next question is "who is to benefit?"

So here we are. Most importantly, what distinguishes the skeptics from the GM partisan? The skeptics try to speak to the facts. And it is therefore extremely important that the facts ought to be really facts, so that there are no mistakes. If I can help in any sense with this, then I shall do my best.

4

Glypho-Gate

By Sheldon Krimsky, with Gilles-Eric Séralini, Robin Mesnage, and Benoît Bernay

This article refers to the study "Ethoxylated Adjuvants of Glyphosate-Based Herbicides Are Active Principles of Human Cell Toxicity," which was conducted at the University of Caen with the structural support of CRIIGEN (www.criigen.org) in the European Network of Scientists for Social and Environmental Responsibility. This article originally appeared in GeneWatch, *volume 26, number 1, January–March 2013.*

Background

Gilles-Eric Séralini is a professor of molecular biology at Caen University, located in the town of Caen in Normandy, France. Professor Séralini was the lead author among a team of eight scientists who submitted a paper to the peer-reviewed journal Food and Chemical Toxicology on the long term toxicity of Roundup herbicide and Roundup-tolerant genetically modified maize. The paper was received by the journal on April 11, 2012, sent out for review, accepted for publication on August 2, 2012, became available online September 19, 2012 and appeared in print in the Elsevier journal November 2012.[1]

Séralini and his team exposed rats to GM maize and Glyphosate and studied them for two years. They found that female rats died at a rate two to three times greater than controls. Female rats developed large mammary tumors more often and earlier in life than the control groups. In treated male rats, liver congestion and necrosis were observed 2.5 to 5.5 more frequently than controls; severe kidney disease was found to be 1.3 to 2.3 times greater with large palpable tumors occurring four times more than controls.

The paper was met with a firestorm of reaction. Some scientists and regulatory bodies found the study inconclusive, citing methodological flaws or limitations in the study design or statistical analysis, and recommended that the study be repeated. Others dismissed the study outright as biased and requested that the journal withdraw it. However, more than a hundred scientists from universities and institutes throughout the world signed on to an open letter supporting Séralini and his team against what they viewed as corporate influence over the science of GM crops.

Many media outlets dismissed the study without even waiting for the paper to be fully aired in the scientific community. Faced with an unprecedented reaction to their journal publication, the editors of *Food and Chemical Toxicology* wrote: "The editors and publishers wish to make clear that the normal thorough peer review process was applied to the Séralini et al. paper. The paper was published after being objectively and anonymously peer reviewed with a series of revisions made by the authors and the corrected paper then accepted by the editor."

Séralini's group issued an eight-page response to critics where they provided a table of criticisms and responses.[2]

Professor Séralini sent *GeneWatch* an announcement in which his research group identifies the most toxic chemical in Ready Roundup—the most widely used herbicide in the world—which is not the active ingredient glyphosate but a substance called POE-15. This substance is an adjuvant added to glyphosate, which the authors state is toxic to human cells. Adjuvant chemicals often escape the rigorous testing of active ingredients in pesticides. Another example of an adjuvant in synthetic pyrethroids is piperonyl butoxide, which is a potential carcinogen.

Sheldon Krimsky
Chair, Board of Directors
Council for Responsible Genetics

Glyphosate Not the Most Toxic Chemical in Roundup

In a new published paper in the scientific journal *Toxicology*, Robin Mesnage, Benoît Bernay and Gilles-Eric Séralini from the University of Caen, France, have proven from a the first of nine Roundup-like herbicides that their most toxic compound is not glyphosate—the substance the most assessed by regulatory authorities—but a compound that is not always listed on the label, called POE-15. Modern methods were applied at the cellular level (on three human cell lines), and mass spectrometry (studies on the nature of molecules). This allowed the researchers to identify and analyze the effects of these compounds.

Glyphosate is supposed to be the "active ingredient" of Roundup, the most widely used herbicide in the world, and it is present in a large group of Roundup-like herbicides. It has been safety tested on mammals for the purposes of regulatory risk assessment. But the commercial formulations of these pesticides as they are sold and used contain added ingredients, or adjuvants. These are often classified as confidential and described as "inerts." However, they help to stabilize the chemical compound glyphosate and help it to penetrate plants, in the manner of corrosive detergents. The formulated herbicides (including Roundup) can affect all living cells, especially human cells. This danger is overlooked because glyphosate and Roundup are treated as the same by industry and regulators on long-term studies. The supposed non-toxicity of glyphosate serves as a basis for the commercial release of Roundup. The health and environmental agencies and pesticide companies assess the long-term effects on mammals of glyphosate alone, and not the full formulation. The details of this regulatory assessment are kept confidential by companies like Monsanto and by health and environmental agencies.

This study demonstrates that all the glyphosate-based herbicides tested are more toxic than glyphosate alone, and explains why. Thus their regulatory assessments and the maximum residue levels authorized in the environment, food, and feed, are erroneous. A drink (such as tap water contaminated by Roundup residues) or a food made with a Roundup tolerant GMO (like a transgenic soybean or corn) was already demonstrated as toxic in the recent rat feeding study[1] from Prof. Séralini's team. The researchers have also published responses to critics of the study.[2] This new research explains and confirms the scientific results of the rat feeding study.

Overall, it is a great matter of concern for public health. First, all authorizations of Roundup-type herbicides have to be questioned urgently. Second, the regulatory assessment rules have to be fully revised. They should be analyzed in a transparent and contradictory manner by the scientific community. Agencies that give opinions to government authorities, in common with the pesticide companies, generally conclude safety. The agencies' opinions are wrong because they are made on the basis of lax assessments and much of the industry data is kept confidential, meaning that a full and transparent assessment cannot be carried out. These assessments are therefore neither neutral nor independent. They should, as a first step, make public on the Internet all the data that underpin the commercial release and positive opinions on the use of Roundup and similar products. The industry toxicological data must be legally made public.

Adjuvants of the POE-15 family (polyethoxylated tallowamine) have now been revealed as actively toxic to human cells, and must be regulated as such. The complete formulations must be tested in long-term toxicity studies and the results taken into account in regulatory assessments. The regulatory authorization process for pesticides released into the environment and sold in stores must be revised. Moreover, since the toxic confidential adjuvants are in general use in pesticide formulations, we fear according to these discoveries that the toxicity of all pesticides has been very significantly underestimated.

5

GM Alfalfa: An Uncalculated Risk

By Philip Bereano

Philip Bereano, JD, PhD, *is professor emeritus in the field of Technology and Public Policy at the University of Washington a cofounder of AGRA Watch, the Washington Biotechnology Action Council and the Council for Responsible Genetics. This article originally appeared in* GeneWatch, *volume 24, number 1, February–March 2011. Reproduced with the permission of the AGRA Watch project of the Community Alliance for Global Justice, Seattle (www.seattleglobaljustice.org/agrawatch).*

At the end of January, the Animal and Plant Health Inspection Service (APHIS) unit of the US Department of Agriculture announced that it would fully deregulate the planting of GE alfalfa, despite its Environmental Impact Statement (EIS) conclusion that such a course of action might lead to genetic contamination. To many observers, this appears to be in direct contravention to its obligations under law and court decisions.

In response to a law suit brought by the Center for Food Safety, a 2007 trial judge found that the Department had not done a proper EIS; included was a finding that alfalfa farmers had established a reasonable probability that their conventional alfalfa crops would be contaminated with the engineered Roundup Ready gene if USDA were allowed to fully deregulate GE alfalfa. The decision recognized that the substantial risk of such contamination was damage that would support a legal action, and the judge issued an order prohibiting the planting or deregulation of genetically engineered alfalfa. It directed the USDA to do a complete EIS and to adopt a course of action which would minimize injuries. The Supreme Court in 2010 overturned the planting ban, but did not restore the Department's approval of GE alfalfa; thus, planting was still not legally allowable.

Congress held two hearings in 2010, in which the agency was criticized for these failures, and over 200,000 citizen comments were filed, mostly objecting to the Department's plans. The Department's final environmental review, issued on Dec. 23, 2010, again concludes, in effect, that there will be no substantial harm from biotech alfalfa! As a part of this environmental impact analysis, the USDA proposed three options for action: 1) No deregulation of GE alfalfa; 2) complete deregulation of GE alfalfa; or 3) partial deregulation of alfalfa with certain government mandated measures to segregate GE production from organic production. Under the National Environmental Policy Act (NEPA), the public is granted a thirty-day period of public review, which ended in late January. It was clear that the Department would reject option (1), since no GE crops have been subjected to regulation.

Most critics believed that the USDA would seek "partial de-regulation," including mandatory conditions such as prohibiting the planting of GE alfalfa in certain parts of the country, and establishing buffer zones between GMO and organic crops, which would, in reality, still allow contamination. And contamination has long been an industry-government strategy for forcing acceptance of GMO. (As Emmy Simmons, assistant administrator of the US Agency for International Development, said to me after the cameras stopped rolling on a vigorous debate we had on South Africa TV in 2002, "In four years, enough GE crops will have been planted in South Africa that the pollen will have contaminated the entire continent.")

There is no such concept in US law as "partial de-regulation." Either the crop is regulated—according to an assessment in a full Environmental Impact Statement as ordered by the Court—or it is not. As the agency itself notes, "The supplemental request that APHIS received from Monsanto/KWS did not clearly explain what the petitioners mean or envision by a 'partial deregulation.'" In other words, this would have been a wholly ad-hoc and fictitious approach to fulfilling the agency's regulatory responsibilities.

"Partial de-regulation" is a faulty and misleading concept as regards the ecology of bioengineered plants. It does not prevent any of the potential harms from GE crops but seems to suggest that on a small enough scale they are tolerable. Tolerable to whom? Small environmental perturbations can lead to large impacts. In international meetings the US repeatedly says it supports "sound science" as the basis for regulation; the proceeding illustrates just how farcical such claims actually are, since there is no science supporting a notion of "partial de-regulation."

Apparently this "partial" proposal ran into considerable opposition in Congress and from some farm groups and biotechnology companies. Claiming that the introduction of restrictions based on economic consequences of pollen drift "politicizes the regulatory process and goes beyond your statutory authority," Representative Frank D. Lucas, Republican of Oklahoma, the new GOP chair of the House Agriculture Committee wrote to Secretary Vilsack on Jan. 19, and held a hearing on the proposals the next day. The letter was co-signed by Republican Senators Saxby Chambliss of Georgia and Pat Roberts of Kansas. Of course, an EIS is supposed to look at economic consequences of major federal actions, but it may be too much to expect these legislators to know what the law requires. Instead, they argued against any restrictions since the Department's environmental impact statement had concluded that growing GE alfalfa would be okay. They seem to have won, at least for now.

Restricting the growing of alfalfa would undermine Washington's repeated position at international meetings that GE is completely safe and would run counter to its efforts to pressure other countries to accept genetically modified crops. And the Obama White House is not going to alter its business-friendly policies—according to Maureen Dowd in *The New York Times* (Jan 30, 2011), chief advisor David Axelrod recently punned that everyone should "'plow forward' on a plan for genetically produced alfalfa."

The press has depicted the debate over GM alfalfa as biotech vs. organic but in reality organic is a small percentage of the alfalfa production segment that is threatened by the introduction of GM alfalfa (it is used as feed for cows whose milk is labeled "organic"). But conventional alfalfa is also threatened. Alfalfa seed companies have huge export markets to GMO-sensitive regions such as the EU, Middle East, and Asia, particularly Japan, and they don't want to jeopardize those markets. Two of the plaintiffs in the alfalfa suit are conventional alfalfa seed companies. Secretary Vilsack and the USDA have proposed a plan for the "coexistence" of GMOs alongside organic and conventional crops. Unfortunately this will result in genetic contamination.

The EIS inadequately assessed the likelihood of contamination injuries and the need for redress (if not, indeed, prevention). Over 200 past contamination episodes have cost farmers hundreds of millions of dollars in lost sales, not always compensated by crop developers. (It is interesting to note that the 160 members of the Cartagena Biosafety Protocol, which does not include the US, agreed a few months ago to a treaty to redress such damages if they occur internationally.)

The EIS misrepresented the situation regarding the inevitable increase in herbicidal chemicals, perhaps up to 23 million pounds per year. And it ignored the likely increase in herbicide-resistant "superweeds," already becoming an important US agricultural concern. It used a short-term and short-sighted approach.

Although the final EIS noted risks to organic and conventional farmers (concerns surrounding purity and access to non-GE seed), the decision still places the entire burden for preventing contamination on non-GE farmers, with no protections for food producers, consumers and exporters. The USDA must take a more proactive role to ensure that these risks are minimalized and that they are not thrown on innocent third parties. "We appreciate the measures that the Secretary has announced to explore ways to develop the science to protect organic and other non-GE alfalfa farmers from contamination. However, to institute these measures after the GE alfalfa is deregulated defies commonsense," said Michael Sligh, founding member of the National Organics Coalition. "Logically, efforts to develop the science of preventing GMO contamination should precede, not follow, any decision to deregulate GE crops."

The underlying problem in this proceeding is that APHIS refuses to follow full risk assessment procedures established for GE food plants, such as those specified in the UN's Codex Alimentarius (although the US was one of the 168 countries which approved their adoption at the meeting of the Codex Commission about a decade ago). Nor would it accord with the norms of the Cartagena Protocol on Biosafety (which the US has not adopted, but are followed by 170 other countries). Too often, APHIS has relied on information and analyses provided by the industry as a rubber stamp without any independent assessment of its own—an actual situation of industry "self-regulation" which has been repeatedly (in all areas of environmental and consumer concern) shown to be a farce.

There have been about 200 incidents of GE crops contaminating non-GE produce, resulting in hundreds of millions (if not billions) of dollars in damages; contamination is a real risk and one of very significant magnitude. Indeed, the trial court in the original lawsuit found that contamination by GE alfalfa has already occurred. Thus, the Department cannot dismiss it as insignificant or rest on Monsanto's assurances that its practices render contamination unlikely. (Monsanto's documented history of lying to governmental bodies and distorting evidence in submissions reduces its credibility to nil, anyway.) APHIS must surely be aware that the US government's definition of "organic" (by the

USDA) contains no threshold for the presence of GE contamination. More than a quarter of a million commentators vigorously objected to the original version of the rule which would have allowed GE components in "organic" foods. APHIS must proceed in a manner which guarantees that contamination will not occur, even if this means denying permission to plant GE alfalfa. Contamination by GE alfalfa violates the basic tort ideas of nuisance and trespass (although most farmers are not economically able to challenge a giant corporation such as Monsanto).

The Department suggests that consumers will forgive unintentional contamination, but intention is irrelevant to the National Organic Standards and to the protection of human health. Consumers have a legal right to demand that products live up to their labeling. Additionally, the claim that consumers will forgive unintentional contamination is unsubstantiated. Most surveys of US consumers indicate that they want to know that their food is free of any kind of contamination; further, most surveys point out the vast majority of US consumers do not want to have unlabeled GMO food in their grocery stores.

The Center for Food Safety is recommencing the litigation. Hopefully, the court will enjoin any planting this spring, so that contamination doesn't lead to a fait accompli and the insidious presence of more GMO in our food supply, untested for its effects on human health and the environment.

6

The Next Generation of Biohazard? Engineering Plants to Manufacture Pharmaceuticals

BY BRIAN TOKAR

Brian Tokar *is the director of the Institute for Social Ecology and a lecturer in environmental studies at the University of Vermont. He is the author of several books, including* Agriculture and Food in Crisis: Conflict, Resistance and Renewal (Monthly Review Press, 2010). *A version of this article originally appeared in* Green Social Thought *(formerly* Synthesis/Regeneration) *and in* GeneWatch, *volume 14, number 5, September 2001.*

With the worldwide rejection of genetically engineered foods, the biotechnology industry is scrambling to develop a new generation of products that might someday be seen as advantageous for consumers and beneficial to humanity. This is the primary motivation, of course, behind the massive PR campaign to sell the benefits of so-called "golden" vitamin A rice. Even though the claimed health benefits of this invention have been widely discredited—and activists in the global South have been in the forefront of pointing out that such inventions will do nothing to help people reclaim the ability to feed themselves—the mainstream press continues to tout this rice as evidence that biotechnology will someday feed the world. This is only the first of many new-generation biotech products designed to hide the industry behind a cloak of "humanitarian" concern.

The widely touted "next generation" of genetically engineered products is quite diverse in nature. There are salmon that can reportedly grow twice as fast as non-engineered varieties, with serious consequences for native ecosystems when these "superfish" escape from coastal fish farms. Also poplar, eucalyptus, and pine trees are being genetically engineered to grow faster and more uniformly, tolerate high doses of herbicides, and become more suitable for chemical processing into paper pulp. Here, the potential ecological consequences are magnified many-fold compared to the already well-known hazards of GE varieties of annual food crops, due to trees' longer lifespans, the more persistent spread of their pollen, and effects on countless other forest-dependent species. Researchers are even claiming to be ready to release genetically engineered insects on an experimental basis. Whether the insects have been engineered to administer vaccines, or are weakened strains intended to compete against pathological insects and crop pests, it is extremely unlikely that such creatures could ever be satisfactorily controlled. The potential problems are reminiscent of the genetically engineered Australian mice that ended up annihilating an entire population, rather than merely shrinking it by competing with them in reproductive ability.

Another relatively recent development is the engineering of plants to produce a variety of pharmaceuticals and industrial chemicals. Nearly everyone has read of efforts to engineer bananas that might someday be used to deliver vaccines. As with countless other applications of genetic engineering, the hype is far more convincing than the reality. Still, companies like Monsanto, DuPont, and Dow have been actively exploring experimental methods for producing vaccine components, human antibodies, and various industrially useful proteins in tobacco, corn, and potato plants. One company, the Texas-based ProdiGene, has been collaborating with Stauffer Seeds to produce eleven different proteins in genetically engineered plants on a commercial scale. Along with the efforts of companies like Massachusetts-based Genzyme to engineer animals as "bioreactors" for drug production, this represents a whole new sphere of biotechnology applications. The resulting health and environmental consequences of these new biotech creations could far exceed those of today's herbicide tolerant and Bt pesticidal crops.

On one level, the new "bioreactor crops" present many of the same potential environmental problems as other genetically engineered crop varieties, particularly if they are to be grown outdoors and on a large scale. Most noteworthy are problems of cross-pollination and unknown deleterious effects on beneficial insects, soil microbes, and other native organisms. But additionally, we may soon see biologically active enzymes and pharmaceuticals, usually only found in nature in minute quantities—and usually biochemically sequestered in very specialized regions of living tissues and cells—secreted by various and unpredictable plant tissues on a widespread commercial scale. The consequences may be even more difficult to detect and measure than those associated with more familiar GE crop varieties, and could escalate to the point where those now familiar problems would begin to pale by comparison.

There are also potentially severe public health consequences. As commercial grain distributors have proved unable to reliably sequester such a relatively well-characterized product as Aventis' Starlink corn, what steps could be taken to prevent the accidental commingling of crops engineered for chemical production into the rest of the food supply? British proponents of this technology have already proposed balancing the high cost of purifying specific proteins from plants with income obtained by extracting food products such as oils, starches and flours from these same crops. Anyone want some pharmaceutical residues or industrial enzymes in their corn flakes or taco shells?

Concerns about the public health and environmental consequences of these crops are exacerbated by their wide range of very high-level biological activities. Products being actively researched for plant-based production include blood coagulants, proteases and protease inhibitors, growth promoters, neurologically active proteins, and enzymes that modify the structure and function of other biologically important compounds, as well as monoclonal antibodies and viral surface proteins potentially useful for vaccination. Among the hazards of such activities is that large-scale releases of antibodies and viral antigens may trigger unexpected allergic or autoimmune reactions in some people.

Even the purported benefits of plant-produced vaccines are cast in doubt by the evidence. One problem is the well-documented phenomenon of oral tolerance: a concerted loss in vaccine efficacy that often follows the administration of antigens through mucous membranes. Hazardous chemicals, such as cholera toxin, are often needed as cofactors to increase the effectiveness of oral vaccines. Even the proponents of this technology have cited the contamination of pharmaceuticals with pesticide residues as a significant problem.

The active collaboration between ProdiGene and Stauffer Seeds has already brought several products of this technology to market in the US, all of which serve to highlight the potential hazards of plants engineered to produce commercial proteins. Stauffer is actively contracting with farmers to grow corn containing the genes involved in the formation of several specific enzymes, three vaccines, a protein-based sweetener, a proprietary "Therapeutic Agent," and two other biologically active chemicals. Three of its products, avidin, beta-glucuronidase, and aprotinin (a protease inhibitor commonly used by surgeons), have been produced in sufficient quantities to be sold through a commercial chemical supplier, the St. Louis-based Sigma Chemical Company.

Avidin is a protein that naturally occurs in raw egg whites. While Sigma markets it for use in medical diagnostic kits, it is also used as an insect growth inhibitor and is being investigated as a next-generation biopesticide. Avidin binds to biotin, an important B-vitamin, and prevents its absorption across the intestinal mucosa. It is known to cause a type of vitamin B deficiency in some people who consume raw egg whites.

There are contradictory reports as to whether beta-glucuronidase is still being produced by Stauffer from plant "bioreactors," but it appears to have been available in this form for a number of years. This enzyme reverses a biochemical reaction that helps render irritant molecules soluble. This added solubility helps to facilitate the detoxification and elimination of compounds as diverse as some hormones, antibiotics, and opiates. In the presence of this enzyme, potential toxins are freed from the molecular complex that facilitates their proper excretion. One can only speculate on the consequences when elevated levels of such compounds are being released into the environment.

Stauffer's professed goal is to maximize distribution of these and other compounds via both foreign and domestic production of transgenic corn, allowing for three growing cycles per year. According to Stauffer's web site, production is currently under way in South America, the South Pacific, and the Caribbean, as well as within the continental US. As South America is the center of biodiversity for maize, the potential for widespread disruptions of indigenous relatives may be quite severe.

Other companies at the forefront of turning plants into chemical factories include Virginia-based 'CropTech, which has produced pharmaceuticals and human enzymes in tobacco, with several products already in clinical trials. The San Diego-based EPIcyte has partnered with Dow Chemical to develop and produce experimental human antibodies in plants, as well as a topical contraceptive and a microbicide that is purported to be active against HIV. Monsanto's Integrated Protein Technologies subsidiary is seeking contracts with a number of clients to produce commercial quantities of various proteins in corn, tobacco, and soybean plants. Monsanto claims to be able to produce several metric tons of any appropriate protein within a three-year period. Several other companies in the US, Canada, and France are also actively exploring these techniques.

The promises offered by many of these companies may seem impressive in that they suggest that therapeutically useful agents could become more widely available at considerably lower cost. If the plants are grown solely in isolated greenhouses, with pollen entirely contained, and byproducts of this process completely isolated from the food supply, the advantages might someday outweigh the hazards in several instances. But, according to Carole Cramer, the founder of CropTech, some products under consideration would require thousands or even hundreds of thousands of acres planted at densities (in the case of transgenic tobacco) of 50,000 to 100,000 plants per acre in order to supply the current market for these proteins. Given the recent track record of the biotechnology industry in aggressively promoting its products at all costs, while denying all potential hazards and refusing to sequester potentially harmful crops, the likelihood of the best-case scenario coming to pass appears extremely slim. If this technology proceeds unchecked, we will see the contamination of the world's food supply with an expanding and increasingly hazardous array of new biotech products.

7

Busting the Big GMO Myths

By John Fagan, Michael Antoniou, and Claire Robinson

John Fagan, PhD, *is founder and, until November 2013, was Chief Scientific Officer of Global ID Group, through which he pioneered the development of innovative tools to verify food purity, quality, and sustainability. These tools include DNA tests for genetically engineered foods, the first certification program for non-GMO foods, and a leading program dedicated to certifying corporate social and environmental responsibility in the food and agricultural sectors. Dr. Fagan currently works primarily through Earth Open Source, a non-profit that focuses on science-policy issues related to GMOs, pesticides, and sustainability in the food and agricultural system and on rural development. Dr. Fagan holds a PhD in molecular biology, biochemistry, and cell biology, from Cornell University.* **Michael Antoniou, PhD,** *is reader in molecular genetics and head of the Gene Expression and Therapy Group at the King's College London School of Medicine. He has twenty-eight years' experience in the use of genetic engineering technology investigating gene organization and control, with more than forty peer-reviewed publications of original work, and holds inventor status on a number of gene expression biotechnology patents. Dr. Antoniou has a large network of collaborators in industry and academia who are making use of his discoveries in gene control mechanisms for the production of research, diagnostic and therapeutic products, and human somatic gene therapies for inherited and acquired genetic disorders.* **Claire Robinson, MPhil,** *is research director at Earth Open Source. She has a background in investigative reporting and the communication of topics relating to public health, science and policy, and the environment. She is an editor at GMWatch (www.gmwatch.org), a public information service on issues relating to genetic modification, and was formerly managing editor at SpinProfiles (now Powerbase).*

The process of doing the research for the book, *GMO Myths and Truths*[1], gave us the opportunity to step back and see the dynamics of the GMO crop and food debate from a broader perspective. It became clear that in addition to the specific myths that we had identified, the chemical-biotech industry was broadcasting certain over-arching messages that were, perhaps, even more important to understand and debunk. This chapter attempts to deal with these systematically by summarizing and discussing the major myths promulgated by GMO proponents.

The biggest myth of all is that GMOs have swept the world and dominate agriculture as it is practiced today. The corollary of this myth is that if your country doesn't accept GMOs it will fall behind. GMO proponents proudly announce that there are seventeen million farmers growing GMOs in twenty-eight countries around the world, and that GMO crop production has increased a hundred-fold since 1996.

On the surface these statistics sound impressive, but a deeper examination reveals that they amount to a story of monumental failure. Despite gargantuan marketing and lobbying efforts costing billions of dollars and reaching every corner of the world during the last twenty years, GMOs have failed to penetrate the world agricultural market significantly.

There are roughly thirty-five crops that feed humanity. If GMOs were a great success, we would expect that most of these would be genetically engineered by now. But in fact only four—soy, corn, canola, and cotton—represent 95 percent of all genetically modified crops commercialized around the world today.

That GMOs are grown in twenty-eight countries sounds extraordinary until one discovers that more than 90 percent of all GMOs are grown in only five of those twenty-eight countries. Another five grow 8 percent, meaning that each grows only 1.6 percent of the world's GMOs on average. The other eighteen countries grow only token amounts of GMOs, on average only 0.11 percent of the global total. Finally, don't forget that there are 195 countries in all. The vast majority—167—have not accepted GMOs at all.

What about those seventeen million farmers? They represent only 1.7 percent of the world's farmers (1 billion[2]). In fact, the actual number of GMO farmers is probably closer to 3.8 million[3], or only 0.38 percent of global farmers. Finally, a hundred-fold increase in the production since 1996 is not so impressive when one realizes how tiny the production was then. GMOs were a tiny, tiny drop in the bucket in 1996, and today they are still just a drop.

The second big myth about GMOs is that genetic engineering of crops is the most advanced agricultural technology ever invented, that it is the cutting edge of agricultural and breeding technology, that it offers tremendous power and promise for improving agricultural performance around the world, and that it will help stem the tide of world poverty. In fact, the main reason that GMOs have failed to penetrate the market and have failed to sweep the world is because this technology is not powerful, it is not successful, and it is not at the cutting edge of agricultural technology today. In fact, quite the opposite is true. Genetic engineering of crops is a cumbersome, inflexible, outmoded, and old-fashioned technology. It is imprecise and poorly controlled; it is like trying to do heart surgery with a shovel. As a result, it is simply not up to the task of creating the kinds of crops that GMO proponents have been boldly promising for twenty years.

A third myth about GMOs is that they are perfectly safe for human consumption and for the environment. Twenty years ago, in the early days of the debate about genetically engineered crops, we could only say that there were *potential* risks. Back then, we would point out that the imprecision and lack of control in the process had the potential to create three basic classes of hazards for human safety. First, GMOs could be allergenic; genetic modification could alter foods so that they cause allergic reactions in consumers. Second, GMOs could be toxic; they could cause harm to specific cellular or organ targets. Third, genetic modifications could reduce the nutritional value of a food.

Today there is strong scientific evidence that each of these potential hazards is real. There are peer-reviewed papers published in the scientific literature documenting that certain GMOs are, in fact, allergenic. Similarly, the toxicity of certain GMOs and the reduced nutritional value of other GMOs have been scientifically demonstrated. More and more evidence is accumulating, showing that GMOs can be harmful to health and the environment, as we will see below. But GMO proponents continue to perpetuate the myth that the genetic engineering process is precise and scientifically controlled, and therefore that genetic engineers can be trusted to consistently deliver GMOs that are safe for human consumption and the environment.

Harm to Health: The Séralini Study

One of the most important papers showing that GMOs harm health was published in 2012 by Gilles-Eric Séralini et al.[4] The focus of this study was a genetically modified corn variety called NK603, which is engineered to survive application of the weedkiller Roundup.

These researchers found that the NK603 corn, either with or without Roundup, and Roundup alone, all resulted in toxicity to multiple organs, leading to severe liver and kidney damage and premature death especially in males. Unexpectedly, they also found an increase in tumor incidence, especially mammary (breast) tumors in females in all treatment groups, which appeared earlier in life and grew more rapidly.

This was a very significant study, because it used quite low levels of GM corn in the diet, and very low levels of the herbicide Roundup, which is always used in conjunction with NK603 because this GM corn variety has been engineered to be resistant to this particular herbicide. Thus, this study was done under conditions of exposure that would be expected, for instance, in livestock that are fed NK603 corn as part of their diet, and even at these low doses, liver and kidney toxicity were observed, as were increased tumor incidence.

This study also highlights the inadequacy of the safety assessment regimes for GMOs. Assessment must be inadequate, because, somehow, this corn variety, which Séralini demonstrated to be hazardous, slipped through the GMO approval procedures in the US, Canada, Europe, and many countries around the world.

How could this happen? The answer is that a safety assessment procedure can only detect the hazards that it is designed to look for, and the biotech industry has lobbied regulators to put in place procedures that, in fact, are fundamentally flawed. They are designed to speed GMOs to market, not to protect the public. As a result, genuine hazards are overlooked.

This study is important not only because it exposes the dangers linked to this GMO, which is widely cultivated in both North and South America, and is used for animal feed and enters processed human food internationally, but also because it highlights the fact that the regulatory oversight and safety assessment of GMOs are inadequate in the US, Canada, Europe, and literally every country around the world. Oversight must be inadequate, because, somehow, inexplicably, this corn variety, which Séralini demonstrates to cause serious kidney and liver toxicity, slipped through the GMO approval processes that have been operating in all of these countries.

In safety assessment, the devil is in the details. For instance, Séralini's study used a protocol that was quite similar to the one that Monsanto used to demonstrate to regulators that NK603 was safe. The main difference between the studies was that Monsanto ended its rat feeding study after ninety days—a little less than three months—whereas Séralini extended the study to two years—the lifetime of a rat. What he found was that tumors and toxicity only began to show up after four months—conveniently too late to be detected in Monsanto's study, since it was terminated after three months.

Given that consumers will be exposed to GMOs during their entire lifetimes it is only logical that long-term studies, like Séralini's, need to be done with all GMOs in order to accurately assess safety.

Needless to say, Séralini's study was heavily attacked by the chemical-biotech multinationals that create and sell GMOs, because it threatens their financial interests. After fourteen months of heavy attacks, the journal that published the article crumbled and retracted the article. This drew a huge amount of criticism from the scientific community.[5] Predictably, the withdrawal of the article was used widely by GMO proponents to further dismiss Séralini's research, but the reality is that the study is sound. It clearly and definitively demonstrates that NK603 and very low doses of Roundup can be harmful to organisms that consume them.

The normal way that the scientific process works is that if a scientist doubts the results published by others, he does another research study and reports his results, which add to the body of scientific understanding of the phenomena of interest. This is not how the chemical-agricultural giants operate. Instead of doing more research to verify or disprove the findings of Séralini, they did "science by press release." They used their huge financial resources to mount a media campaign attacking not only the research but also Séralini personally. These *ad hominem* attacks are standard practice among the GMO proponents and are a serious deterrent to any scientist who considers entering this area of research.

Séralini's is one of many studies that casts doubt on the safety of GMOs. The evidence is strong and more studies are coming out regularly. Séralini's study is particularly informative, because it is in-depth and long-term and illustrates the ends to which GMO proponents will go to protect their financial interests.

Examples of Environmental Harm

There are also environmental hazards related to GMOs. The first is that use of GMO seeds results, long-term, in increased use of toxic carcinogenic and teratogenic (birth-defect-causing) compounds, such as Roundup. This leads to pollution of water, air, and soil and increased incidence of cancer, birth defects, and other diseases. This has been observed in countries such as Argentina, where increases in the cultivation of Roundup Ready genetically engineered soy have been reported to be paralleled by increases in incidence of both cancer (a 300 percent increase) and birth defects (a 400 percent increase).[6]

Super pests are also a problem in the case of GMO cotton in the US, China, Australia, and Spain, where the boll worm has developed resistance to Bt toxins.[7] Boll worm resistance has also become a significant problem in India.[8] Secondary pests are another problem that has been encountered around the world. When first introduced, GM cotton performs well, controlling the predominant pest, the boll worm. But nature abhors a vacuum and as a result, within two to three years secondary pests, such as mealy bugs, mirids, and thrips, enter to fill the vacuum left by the boll worm.

Feeding the World

What about the claim that GMOs increase crop productivity and that the increases they offer are essential to feeding the world's growing population? Proponents claim that genetically engineered crops promise better yields, as well as all of the other promises that we debunked earlier. They say it is the only way to feed the nine billion people of the future, and that rejecting GMOs is rejecting the one technology that will save the starving billions around the world. Rejection is therefore not only anti-scientific, GMO-proponents argue, but also cruel and inhuman because it deprives the hungry and destitute.

This argument is gripping emotionally, but when we look more deeply it becomes apparent that the proponents of biotechnology themselves are a bit cruel; they are cynically exploiting the circumstances of the poor and hungry to promote themselves. This is well expressed in the following statement signed by twenty-four delegates from eighteen African countries to the United Nations Food and Agricultural Organization, 1998:

> We strongly object that the image of the poor and hungry from our countries is being used by giant multinational corporations to push a technology that is neither safe, environmentally friendly nor economically beneficial to us. We do not believe that such companies or gene technologies will help our farmers to produce the food that is needed in the twenty-first century. On the contrary, we think it

> will destroy the diversity, the local knowledge and the sustainable agricultural systems that our farmers have developed for millennia, and that it will thus undermine our capacity to feed ourselves.

It is widely recognized that hunger is not a technical problem that can be solved by increasing production. Production is not the problem. There is plenty of food in the world for everyone today. Hunger is not a production problem; it's a problem of lack of access to food or the resources for producing food. Hunger is fundamentally a social and political problem, not a technical production problem.

The problem of global hunger must be solved in the hearts of humankind, not in laboratories or agricultural fields. It is from there that the solution must come, not from improved production methods.

Even if hunger were a production problem, genetic engineering is not capable of solving that problem. Genetic engineers have failed to create GMO crops with increased intrinsic yields or improved nutrition. Increasing productivity through GMOs is only a promise, and it is an old, re-heated, leftover promise that the GMO proponents have failed to keep and have used far too often in the last twenty years to justify their existence while profiting handsomely on herbicide-resistant crops that amplify the sales of their chemicals. As we saw earlier, these promises are inevitably destined to fail because of the technological limitations of genetic engineering.

Let us emphasize: To date there is not a single GMO on the market that delivers consistently and significantly higher intrinsic yields than the natural crops from which they are derived. In his report, "Failure to Yield," Doug Gurian-Sherman states, "Commercial GE crops have made no inroads into raising the intrinsic or potential yield of any crop. By contrast traditional breeding has been spectacularly successful in this regard; it can be solely credited with the intrinsic yield increases in the United States and other parts of the world that have characterized the agriculture of the twentieth century."[9]

If Not GMOs, Then What?

In closing we can ask the question, if not GMOs, then what? We are not condemning biotechnology or science, we are only questioning the use of genetically modified seeds. GMOs are just one of many biotechnologies. Many dimensions of biotechnology are very safe, very useful, and fully compatible with agro-ecology. These "good" biotechnologies offer great promise for sustainable agriculture.

Examples of safe and beneficial biotechnologies and beneficial uses of advanced genetics and cell biology include Marker Assisted Selection (MAS), which makes use of the most advanced knowledge of crop genetics to guide the process of conventional breeding. MAS enables breeders to rapidly and effectively create crop varieties that deliver high quality, high yields, and resistance to pests and stress. Cell culture and tissue culture are also powerful biotechnologies. These are tools that are already widely used. The food system can benefit greatly from using these tools even more broadly to propagate elite crop varieties that have highest yields, greatest resistance to pests, highest nutritional content, greatest resistance to drought, heat stress and a range of other problems.

It should also be pointed out that genetic engineering itself is a powerful and extremely useful research tool. For example, it can be used to identify genetic markers that can then be used in MAS to improve crops. It can also be used to identify specific genes that have useful functions. Snorkel rice[10] is a good example of how genetic engineering can be used to great benefit as a research tool. The snorkel rice story has demonstrated the limitations of genetic engineering, by showing how genetic engineering was a very useful research tool for helping identify potentially useful genes, but that the engineered versions of those genes were unable, themselves, to be used to create a flood-resistant rice variety. The researchers had to resort to natural breeding augmented with MAS to achieve their goal.

There are many examples of "good" biotechnology; both "high-tech" and "low-tech". Bread, beer, and wine fermentation are examples, as is cheese-making. Use of bio-control agents to deal with agricultural pests, and composting techniques to generate renewable fertilizers are other examples of traditional biotechnologies. Even agriculture, itself, can be considered a biotechnology. These "good" biotechnologies have several characteristics in common. First, these methods are adaptable and therefore can be made compatible with the geography, climate, and culture where they're implemented. They employ and integrate local knowledge. They foster crop diversity, use inter-cropping, use indigenous crop varieties that are naturally adapted to local environments and employ natural indigenous renewable pest management strategies. They use bio-control strategies. They are low input, use renewable inputs, and often those inputs are produced right on the farm. Good biotech conserves energy, and uses renewable energy; it conserves and builds the soil and conserves water. Good biotech uses region-adapted genetics, enhances natural pest resistance and enhances natural resilience of crops.

There is a lot of research demonstrating the advantages of the agro-ecological approach in contrast to the biotechnological approach. For instance, the International Assessment of Agricultural Knowledge Science and Technology for Development (IAASTD) was a four-year study conducted by four hundred scientists from eighty countries around the world.[11] The final, extensive report was peer-reviewed by hundreds of scientists and was finally endorsed by sixty-two governments. This study concluded that the key to food security lay in agro-ecological farming methods. It specifically did not endorse genetic engineering or GMO crops, noting that yields were variable and that better solutions were available.

There are scores of peer-reviewed research reports describing dramatic increases in yields and improvements in food security in many countries around the world resulting from the application of agro-ecological and organic agricultural methods. For instance, a review published by the UN in 2008 described 114 farming projects in twenty-four African countries in which organic practices resulted in yield increases averaging more than 100 percent. The report concluded that organic agriculture can be more conducive to food security in Africa than chemically-based production systems, and that it is more likely to be sustainable in the long term.[12]

Oliver De Schutter, the UN special rapporteur on the right to food, summarizes other studies, saying, "yields went up 214 percent in forty-four projects in twenty countries in sub-Saharan Africa using agro-ecological farming techniques over a period of three to ten years, far more improvement than any GM crop has ever done."[13] In the same article he also says "agroecology mimics nature, not industrial processes, it replaces the external inputs like fertilizer with knowledge of how a combination of plants, trees and animals can enhance productivity of the land."[14]

The story of GMOs is the story of world agriculture seduced into a twenty-five-year, multi-billion dollar wild goose chase. Agriculture needs to get back to practicalities and come to grips with the realities of humanity's food system. We have been worrying too much about corporate profits and too little about the hungry bottom billion of humanity. When we get back to practicalities, we will find much more profitable and beneficial ways forward. We need to adapt approaches that have integrity, that are good for all of humanity, not just for that small subset of humanity that has invested in the chemical-biotech giants.

These sustainable approaches are in easy reach. We just need to employ them. Humanity is not dependent on genetic engineering or GMOs or on any company's patented technology to secure its future. Mother Nature has what we need, and we have the freedom and ability to choose the better approach. What is key to humanity's future is people who are dedicated, not to a technology or to a short-term, personal objective like improving the profits of the company they work for, but to creating a better world, a world of abundance, harmony, and freedom for all. This is more than a better food system, but creating a better food system will take us a long way toward that bigger goal.

PART 2

Labeling and Consumer Activism

The US government does not require that genetically engineered foods be labeled as such unless it determines that the genetically engineered food is no longer "substantially equivalent" to the unmodified version. This is a vague and misleading principle that encompasses most genetically engineered foods on the market today.

In light of the uncertainty regarding the safety and environmental impact of genetically engineered foods, many consumers want to take a precautionary approach. They have demanded mandatory labeling of foods produced by or containing genetically engineered organisms (GMOs).

It's a demand with much precedent. The US already allows "process" labels on other products. Kosher foods, for example, are equivalent in nutritional value and taste to non-Kosher foods. The Kosher label refers to the process by which livestock is slaughtered or foods are prepared. Similarly, "dolphin-safe" tuna is equivalent in nutritional value and taste to other types of tuna. The "dolphin-safe" process label indicates that special nets have been used that do not entrap dolphins.

Corporations promoting GMOs have spent untold millions of dollars promoting their products and suing their detractors. And yet, dedicated activists for consumer and environmental rights have continued the fight to educate Americans and promote regulations and legislation to govern GMOs. Until such time as the legal, regulatory, and ethical structures are put in place to more adequately deal with the implications of genetically engineered food, these activists are seeking to reassign the risks of GMOs from consumers to the industry that produces them.

The following essays detail their struggle.

—Jeremy Gruber

8

Codex Food Labeling Committee Debates International Guidelines

By Diane McCrea

Diane McCrea *is a London-based consultant who is particularly interested in the issues of labeling and consumer choice. She serves as spokesperson for Consumers International at Codex Alimentarius food-labeling committee meetings. This article originally appeared in* GeneWatch, *volume 11, numbers 1–2, April 1998.*

An ongoing debate among representatives to a little-known United Nations committee could have far-reaching impacts on the choices shoppers get to make around the world.

The Codex Committee on Food Labeling (CCFL), which comes under the Codex Alimentarius Commission, will meet next in Ottawa, Canada, at the end of May to continue discussion of international labeling guidelines for genetically modified foods.

The Codex Alimentarius Commission, referred to as Codex, was established in 1962 by the Food and Agriculture Organization of the United Nations and the World Health Organization. Codex is designated by the World Trade Organization as the rule-making body for international trade issues related to food. Under the General Agreement on Trade and Tariffs, Codex decisions carry the weight of international law. Codex Alimentarius means "food code" in Latin.

The mission of the Codex Alimentarius Commission, as spelled out in its procedural manual, is to "guide and promote the elaboration and establishment of definitions and requirements for foods, to assist in their harmonization and, in doing so, to facilitate international trade." Although Codex is also mandated to come up with "requirements for food aimed at ensuring the consumer a sound, wholesome product free from adulteration, correctly labeled and presented," the unfortunate reality is a commission dominated by industry.

Governments elect to become members of Codex, and only government representatives can vote at official proceedings. Each member nation has one vote, regardless of size or population. Selected others are invited to participate in Codex proceedings and attend meetings as "observers." Those granted observer status are generally international nongovernmental organizations (NGOs).

A fundamental concern is the balance of representation between industry-funded groups and the public interest sector. This representation is dramatically skewed to those with the financial resources to send delegations to meetings. The current list of 111 approved organizations with "observer" status stands at 104 industry-funded groups, six health and nutrition foundations, and one broad-based global consumer group—Consumers International.

A 1993 request from the Codex Commission stated that the CCFL "should provide guidance on the possibilities to inform the consumer that a food had been produced through modern biotechnology." Consumers International believes that the precautionary principle should be paramount when it comes to food safety concerns, but Codex standards are too narrow.

At the committee's 1997 meeting the Codex secretariat presented a paper with recommendations for the labeling of genetically modified foods. Based on the principle of "substantial equivalence" (the same principle currently guiding US government labeling policy), the paper called for a label only if a genetically modified food is no longer substantially equivalent to the unmodified version of the food as regards composition, nutritional value or intended use. In other words, labels would be required on only a fraction of the many genetically engineered foods entering the market.

Consumers International, along with Norway, lobbied for broadened criteria to take full account of consumer needs. In addition to and aside from science-based safety concerns, consumers have the absolute right to know whether their food has been genetically modified. For example, many consumers have serious ethical and environmental concerns about these new foods. There are certainly valid concerns about the short-term and long-term health effects of genetically modified foods, but human health is not the only issue.

Several other nations also expressed doubts about the secretariat's recommendations, and further consultations and negotiations will ensue. Given the slow pace of Codex procedures and the desire of committee members to make decisions by consensus, it may be another ten years before any agreement is reached on the labeling issue. Of course this works to the advantage of the biotech industry; the lack of international regulations allows genetically modified foods to continue entering the market unlabeled.

Consumers need to become more informed about Codex and the way in which it decides on international food standards. A public debate on Codex and the way in which it sets standards is long overdue. While ensuring fair trade, Codex needs to acknowledge its role and duty to protect the health of consumers and at the same time meet their other needs and concerns.

Consumers Call on FDA to Label GMO Foods

By Colin O'Neil

Colin O'Neil *is director of government affairs at the Center for Food Safety. This article originally appeared in* GeneWatch, *volume 24, number 6, October–November 2011.*

Americans cherish their freedom of choice. If you want to choose food that doesn't contain gluten, aspartame, high fructose corn syrup, transfats or MSG, you simply read the ingredients label. But one choice Americans are not free to make is whether their food contains genetically engineered ingredients. Unlike most other developed countries—including fifteen European Union nations, Japan, Australia, Brazil, Russia and China—the US has no laws requiring the labeling of genetically engineered foods. Yet polls have repeatedly shown that the overwhelming majority of Americans—more than 90 percent in most polls—believe GE foods should be labeled.

Citing this overwhelming support, last month the Center for Food Safety (CFS) filed a groundbreaking legal petition with the US Food and Drug Administration (FDA) demanding that the agency require the labeling of all food produced using genetic engineering. The CFS prepared the legal action on behalf of the "Just Label It" campaign and a number of health, consumer, environmental and farming organizations, and food companies are also signatories to the petition.

A Choice Deferred

In 1992, the FDA issued a policy statement that GE foods were not "materially" different from non-GE foods and thus did not need to be labeled. The agency severely constricted what it called "material," limiting it to changes that could be tasted, smelled, or detected through the other human senses. Because GE foods cannot be "sensed" in this way, the FDA declared them to be "substantially equivalent" to conventionally produced foods, and no labeling was required.

The FDA adopted this stance despite a lack of scientific studies and data to support its underlying assumption that genetically engineered foods were "substantially equivalent" to conventional foods. It was a political, not scientific, decision to apply nineteenth-century logic to a twentieth-century food technology, and in the process left all consumers in the dark to hidden changes to their food. We as consumers no longer base our decisions solely on what we can see or taste or smell, so why should the FDA continue to do so?

The FDA Should Prevent Consumer Deception, Not Create It

The FDA's authority to require labeling goes well beyond the agency's outdated definition of "material." Rather, the law authorizes the FDA to require labeling for GE foods in order to prevent consumer deception. Because the FDA allows these facts to go unlabeled, consumers believe they are purchasing something different than what they actually are.

To be patentable, a genetically engineered food must be "new" and "novel." Thus, a product or process that is patentable cannot be both "novel" for patent purposes yet "substantially equivalent" to existing technology in other contexts. Continuing to treat GE foods as novel for patenting purposes but traditional for labeling purposes is a clear error in judgment by the FDA and abuse of the public's trust.

Polls consistently show[1] that more than 90 percent of Americans want GE foods to be labeled and consumers do not expect food to be genetically engineered absent labeling. The FDA's continued failure to mandate labeling is an abdication of its duty to protect consumers from deception.

Unlabeled, Untested, and You're Eating It

Unlabeled GE foods are misleading not only because they contain unperceivable genetic and molecular changes to food, but also because they have unknown and undisclosed risks. The FDA has never conducted a single safety assessment for GE foods and does not affirm their safety. There have been very few independent, peer-reviewed, comprehensive studies of their long-term human health and environmental impacts, and the few that exist give cause for concern. In fact, scientists both within the FDA and outside the agency agreed that there are profound differences between genetically engineered foods and those produced through traditional breeding practices.

Yet, rather than requiring the necessary safety assessment, the FDA explicitly places responsibility for determining the safety of GE foods and crops back in the hands of their makers the biotechnology companies, and uses what it calls a "voluntary consultation" process. Companies that develop a GE crop are encouraged, but not required, to share the conclusions (but not the raw data or methodology) of any studies they may have conducted on their GE crop. This system does not favor health, safety or transparency.

A recent independent Canadian study found that a toxin from the soil bacterium *Bacillus thuringiensis* (Bt), which has been engineered into Bt corn, was present in the bloodstream of 93 percent of pregnant women, as well as in the fetal cord blood of 80 percent of the pregnant women.[2] These findings cast grave doubt on the biotechnology industry's assurances—accepted at face value by federal agencies, including FDA—that the genetically engineered Bt toxin would be broken down by human digestive systems before entering the bloodstream. This study not only underscores the scientific uncertainty surrounding the health impacts of GE crops, but also casts doubt on the wisdom of federal agencies' practice of relying excessively on crop developers' own safety assessments rather than on independent studies.

The FDA's Looming Decision on GE Salmon Labeling

One issue related to GE food labeling currently sitting at the FDA is the pending approval of AquAdvantage transgenic salmon, the first GE animal intended for human consumption. The genetically engineered Atlantic salmon was developed by AquaBounty Technologies, which artificially combined growth hormone genes from an unrelated Pacific salmon (*Oncorhynchus tshawytscha*) with DNA from the anti-freeze genes of an eelpout (*Zoarces americanus*).

This genetic modification causes the AquAdvantage salmon to produce growth hormone year-round, creating a fish the company claims grows at twice the normal rate. This GE salmon poses a number of health, environmental, economic, and animal welfare concerns and is only made worse by the FDA's acknowledgment that it would likely not require labeling despite these concerns. Yet a careful look at this fish reveals that it is not the safe and healthy fish that its proponents would lead you to believe.

According to the company's own data, its GE salmon contains less healthy fatty acids than other farmed salmon and far less healthy fatty acids than wild salmon. FDA claims that the omega-3 to omega-6 fatty acid ratio in the AquAdvantage salmon is similar to the ratios found in scientific literature for farmed Atlantic salmon. In fact, the ratio for the AquAdvantage salmon is nearly 15 percent less than the recorded ratio for conventionally farmed Atlantic salmon and 65 percent less than wild salmon.

GE salmon also contain levels of healthy vitamins and minerals inferior to the levels present in other farmed salmon. The company study provided to the FDA identified a number of vitamins and essential nutrients for which the levels present in the AquAdvantage salmon differed from non-GE salmon by more than 10 percent. The AquAdvantage salmon has lower levels of every essential amino acid tested and nearly 25 percent less folic acid and vitamin C. As a result of the genetic modification, this fish is fattier, less nutritious, and at higher risk for physical deformities than other salmon.

With regard to food allergies, the FDA stated: "the technical flaws in this [AquaBounty's allergy] study so limit its interpretation that we cannot rely on its results." It's no wonder a 2008 Consumers Union nationwide poll found that 95 percent of respondents said they thought food from genetically engineered animals should be labeled.[3] And while you're not going to see this type of comparison on a nutrition label, absent mandatory labeling for GE foods, you will not be able to choose between a GE fish and regular farmed salmon.

In the US, we pride ourselves on having choices and making informed decisions. The longer the US clings to its antiquated policy on GE food labeling, the more its standing as a leader in scientific integrity will be compromised. It is long overdue that FDA acknowledges the myriad reasons GE foods should be labeled and rewrites its outdated policy, lest it continue to foster consumer deception.

10

Genetically Engineered Foods: A Right to Know What You Eat

By Phil Bereano

Philip L. Bereano, JD, PhD, *is professor emeritus in the field of Technology and Public Policy at the University of Washington, a cofounder of AGRA Watch, the Washington Biotechnology Action Council, and the Council for Responsible Genetics. This article originally appeared in* GeneWatch, *volume 12, number 1, February 1999.*

Genetic engineering consists of a set of new techniques for altering the basic makeup of plants and animals. Genes from insects, animals, and humans have been added to crop plants; human genes have been added to pigs and cattle. Most genetic engineering is designed to meet corporate—rather than consumer—needs. Foods are engineered, for instance, to produce "counterfeit freshness." Consumers believe that the engineered physical characteristics, such as color and texture, indicate freshness, flavor, and nutritional quality. Actually, the produce is aging and growing stale, and nutritional value is being depleted. Genetic engineering techniques are biologically novel, but the industry is so eager to achieve financial success that it argues that the products of the technologies are the same, or "substantially equivalent" to normal crops. Despite the gene tinkering, the new products are not being tested extensively to find out how they differ from normal food crops, and whether they present unacceptable hazards.

Genetically engineered foods are now appearing in the supermarkets and on our dinner plates, despite consumer attempts to label these "novel foods" in order to distinguish them from more traditional ones.

The failure of the US government to require that genetically engineered foods be labeled presents consumers with a number of quandaries: issues of free speech and consumers' right to know, religious rights for those with dietary restrictions, and cultural rights for people, such as vegetarians, who choose to avoid consuming foods of uncertain origins. Some genetic recombinations can lead to allergic or auto-immune reactions. The products of some genes which are used as plant pesticides have been implicated in skin diseases in farm and food market workers.

Product labels perform an important social function, namely communication between a seller and a would-be buyer. The struggle over labeling is occurring because industry knows that consumers do not want to eat GEFs, and that labeled products will likely fail in the marketplace. However, as the British publication *The Economist* noted, "if Monsanto can not persuade us it certainly has no right to foist its products on us." Labels would counter "foisting" and are legally justifiable.

The Government's Rationale

In 1992, the government abdicated any supervision over GEFs. Under the Food and Drug Administration's rules, the agency does not even have access to any industry information about a GEF unless the company decides voluntarily to submit it. Moreover, important information bearing on questions of assessment of risks is often withheld from the public as being proprietary, "confidential business information." So "safety" cannot be judged in a precautionary way; we must await the inevitable hazardous event.

James Maryanski, the biotechnology coordinator of the FDA, claims that whether a food has been genetically engineered is not a "material fact" and the agency would not "require things to be on the label just because a consumer might want to know them."

Yet a standard law dictionary defines "material" as "important," "going to the merits," and "relevant." Since labeling is a form of speech from growers and processors to purchasers, it is reasonable, therefore, to interpret "material" as including whatever issues a substantial portion of the consuming public defines as "important."

Consumers' Point of View

Numerous public opinion polls, in the US and abroad over the past decade, have shown great skepticism about GEFs; a large proportion of respondents, usually majorities, are reluctant to purchase and consume such products. Regardless of whether they would consume GEFs, consumers feel even more strongly that such foods should be labeled. In a 1998 *Toronto Star* poll reported on June 2,

98 percent favored labeling. Bioindustry giant Novartis surveyed US consumers and found that 93 percent of them wanted information about genetic engineering of their food products.

However, the US government has been resisting attempts to label genetically engineered food. Despite the supposed environmentalist and consumer sympathies of the Clinton-Gore Administration, the government holds an ideological belief that nothing should impede the profitability of biotech as a mainstay of the future US economy. In addition, the Clinton Administration's hostility to labeling may be attributed to documented political contributions made to it by the biotechnology industry.

Last May, several religious leaders and citizen groups, including members of the Council for Responsible Genetics Board of Directors, sued the FDA to change its position and to require that GEFs be labeled. The lawsuit is based on scientific and religious grounds; the less explored issue of free speech and the right to know will be discussed in this article.

Process Labels

Some government officials have said that labeling should contain information only about the food product itself, not the process by which it is manufactured. Yet, the US has many process food labels: kosher, dolphin-free, Made in America, union-made, free-range (as applied to chickens, for example), irradiated, and a number of "green" terms such as "organic." For many of these products, the scientific difference between an item which can carry the label and one that cannot is negligible or non-existent. Kosher pastrami is chemically identical to non-kosher meat. Dolphin-free tuna and tuna caught by methods which result in the killing of dolphins are the same, as are many products which are "made in America" when compared to those made abroad, or those made by unionized labor as opposed to those made by workers who are not organized.

These labeling rules recognize that consumers are interested in the processes by which their purchases are made and have a legal right to such knowledge. In none of these labeling situations has the argument been made that if the products are "substantially equivalent" no label differentiation is permissible. It is constitutionally permissible for government to require that food producers provide certain information, even when it intrudes slightly on the commercial speech of producers, in order to expand the First Amendment rights of consumers to know what is of significant interest to them.

"The Precautionary Principle"

Consumers International, a global alliance of more than two hundred consumer groups, has suggested that "because the effects [of GEFs] are so difficult to predict, it is vital to have internationally agreed upon and enforceable rules for research protocols, field trials, and post-marketing surveillance." In a movement towards the establishment of such rules, the "Precautionary Principle," has actually entered into the regulatory processes of the European Union. The "Precautionary Principle" reflects the oldest commonsense aphorisms such as "look before you leap," "better safe than sorry," or "an ounce of prevention is worth a pound of cure." It rests on the notion that parties who wish to change the existing social order (often to profit or increase their power and influence) ought not to be able to slough off the costs and risks onto others. The proponents of GEFs ought to bear the burden of proof that the new procedures are safe and socially acceptable rather than forcing regulators or the general citizenry to prove a lack of risk.

The process of applying the Precautionary Principle must be open, informed, and democratic, and must include potentially affected parties. It must also involve an examination of the full range of alternatives. For GEFs, labeling performs important functions in carrying out the Precautionary Principle. It places a burden on industry to show that genetic manipulations are socially beneficial and provides a financial incentive for them to conduct research to reduce uncertainty about the consequences of GEFs.

Look Before You Eat

Democratic notions of free speech include the right to receive information as well as to disseminate it. It is fundamental to capitalist market theory that for transactions to be maximally efficient all parties must have "perfect information." The realities of modern food production create a tremendous imbalance of knowledge between producer and purchaser. Our society has relied on government intervention to redress this imbalance in order to make supermarket shopping a fairer and more efficient—as well as safer—activity. The notion of "consumer right to know" articulates the perspective that in an economic democracy, choice is the fundamental prerogative of the purchaser.

As some biologists have put it, "the category of risk associated with genetically engineered foods is derived from the fact that, although genetic engineers can cut and splice DNA molecules with . . . precision in the test tube, when those altered DNA molecules are introduced into a living organism, the full range of effects on the functioning of that organism cannot be predicted or known before commercialization . . . [T]he introduced DNA may bring about other unintended changes, some of which may be damaging to health." [1]

Government Regulation and Free Speech

Can the government *prohibit* certain commercial speech, i.e., such as barring a label "this product does not contain genetically engineered components"? The government is constrained by the First Amendment from limiting or regulating speech in the content of labels except for the historic functions of protecting the public health and safety and eliminating fraud or misrepresentation.

In several recent cases, the US Supreme Court has restricted government regulation of commercial speech, in effect allowing more communication. The First Amendment directs us to be especially skeptical of regulations that seek to keep the public in the dark. Thus, it would seem difficult for the government to sustain an effort to prohibit labeling GE foods.

In 1995, James Maryanski of the FDA stated that "the FDA is not saying that people don't have a right to know how their food is produced. But the food label is not always the most appropriate method for conveying that information." Is it acceptable for a government bureaucrat to make decisions about what are appropriate methods of information exchange among citizens?

The first food product bearing a label "No GE Ingredients," a brand of corn chips, made its appearance this summer. A trend may be developing as other brands, such as the RAIN food company, announced recently that they too would begin labeling their foods as "free from genetically engineered organisms."

In Support of Mandatory Labeling

Can the government *mandate* commercial speech, requiring that GEFs bear a label? The government does require some labeling information which goes beyond health effects, and not every consumer must want or need mandated label information in order for it to be required by law. Such requirements have never been judged an infringement of the producers' constitutional rights. For example:

- Very few consumers are sensitive to sulfites, although all wine must be labeled.

- The burden is put on tobacco manufacturers to carry the Surgeon General's warning, even though the majority of cigarette smokers will not develop lung cancer and an intended effect of the label is to reinforce the resolve of non-consumers to refrain from smoking.
- The burden of labeling virtually every processed food with its fat analysis (saturated and unsaturated) and caloric values is mandated even though vast numbers of Americans are not overweight nor suffering from heart disease, and are not interested in this information.
- Irradiated foods (other than spices) must carry a specific warning.
- Finally, the source of hydrolyzed proteins in foods must be on a label to accommodate vegetarian cultural practices and certain religious beliefs.

These legal requirements are in place because large numbers of citizens want such information, and a specific fraction needs it. An identifiable fraction of consumers actually needs information about genetic modifications—for example, regarding allergenicity—as the FDA itself has recognized in the Federal Register, and almost all consumers want it. Foods which are comprised, to any extent, of genetically altered components or products should be required to be labeled. Indeed, such labeling is necessary to prevent fraud and misrepresentation, so that consumers do not mistakenly purchase GEFs when they wish regular foods.

Consumers' right to know is an expression of an ethical position which acknowledges individual autonomy; it is also a social approach which helps to rectify the substantial imbalance of power which exists in a modern society where commercial transactions occur between highly integrated and well-to-do corporations, on the one hand, and consumers on the other. We should be willing to let genetic engineering run the test of the market place; if industry and government officials believe in it so strongly they should fly their genetic engineering flags proudly in the sunshine rather than seeking to obtain profits and economic advantage through concealment.

11

Latina/o Farmers and Biotechnology

By Devon Peña

Devon Peña *is a former CRG board member. He is Professor of Anthropology and Ethnic Studies at the University of Washington and is coordinator of the Ph.D. Program in Environmental Anthropology. Professor Peña serves on the National Planning Committee for the Second National Indigenous and People of Color Environmental Leadership Summit. He is the author of* Mexican Americans and the Environment: Tierra y Vida; The Terror of the Machine: Technology, Work, Gender, and Ecology on the US-Mexico Border; *and coauthor of the* Citizen's Guide to Colorado's Water Heritage. *This article originally appeared in* GeneWatch, *volume 14, number 5, September 2001.*

It is well known that the American family farm has suffered a dramatic decline over the past fifty years, a period defined by a tumultuous shift toward large-scale industrial monoculture- and corporate-dominated agriculture. The rise of commercial agriculture biotechnology is the most recent expression of this industrial, anti-nature paradigm.

A lesser-known trend is that the past thirty years also have marked a steady increase in the number of Latina/o farm owners and operators. We may be seeing a new phase involving the "Mexicanization" and not just the "mechanization" of agriculture.

The latest USDA data documents an increase of 40 percent between 1987 and 1997 in the number of Latina/o owned and operated farms in 1997. If current trends continue, in twenty years the number will increase to more than 50,000. Similar trends are evident among some Asian immigrant groups.

The number of family farmers among all other ethnic and national-origin groups, including Euro-Americans, continues a pattern of steady decline. Together, these trends suggest that by 2040 close to 30 percent of family farms will be owned or operated by Latina/os and Asians. This will constitute a major demographic shift in American agriculture.

A question of concern for environmental justice activists is whether these farmers of color will use genetically modified (GM) crops or can they be persuaded to adopt more sustainable and community-oriented practices?

Most Latina/o farmers in the US are of Mexican origin. Some are tenth generation Spanish-Indian *mestizos* farming ancestral lands in New Mexico and Colorado; these are the oldest family farmers in the United States and many have pre-Hispanic Pueblo Indian roots. Others include recent immigrants, among them Mixtec, Zapotec, and other indigenous groups. These new settlers are buying farms, ranches, and orchard lands at a remarkable pace. Latina/o farmers now own some eight to ten million acres of farms, ranches, orchards, and open space lands. By 2040, they will own close to 20 million acres of land.

Why should Latina/o farmers and other farmers of color be concerned about biotechnology? We must first recall that aboriginal Mexican farmers developed some of the world's most important crops like corn, common bean, scarlet runner bean, squash, chile, peanut, avocado, amaranth, sweet potato, tomato, cassava, yam bean, and vanilla bean.

We recently witnessed a struggle over the patenting of the Mexican yellow bean, a locally adapted native crop grown for centuries by Indian and mestizo peasants. Larry Proctor, the president of an American seed company, PODNERS, brought a yellow bean that is commonly farmed in Mexico to the US. After a few years of planting and selecting for an even size and shade of yellow, he applied for and received a patent for the seed, despite the fact it has been grown for centuries in Mexico. The Mexican government has challenged the patent because PODNERS is attempting to ban exports of the beans from Mexico and because they are charging Mexican farmers royalties. The "Enola bean" patent conflict illustrates the threat posed by commercial agricultural biotechnology to the traditional crops of Mexican and Mexican American farmers.

These farmers should become more concerned with the efforts by biotechnology corporations to appropriate locally stewarded germplasm. How many of the traditional Mexican crops will be collected, genetically modified, and patented? How will these practices affect the autonomy and integrity of Latina/o farmers as plant breeders and seed savers?

There is a saying among Mexican American farmers in Colorado: "We have always been organic. We were just too poor to call it that." Biotechnology poses additional threats to traditional organic practices among Latina/o farmers. The dangers are primarily posed by the threat of horizontal gene transfer, in which GM crops exchange genes with non-GMO crops and their wild, weedy relatives. This would clearly undermine the increasing number of Latina/o farmers who are working to protect traditional organic crop varieties and farming practices.

Small farmers have no use for biotechnology. Anything that can be done with genetic modification can also be done naturally, with fewer, if any, environmental consequences. In addition, the political and economic interests of small farmers are not served by the contractual obligations created by the use of biotechnology. They, and the environmental justice movement, are far better protected with traditional methods of farming.

The environmental justice movement must articulate a coherent critical analysis of the threats posed by commercial agricultural biotechnology to farmers of color. Moreover, the environmental justice movement must support the efforts of farmers of color to preserve or adopt sustainable and regenerative farming practices that are grounded in local and regional economics.

Resistance to biotechnology dovetails with resistance to globalization. The biodiversity-based livelihoods of farmers of color will prove to be an important battlefront in the movements for global environmental justice and a sustainable and equitable future.

12

Labeling Genetically Engineered Food in California

By Pamm Larry and CRG Staff

Pamm Larry *is founder of LabelGMOs.org. This article originally appeared in* GeneWatch, *volume 25, number 1, January–February 2012.*

Less than an hour ago, I got word that AB 88, a California Bill that would require labeling of genetically engineered fish, got voted down in the Assembly Appropriations Committee . . . again. California has tried to get GE foods labeling regulations a number of times before this. The last time was in 2010 when the California State Grange "shopped" a version and no legislator would touch it.

Because our elected officials will not enact laws to give us the right to know what's in our foods, a year ago this month, I, a grandmother with no managerial campaign experience, decided that it was my job to get this issue on the ballot so the people of the State of California could vote on it. I started out with no knowledge of the logistics of this process. I had no funding, no support from the leading GMO organizations (aside from the Organic Consumers Association) and no support from the organic industry. The only people who lit up were the people I started to share my crazy idea with. They all *knew* that this was the game changer that would get us the labeling that 80 percent of the population repeatedly say they want in poll after poll.

I am happy to say that through our tenacity and commitment, we have grown from one person to more than 115 leaders throughout the state, all committed to organizing and educating their communities.

Although we started as a grassroots movement and continue to have that as a crucial arm of our campaign, we realize that for us to win, we need everyone onboard, large and small. We have been joined by major organizations, health groups, environmental groups, farmers, activist organic companies, parent groups and faith based groups to create a solid, broad base coalition that continues to grow exponentially. We now have a professional campaign manager and are gearing up to gather 850,000 signatures mid-February to Earth Day in April. We are confident we will get this on the ballot, and then win in November.

We have other bright spots on the GE labeling front. In November 2011, a court ruled that GE canola could not be labeled "natural" without the possibility of the company being sued. Within the last few months, Connecticut and Washington have newly introduced labeling legislation. Dennis Kucinich (D-Ohio), re-introduced three GE bills: H.R. 6636, the Genetically Engineered Food Right to Know Act, H.R. 6635, the Genetically Engineered Food Safety Act, and H.R. 6637, the Genetically Engineered Technology Farmer Protection Act.

Things look promising, but in order for anything to be enacted, we need all hands on deck. One easy yet powerful thing to do is to leave a comment for the national formal petition to the FDA written by the Center for Food Safety, at www.justlabelit.org. It's clear that in order for us to get labeling, voting with our dollars, although vital, is not enough. There are increasing numbers of GE foods up for deregulation. The time for labeling is now. Please join us!

13

Lax Labeling Policies Betray Public Trust

BY JOSEPH MENDELSON

Joseph Mendelson III *is the Chief Climate Counsel at the US Senate Committee on Environment and Public Works and the former legal director of the International Center for Technology Assessment. This article originally appeared in* GeneWatch, *volume 13, number 1, February 2000.*

As the latest scandal rocks Washington, political pundits of all stripes have rushed on camera to debate whether the president has betrayed the trust of the American public. While the media titillates the public over an alleged presidential affair, the everyday truth is that our government has betrayed our trust in many ways. There is no better example than the public's increasing inability to know exactly what it is eating.

After you turn off the latest news from Washington and sit down to supper tonight, you may want to take a moment to ponder your plate. Your lamb chops, tomatoes, and corn on the cob may look the same as usual, but it is quite possible that the lamb chop contains human genes, the tomato has been genetically altered for a longer shelf life, and the corn has been genetically engineered to resist infestations from the European corn borer. While this may be shocking to you, it certainly isn't news to the government agencies established to protect our food supply.

For more than a decade the US Food and Drug Administration (FDA) and the US Department of Agriculture (USDA) have been eroding the consumer's right to know how food has been produced. Since the mid-1980s the FDA and other government agencies have been considering how to review and regulate foods developed through genetic engineering. By law, these agencies are charged with reviewing the environmental, food safety and ethical issues surrounding these novel foods. Unfortunately, this review so far has resulted in government policies that keep all of us in the dark by allowing genetically engineered foods to enter the marketplace without labeling or significant government oversight.

The slow trickle eroding our right to know became a torrent on May 29, 1992, when the FDA issued a policy for foods derived from new plant varieties. The policy determined that all transferred genetic material and the resulting food products derived from genetically engineered plant varieties were to be "generally regarded as safe." As a result, genetically engineered food products derived from genetically engineered plants could appear in interstate commerce without labeling, without pre-market notification to the FDA, and without specific FDA approval.

After receiving more than 5,000 comments from consumers requesting that the FDA require genetically engineered foods be labeled for health, safety and religious reasons, the agency requested additional information from the public related specifically to the labeling of foods derived from genetically engineered plants. Since seeking this information in 1993, the FDA has yet to finalize any regulations concerning the use and approval of genetically engineered foods. The result has been a de facto approval process for these products without the government addressing such critical safety issues as the genetic stability of these new foods or even requiring a registry of what genetically engineered products we are all unknowingly purchasing.

Currently, at least twenty-seven different genetically engineered foods are known to be on the market. These foods may have already made it onto your plate. In fact, you could be encountering many more than these twenty-seven foods because the government itself has no way of even knowing what is truly on the market.

This situation doesn't just apply to the fruits and vegetables regulated by the FDA. In 1992 the USDA also began finalizing a policy allowing meat from genetically engineered livestock and poultry to enter the food supply unlabeled. Finalized in 1994, the USDA policy allows meat from animals used in genetic engineering experiments to be sold for slaughter in the regular meat market after a cursory notification process.

Most of the animals that are subject to this policy are from research that involves using human genetic material within the animal in an attempt to create novel pharmaceuticals. As a result, consumers could now be unknowingly eating meat and poultry containing human genetic material.

Amazingly, these policies have come into place despite laws, such as the Federal Food, Drug and Cosmetic Act, which mandate the labeling of materially altered foods. The policies ignore public opinion surveys conducted and funded by USDA showing that 90 percent of the public believes placing human genes into animals is "unacceptable" and other polls finding consumers overwhelmingly in favor of the mandatory labeling of all genetically engineered foods. Among these consumers are practitioners of a wide variety of religious denominations that may have a constitutional right to avoid consuming genetically engineered organisms (including meat with human genetic material) based on theological belief or adherence to specific dietary covenants.

Whatever the particular concerns, the FDA and USDA policies have fundamentally betrayed the public's trust by eliminating our ability to know what we are eating. Instead of worrying about sexual behavior in the Oval Office, maybe the political pundits should be talking about how the federal government has shirked its responsibility to ensure the safety of our food.

14

A Conversation with John Fagan

By Sam Anderson for GeneWatch

John Fagan, PhD, *is founder and chief scientific officer of Global ID Group, which includes FoodChain Global Advisors, Genetic ID, and Cert ID. Through these companies he has pioneered the development of innovative tools to verify food purity, quality, and sustainability. These tools include DNA tests for genetically engineered foods, the first certification program for non-GM foods, and the first program dedicated to certifying corporate social and environmental responsibility in the food and agricultural industries (ProTerra Certification). Dr. Fagan holds a Ph.D. in molecular biology, biochemistry, and cell biology from Cornell University. He was interviewed by* GeneWatch *editor* Sam Anderson.

GeneWatch: You've been involved in GMO activism for quite a while. What's different about the issues now compared to a decade or two ago?

Fagan: In 1994, when I started working on this issue, this was the defensible position for somebody who was critical of GMOs: If you look at how GMOs are made, you can expect that this technology will generate problems. We could say it was *likely* to generate foods that are allergenic, or toxic, or reduced in nutritional value, and it's *likely* to generate plants that are going to have impacts on the environment, alternating host-pest relationships, influencing soil fertility, altering the balance of ecosystems in various ways.

But that was all we could say at the time, because there wasn't much research. Today, there is clear, concrete evidence. There is an abundance of research showing that yes, GMOs can be allergenic, they can be toxic, they can be reduced in nutritional value, and they cause a host of environmental and agro ecological problems. So the big change is that the science has actually demonstrated the problems that in 1994 we were saying *might* happen.

That's a big change, but there's actually something bigger. These fifteen or twenty years have shown that although there are health and environmental problems, the biggest problems with GMOs have to do with their social and cultural impacts. They pose threats to food security for many nations, and, as a result of food security, national security. Any country that doesn't control its food system and cannot assure access of food to every citizen of the nation is not a sovereign nation. Whoever is controlling the food system is really the controller of that nation.

In America, to a certain extent, these multinational biotechnology and chemical companies are controllers of the food system today. The genetics of the major crops grown in the US are now controlled by just five or six players. So the germplasm that is the basis for our food system is in the hands not of a bunch of family companies that have a diversity of concerns, including concerns for the local farmers, but in the hands of huge multinationals which answer only to their shareholders. And that consolidation of genetic resources into those hands is one of the most dangerous things to happen to humanity in a long time.

GeneWatch: Despite that, it seems like anti-GMO activists' main talking points—at least in the US—usually center around things like health concerns rather than food sovereignty. At least that's my impression—are you finding that to be true?

Fagan: It depends on where you are. In the US, health and environmental issues do hold sway. In Europe, too, although the food security and sovereignty issues are also important there. But if you go to a country like India, or if you saw the discourse that unfolded in Turkey, where the GMO issue is being debated strongly, there, it has a lot to do with national security and national sovereignty. They are absolutely deeply concerned about the idea of an American multinational controlling the seed supply. The pushes in India and Russia to ban GMOs have had to do with concerns about who is controlling the technology.

If you look at the US dialogue right now, there has been a tendency to not even talk about the health and environmental issues, and just focus on "right to know." Personally, I feel that has restricted the discussion way too much. The goal was to get as broad an audience as possible—and yes, that has been the case, "right to know" is a broad issue—but it has not had the visceral impact that food safety has. My feeling is that in the US we need to emphasize the food safety and environmental aspects more.

But the reality is that the chemical and biotech companies love it when the focus is on those things, because then issues like food sovereignty are not really in sight. If you look at it, what they have done is translate property rights into a strategy for controlling the entire seed industry, and therefore controlling the base of the food system. And they've done that successfully in the US They have been less successful in Argentina and Brazil, and in India they were successful in the cottonseed area but haven't really reached out beyond that because other GMOs haven't been commercialized. In India right now, one of the major pressures that keeps farmers using GM cottonseed is that they can't find anything else to grow. I've heard that over and over. But what you hear from the biotechnology industry is that this is a scientific issue, and we need to debate it on the merits of the science. And that means essentially trying to keep it to a discussion of whether GMOs are safe or not. And they've been able, as we saw with the Séralini situation, to control the discussion in a way that, despite the scientific evidence that GMOs are a real problem, they've been able to drown that out with a large amount of public relations. It's been science by press release rather than laboratory science.

I don't want to sound like a conspiracy theorist, but when the next generation looks back at these issues fifty years from now, they're going to see the food safety and sovereignty issues as being the primary drivers. Anyway, that's my perspective on it.

GeneWatch: You mentioned farmers in India not being able to find non-GM cottonseed. Can you say more about that?

Fagan: It's a common situation. Often farmers don't have a choice of what they buy because the cottonseed industry is controlled by the biotechnology companies. You've heard the stories of how they've done it in the US, and they've done the same thing in India.

It's so funny. The Brazilian farmers wouldn't stand for the shenanigans that most of the US farmers didn't even blink at. Monsanto got sufficient control of the Brazilian seed sector so that they started setting quotas. This meant they could tell seed dealers, "If you want to sell our GM soy seed, we will give you a license to do that, but you have to agree to sell no more than 15 percent non-GM seed." In other words, for every fifteen bags of non-GM seed you sell, you have to sell 85 bags of GM seed. Now, they've been doing this in the US for twelve years; it's standard practice now. The situation in the US was that Monsanto had marketed its GM soy to the point that there was huge demand for it, and seed companies could not stay in business unless they were offering GM soy. So Monsanto would come in and say, "Yes, you can sell our seeds, but we're going to write a five-year contract with you. There are two parts to it: Firstly, you're going to buy this much of our GM soy every year for the next five years.

In parallel with that, you aren't going to sell more than X amount of non-GM soy seed." So Monsanto was able to restrict farmers' access to non-GM seeds.

This put seed companies into a position where they were committed to meeting these quotas, which kept getting bigger, so they were obligated to buy a certain amount of Monsanto seed every year, whether they sold it all or not. What would happen is that over the length of the contract, some of the smaller companies would end up being unable to sell all of the Monsanto seed, and they would eventually accrue a large debt to Monsanto. And at some point, Monsanto would come in and say, "You know, you owe us enough that it puts you into a position of bankruptcy. But we're prepared to be really kind to you in this situation. All you need to do is turn the company over to us and we'll erase the debt. You can still work here, we'll give you a salary, but it's our company from now on."

In Brazil, the farmers objected to this. They saw these restrictions being placed on non-GM seeds and they went to the government and put pressure on them to do something about it. They also saw that Monsanto had been collaborating with Embrapa, the research arm of the Brazilian agricultural industry. Monsanto had gotten Embrapa to withdraw some of the best non-GM soy seed varieties available. The farmers pushed back, and essentially this forced Monsanto to back off of trying to control the markets in Brazil. And all of this resulted in increased awareness among the agricultural sector—and, I think, a number of government officials—about the risks to food security and national sovereignty that can come from this kind of situation.

GeneWatch: What's one change you would like to see in GMO activism today?

Fagan: It seems many activists don't realize that the science has caught up with the issue. I still hear a lot of people talking today the same way that we were talking back in 1994, that it *might* be possible that there are health or safety issues; but today we can take a much stronger position. We can say it's well established that genetic engineering causes foods to be allergenic, toxic, reduced in nutritional value, damaging to the environment, damaging to agro ecology. There's evidence for all of these things. There's also evidence that GMOs do not give better yields, they don't feed the hungry, et cetera. The evidence is there. I would really encourage those who are working on this issue to stand strongly behind that; and if they're not comfortable doing that, it probably means they need to look more closely at the science that has been done. Sometimes people are way too conservative on this.

PART 3

GMOs in the Developing World

One of the most often repeated claims in support of the development and proliferation of GM crops has been their ability to "feed the world." That claim belies for what purpose the biotech companies have engineered GM crops. There is not a single GM crop on the market engineered for increased yield, drought-tolerance, salt-tolerance, enhanced nutrition, or any other appealing trait that is being touted by the industry. Disease-resistant GM crops are practically non-existent. In fact, commercialized GM crops incorporate primarily two "traits"—herbicide tolerance and/or insect resistance. Herbicide-tolerant crops are popular because they simplify and reduce labor needs for weed control, thereby facilitating the concentration of farmland in fewer, ever bigger, farms. Moreover, engineered traits have produced only modest to no gains in yield depending on the crop in question and are actually falling behind conventional productivity improvements in more traditional breeding and crop production methods.

Indeed, Western efforts to aid developing nations by introducing GMOs have too often been shown to cause more harm than good, as plans to aid developing nations by relying on biotechnology have often been formulated without embracing localized knowledge of the agro-ecology. In many cases, the introduction of GMOs in developing countries has led to higher prices at the market to offset the necessary investment costs of expensive seeds and fertilizers.

The tremendous hype surrounding biotechnology has obscured the basic fact that GMO technology, as it exists today, is dominated by multinational firms intent on controlling the world's seed supply, raising seed prices, and eliminating farmer seed-saving. And too often, such top-down interventions are specifically designed to accumulate profit for corporations and their shareholders at the expense of the actual needs of local populations.

Biotechnology is regularly touted as a solution to hunger and food sovereignty when social, political, and economic factors must first be addressed in order to ensure food access and appropriate economic development. Even then, often more conventional agricultural management solutions are preferable. To succeed, any solution must empower citizens to define their own agricultural management system unrestricted by intellectual property rights and GMO patents.

The essays in this section explore the serious implications for local populations when industrial biotechnology replaces local agricultural systems of developing countries.

—Jeremy Gruber

15

The Agrarian Crisis in India

By Indrani Barpujari and Birendra Biru

This article originally appeared as part of "ÃJan Sunwai on the Present Agrarian Crisis: A Report" and appears courtesy of Gene Campaign. Gene Campaign is an Indian non-profit that works on food security and GE issues. This article originally appeared in GeneWatch, *volume 20, number 5, September–October 2007.*

Twenty-first century India has emerged as a major economic power in the world, with the growth rate of the gross domestic product reaching impressive levels and the poverty ratio coming down significantly. In the context of such a scenario, it is indeed very incongruous and difficult to believe that the Indian countryside where the large majority of its people reside is in the grip of a severe agrarian crisis. In the opinion of [economist] Prabhat Patnaik, this crisis in Indian agriculture is "unparalleled since independence and reminiscent only of the agrarian crisis of pre-war and war days."[1]

According to Suman Sahai of GeneCampaign, the most tragic face of India's agrarian crisis is seen in the increasing number of farmer suicides, not just in the hotspot areas of Andhra Pradesh and Vidarbha but in the allegedly prosperous agricultural zones of Punjab and Karnataka.[2] Farmers' suicides are no longer limited to the drought- and poverty-stricken areas of the country. Now farmers in the most productive agricultural regions such as Karnataka, Punjab, West Bengal, Andhra Pradesh, and Maharastra are ending their lives because of their massive indebtedness.

The economist Deepak K. Mishra also expresses a similar view when he says that the conventional notion of agrarian distress being part of the broader landscape of underdeveloped agriculture and backwardness no longer fits to the emerging evidences from rural India.[3] Manifestations of agrarian distress in contemporary India are not confined to the pockets of backwardness; even the regions with a high degree of commercial agriculture, that use relatively better technology and have a relatively diversified cropping pattern have reported high indebtedness and distress of various kinds.

More than 6,000 indebted farmers, mainly cotton farmers, have committed suicide in Andhra Pradesh alone during the period from 1998 to 2005 as its government, which had entered into a state-level Structural Adjustment Programme with the World Bank, raised power tariff five times even as cotton prices fell by half. In Maharashtra, 644 farmers committed suicide across three of its six regions between January 2001 and December 2004.

In Karnataka, 49 suicidal deaths occurred between April and October 2003 in the drought-prone region of Hassan. Over the same period of time, 22 suicides occurred in Mandya, the state's "sugar bowl;" 18 occurred in Shimoga, a heavy rainfall district, and 14 occurred in Heveri, a district that receives average rainfall. While statistics may show Punjab to be India's "breadbasket," claiming that its soils are rich and its five rivers supply abundant water throughout the state, the reality of this image of prosperity is revealed by the increasing number of suicidal deaths among Punjabi farmers. More than a thousand farmer suicides have taken place in Punjab, mainly in the cotton belt. Between 2001 and 2005, over 1,250 suicides took place in Wynaad in Kerala. In Burdwan, the region of West Bengal commonly called the "rice bowl of the East," 1,000 farmers ended their lives in 2003.

Various explanations have been offered for the present agrarian crisis. It has been felt that the present crisis is the result of deflationary public policies and trade liberalization (with falling global prices), which has slowed output growth, contributed to rising unemployment, income deflation for the majority of cultivators and laborers, enmeshing of cultivators in unrepayable debt, and loss of assets (including land) to creditors. According to [economist] Utsa Patnaik, "forty years of successful effort in India to raise foodgrains absorption through Green Revolution and planned expansionary policies, has been wiped out in a single decade of deflationary economic reforms and India is back to the food grains availability level of fifty years ago."[4]

Another explanation given for the agrarian crisis is the drastic reduction in state spending on rural development which has led to loss of purchasing power among rural people. Expenditures in rural development, under which fall agriculture, rural development, special areas programs, irrigation and flood controls, and village and small scale industry, have been slashed to an all-time low of 0.6 percent of NNP in 2004.

An attempt to have a correct appraisal of the crisis afflicting Indian agriculture in recent times has been made by the Tata Institute of Social Sciences (TISS) in Mumbai, which had conducted an investigation into the

Vidarbha agrarian crisis and farmer suicides at the behest of the Bombay High Court. The study found that the main reasons for the crisis are repeated crop failure, inability to meet rising cost of cultivation, and indebtedness. According to Sahai, emergency in agriculture has developed because of the rising cost of agricultural production which is not offset by either the Minimum Support Price offered by government or prices available on the market.[5] The combination of high cost of production (owing to higher input prices and higher cost of labor), low market price and non-availability of easy credit have contributed to an enormous debt burden. This is further compounded by personal loans taken for social needs like marriage and education. The crisis becomes acute when farmers, exhausting their credit with banks, turn to private money lenders who charge usurious rates of up to 60 percent per annum.

The official policy response to the present agrarian crisis has generally been one of denial and insensitivity. The recent initiative of the government regarding the rural employment program has been criticized as "a limited gesture totally inadequate to meet the enormity of the crisis," while the projected enhancement of agricultural credit by the government has been dismissed as "exaggerated" and "inadequate" in the context of the policy environment of withdrawal of reduction of minimum support price programs.[6]

A strange argument has also been advanced in certain quarters to account for the decline in per capita food availability. It is contended that because of a change in the dietary habits of the people, they have diversified their consumption pattern from food grains towards all kinds of less elemental and more sophisticated commodities. Therefore, far from it being a symptom of growing distress, the decline in food availability is actually indicative of an improvement in the conditions of the people, including the rural poor.

Some have even gone to the extent of suggesting that, with the changes occurring in Indian agriculture in terms of the cropping pattern and use of machinery, peasants and workers do not need to put in hard manual labor. Correspondingly, the need for consuming huge amounts of foodgrains no longer arises. This argument is completely untenable in the light of the hard facts of rising unemployment, falling output growth, entrapment of farmers in debt and land loss and especially, when the agrarian crisis has found expression in the acute desperation and hopelessness of the farmers, leaving them with no recourse but to take their own lives. Gene Campaign strongly feels that such policy conclusions which are contrary to realities would have dangerous repercussions if implemented, reducing food security further and impoverishing farmers.

The Debate Surrounding GM Crops

In the backdrop of the severe agrarian crisis with which Indian agriculture is faced, proponents of GM technology perceive GM crops as offering a solution to hunger in the developing countries. The Department of Biotechnology and the Biotechnology industry in India have taken the position at several policy forums that raising agricultural growth from the current 1.7 percent to the desired 4 percent and alleviating the agrarian crisis could be achieved by promoting genetically engineered crops. US-led programs like the secretly concluded and controversial Indo-US deal on agriculture and the ABSP I and ABSP II (Agriculture Biotechnology Support Project) funded by the USAID, led by Cornell University and implemented in India through the Department of Biotechnology, are invoked by the government and the science administration as enabling programs to achieve the goal of uplifting Indian agriculture. Sahai has questioned the desirability of such direct US intervention in India's program on GE crops and foods and also the ridiculously simplistic approach of suggesting that one single technology could address the many factors responsible for decline in agriculture.[7] She further expresses the view that as genetically engineered crops have been developed essentially for the large land holding, mechanized agriculture of industrialized countries, they do not fit the developing country context.

Further, there is little available in the repertoire of genetic engineering today that is geared to address the problems of developing country agriculture. At present, GE technology offers only four major crops: soybean, corn, cotton and canola. Apart from a few virus resistant GE varieties, herbicide tolerance and insect resistance (the Bt trait) are the two traits that dominate the field of genetically engineered crops.

Herbicide tolerant crops contain a gene that makes them resistant to the herbicide that is sprayed to kill weeds. The company that owns the herbicide tolerant crops (in this case Monsanto) is also the company that owns the herbicide that particular crop variety will tolerate. Hence the company promoting herbicide tolerant crops makes a double killing, first on the sale of the herbicide itself and, second, on the sale of the crop varieties which are tolerant to that proprietary herbicide.

Herbicide tolerance was developed for industrial agriculture with its large farms and labor starved conditions, where weed control was possible only by using chemicals like herbicides.

In developing countries, like India, weeds are controlled manually. Weeding is an income source in rural areas, especially for women. Sometimes it is their only source of income. Farm operations like sowing, weeding, harvesting and winnowing are the key sources of rural employment. As the herbicide tolerance trait is essentially a labor saving and hence a labor displacing trait, its introduction will take work away from agriculture labor and destroy income opportunities in rural India.

Bt technology is the second category of genetically engineered crops, like Bt cotton, which is the only GE crop being cultivated in India at present, although many others are in the pipeline. In Bt crops, a toxin producing gene from the soil bacterium *Bacillus thuringiensis* (Bt) is put into plants. The plants that produce the Bt toxin are, in essence, producing their own insecticide. Pests that feed on the plant are supposed to die upon eating the toxin.

Like other forms of insecticide, however, pests will eventually develop resistance to Bt. This resistance is already beginning to develop. Reports are coming in about the collapse of the Bt cotton technology from China and Arkansas. Cotton scientists in India are warning that, with the way in which legal and illegal Bt cotton is spreading everywhere, and without farmers following the recommended crop management practices, it is only a matter of time before local pests become resistant to the toxin and the technology collapses in India as well.

In India, the Bt strategy for pest resistance is likely to collapse earlier than predicted, as, in the absence of any coherent policy, the Department of Biotechnology has sanctioned its use in a large number of crops. Today, about 42 percent of all the research on GE crops in India is based on the Bt gene. Ranging from cotton to potato, rice, eggplant, tomato, cauliflower, cabbage, tobacco and maize, the Bt gene is used everywhere.

Assuming that the crops that are being researched are targeted to reach the fields one day, we are facing a situation where a wide range of crops growing in both the Rabi and Kharif season will contain the Bt gene. So, throughout the year, there will be standing crops containing the Bt toxin. Not only that, in the same season, there will be a number of different Bt crops growing next to each other in small fields, especially in regions where farmers grow a variety of vegetables. When pests, such as the bollworm, are consistently exposed to the toxin in every season, year after year, resistance to the Bt toxin will surface very quickly. All pests ultimately develop a resistance to the poison that is aimed to kill it. That is why a constantly evolving Integrated Pest Management (IPM) approach, using a variety of strategies, is the only approach that can work over the long term to control plant pests and diseases.

Economic Issues

On top of all this, the high cost of Bt technology makes its cultivation economics adverse for small farmers. Bt cotton seeds cost several times the price of successful, local non-Bt seeds. So exorbitant has the pricing been, that the Government of Andhra Pradesh has filed a case against the owner of the Bt technology, the Monsanto Company.

Gene Campaign, which presented the first scientific data from the first harvest of Bt cultivation in Maharashtra and Andhra Pradesh, showed that net profit from Bt cotton was lower per acre compared to non-Bt cotton in all types of soils and that because of the high investment costs and poor performance of Bt cotton, 60 percent of the farmers cultivating Bt cotton were not even able to recover their investment and incurred losses averaging seventy-nine rupees per acre.[8] Research has shown that Bt cotton has been a disaster and in fact responsible for crop failure leading to suicide by victims. The TISS study found that 70 percent of the total number of suicide victims in Vidarbha grew cotton as their primary cash crop; the district records of the region show that seventy percent of the farmers who killed themselves were cultivating Bt cotton.[9]

An independent study was conducted by agricultural scientists Dr. Abdul Qayum and Kiran Sakkhari on Bt cotton in Andhra Pradesh that involved a season-long investigation in eighty-seven villages of the major cotton growing districts—Warangal, Nalgonda, Adilabad and Kurnool.[10] Bt cotton was found to have failed on all counts: it failed miserably for small farmers in terms of yield; non-Bt cotton surpassed Bt by nearly 30 percent and at 10 percent less expense. It did not significantly reduce pesticide use; over the three years, Bt farmers used 571 rupees worth of pesticide on average while the non-Bt farmers used 766 rupees worth of pesticide. It did not bring profit to farmers; over the three years, the non-Bt farmer earned, on average, 60 percent more than the Bt farmer. It did not reduce the cost of cultivation; on average, the Bt farmer had to pay 12 percent more than the non-Bt farmer. It did not result in a healthier environment; researchers found a special kind of root rot spread by Bollgard cotton infecting the soil, preventing other crops from growing.

The Vidarbha Jan Andolan Samiti (VJAS) has also alleged that 170 cotton growers from Western Vidarbha, who had opted to sow Bt cotton from a US-based seeds company, had committed suicide during the period from June to December last year.[11] According to VJAS president Kishore Tiwari, among the 182 suicides in Western Vidarbha, 170 were by Bt cotton growers. According to him, over six lakh farmers from Vidarbha had sown Bt cotton on the assurance that the minimum yield would be twenty quintals per acre. However, the average yield per acre was only two to three quintals per acre (one quintal is equal to 100 kilograms, or about 220 pounds).

Leading farmers' organizations have demanded a ban on Bt cotton and a moratorium on any further approval of genetically modified crops for commercial cultivation. Three varieties of Monsanto's Bt cotton failed miserably in Andhra Pradesh.[12] The Genetic Engineering Approval Committee (GEAC) had to ban its cultivation in Andhra Pradesh on receiving adverse reports from the state government and farmers. The GEAC also banned the cultivation of Monsanto's Mech-12 Bt in all South India. The government also had to concede for the first time that Bt cotton had indeed failed in parts of India, particularly in Rajasthan and Andhra Pradesh.[13]

Complaints of allergic reactions arose from farmers growing genetically modified cotton has come in from Barwani and Dhar Districts of the state of Madhya Pradesh.[14] A report from Nimad district states that Bt cotton is causing allergic reactions in those coming into contact with it, and cattle have perished near Bt cotton fields in another district. Sixteen hundred sheep died in Warangal district after grazing in fields on which Bt cotton had been harvested.[15] This year again, Bt cotton has been found to have raised its ugly head with the deleterious effect of Bt cotton on livestock starting to re-surface in Warangal district.[16]

No comprehensive health and risk assessment of Bt cotton has been done. Thus, in the light of the above facts, it is unrealistic to assume that GM crops, in their current form, could contribute to alleviating the agrarian crisis.

Recommendations

An in-depth analysis of the above issues reveals the severity of the present agrarian crisis. To mitigate the present crisis, the Jan Sunwai came up with a number of recommendations, which are as follows:

- Input costs should be reduced.
- Markets must be made available for agricultural produce.
- A good market price must be provided for agricultural products.
- For farmers, credit should be made available at low interest rates.
- The extension system should be revived to solve problems in the field.
- There should be a proper system to address the issue of water scarcity.
- Adequate water for irrigation should be provided.
- Conserve Agro Bio-Diversity in Gene and Seed banks.
- Increase budget outlay for Agriculture in every Five Year plan of the Government of India.
- Agricultural land should not be given to Special Economic Zones (SEZ).
- The use of Genetically Modified Seeds should be stopped and organic agricultural practices encouraged.
- Farmers' Rights law to be implemented immediately.
- Investments should be made to restore soil health.
- Agriculture should be diversified with introduction of new varieties.

16

Bill Gates's Excellent African Adventure: A Tale of Technocratic AgroIndustrial Philanthrocapitalism

By Phil Bereano

Phil Bereano, JD, PhD, *is a Professor Emeritus at the University of Washington and a co-founder of the Council for Responsible Genetics. This essay is based on work he and other researchers have done for AGRA Watch, a project of Seattle's Community Alliance for Global Justice (http://www.seattleglobaljustice.org). This article originally appeared in* GeneWatch, *volume 26, number 1, January–March 2013.*

From 2009 to 2011, the Bill and Melinda Gates Foundation spent $478,302,627 to influence African agricultural development. Adding in the value of agricultural grants going to multiple regions and those for 2012, the Foundation's outlay to influence African agriculture is around $1 billion. Of course, Gates is not an African, not a scholar of Africa, not a farmer, and not a development expert. But he is a very rich man, and he knows how he wants to remake the world.

Gates's support for agricultural development strategies favors industrial, high-tech, capitalist market approaches. In particular, his support for genetically engineered crops as a solution for world hunger is of concern to those of us—in Africa and the US—involved in promoting sustainable, equitable agricultural policies.

First, his technocratic ideology runs counter to the best informed science. The World Bank and the UN funded 400 scientists, over three years, to compile the International Assessment of Agricultural Knowledge, Science and Technology for Development (IAASTD). Its conclusions in 2009 were diametrically opposed, at both philosophical and practical levels, to those espoused by Gates. It recommended research that "would focus on local priorities identified through participatory and transparent processes, and favor multifunctional solutions to local problems," and it concluded that biotechnology alone will not solve the food needs of Africa.

The IAASTD suggests that rather than pursuing industrial farming models, "agro-ecological" methods provide the most viable, proven, and reliable means to enhance global food security, especially in light of climate change. These include implementing practical scientific research based on traditional ecological approaches, so farmers avoid disrupting the natural carbon, nitrogen and water cycles, as conventional agriculture has done.

Olivier De Shutter, the UN's Special Rapporteur on the Right to Food, reinforces the IAASTD research. He too concludes that agro-ecological farming has far greater potential for fighting hunger, particularly during economic and climatically uncertain times.

Agroecological practices have consistently proven capable of sustainably increasing productivity. Conversely, the present GM crops, based on industrial agriculture, generally have not increased yields over the long run, despite their increased input costs and dependence. The Union of Concerned Scientists details GM crops' underperformance in their 2009 report, "*Failure to Yield*."[1]

Second, Gates funds African front groups whose work with Monsanto and other multinational agricultural corporations directly undermines existing grassroots efforts at improving African agricultural production. Gates has become a stalking horse for corporate proponents promoting industrial agricultural paradigms, which view African hunger simply as a business opportunity. His foundation has referred to the world's poor as presenting "a fast growing consumer market." Referring to the world's poor as "BOP" (the bottom of the pyramid), he insists they must be subsumed into a global capitalist system, one which has done so well to enrich him. His philanthropy is really "philanthrocapitalism."

By and large, Gates' grants do not support locally defined priorities, they do not fit within the holistic approach urged by many development experts, and they do not investigate the long-term effectiveness and risks of genetic modification. The choice of a high-risk, high-tech project over more modest but effective agricultural techniques is problematic, offering no practical solutions for the present and near-future concerns of the people who run small farms.

For example, the Gates Foundation touted a $10 million grant to Conservation International in 2012 as "agroecological," an important concept emerging as a touchstone criterion for assessing development assistance.

Using the guidelines that Miguel Altieri has laid down, it consists of "broad performance criteria which includes properties of ecological sustainability, food security, economic viability, resource conservation and social equity, as well as increased production. . . . To attain this understanding agriculture must be conceived of as an ecological system as well as a human dominated socio-economic system."[2] This goes far beyond the definition used, for example, by the Organization for Economic Co-operation and Development (OECD) as "the study of the relation of agricultural crops and environment." In other words, in addition to embodying the idea of sustainability, agroecology includes principles of democracy.

However, the Conservation International grant is merely a program of monitoring what is happening on the ground in African agriculture. The Foundation's press release describes it as:

> (Providing) tools to ensure that agricultural development does not degrade natural systems and the services they provide, especially for smallholder farmers. It will also fill a critical unmet need for integrating measurements of agriculture, ecosystem services and human well-being by pooling near real-time and multi-scale data into an open-access online dashboard that policy makers will be able to freely use and customize to inform smart decision making. The raw data will be fully accessible and synthesized into six simple holistic indicators that communicate diagnostic information about complex agro-ecosystems, such as: availability of clean water, the resilience of crop production to climate variability or the resilience of ecosystem services and livelihoods to changes in the agricultural system.[3]

This is really a top-down technocratic program, hardly qualifying as agroecological. In fact, while it might be a beneficial activity, it could be used as a perfect illustration of trying to use an appealing label to whitewash its opposite. A Gates's official claims that it will be "for decision-makers," but these users appear to be hierarchical elites, not smallholders, who are unlikely to have "an open-access online dashboard" in their fields.

Genetically modified crops are also supported by the Gates Foundation, although they threaten conventional and organic production as well as the autonomy of African producers and nations. In 2002, Emmy Simmons, then-assistant administrator of the US Agency for International Development, stated that "in four years, enough [genetically engineered] crops will have been planted in South Africa that the pollen will have contaminated the entire continent." Biotechnology cannot coexist with agro-ecological techniques and traditional knowledge.

Mariam Mayet of the African Centre for Biosafety said of the Gates Foundation grant, "[Genetically modified] nitrogen-fixing crops are not the answer to improving the fertility of Africa's soils. African farmers are the last people to be asked about such projects. This often results in the wrong technologies being developed, which many farmers simply cannot afford."

She said farmers need ways to build up resilient soils that are both fertile and adaptable to extreme weather. "We also want our knowledge and skills to be respected and not to have inappropriate solutions imposed on us by distant institutions, charitable bodies or governments," Mayet said.

While successful in his chosen field, Gates has no expertise in the farm field. This is not to say that he and his fellow philanthropists cannot contribute—they certainly can. However, some circumspection and humility would go a long way to heal the rifts they have opened. African farmers never asked to be beaten with the big stick of high-input proprietary technology; doing so continues neo-imperialism and the perpetuation of foreign-imposed African "failure." Africans urge Bill Gates to engage with them in a more broadly consultative, agroecological approach.

17

Bt Brinjal in India: Why It Must Not Be Released

By Aruna Rodrigues

Aruna Rodrigues *is spearheading a legal battle in India seeking a moratorium on release of genetically modified crops, a ban on imports of genetically modified products, and setting up an independent testing facility which meets international standards. This analysis is based on evidence provided to the Supreme Court of India. This article originally appeared in* GeneWatch, *volume 22, number 1, January–February 2009.*

India is the world's second leading producer of brinjal (eggplant), a species which originated there. In 2008, the Indian government approved the production of brinjal genetically modified to internally produce *bacillus thuringiensis*, or Bt, which kills insects. Bt cotton has already been widely commercialized in India, but this would be country's first commercialized GM food crop. The applicants, Monsanto and the Indian biotech company Mahyco, conducted safety studies on the Bt brinjal, and while the regulating agencies accepted the results of those studies, independent scientists and activists have sharply criticized them as severely deficient. This article's author and others have petitioned India's Supreme Court to halt the release of Bt brinjal.

The pressing need for an overhaul of GM foods regulation in India was made clear when India's governmental regulators accepted the conclusions of Mahyco-Monsanto's safety studies of its own Bt brinjal without subjecting the studies to independent scrutiny or oversight or separate tests not funded and carried out by the product's own producers.

It has taken two years for these safety studies to be put in the public domain. The regulator is complicit in having supported the biotech industry, and Monsanto in particular, in their attempts to keep the studies secret by claiming them as "confidential business information" until forced to change their stance by a court order.

Much more serious than Mahyco's "misdemeanours" is the role of the Regulators, the Genetic Engineering Approval Committee (GEAC) and Review Committee on Genetic Modification, who appear to be incapable of conducting a proper safety assessment of Bt brinjal, and therefore possibly of any GM crop.

Independent scientists have examined the studies, and their appraisals provide evidence of badly designed studies, fuzzy data masked by too many controls, no "p" values (a most serious omission), a paucity of raw data, no peer review, and sample sizes which make sheer mockery of good biosafety testing, among other things. The Mahyco-Monsanto studies are a Gold Standard for how bio-safety testing ought not to be conducted.

In short, the studies are a smokescreen. The study defects are long and would fill a dossier on their own demerits. It is difficult to avoid the serious conclusion of intent to mislead, even cover up and commit fraud.

There is no scientific basis for the industry claim that Bt crops are safe to eat. Furthermore, to base such a claim on apparent evidence of the success of Bt cotton misses the point—particularly, that cotton is not a food crop, and that Bt cotton as an animal feed has never been tested for human safety and is seriously implicated in animal toxicity, infertility and deaths.

In nature, *bacillus thuringiensis* (Bt) kills insects. Bt or the Cry class of proteins are toxic, are not eaten by human beings (or other mammals), and are not declared safe for human consumption. When the Bt protein is engineered into a plant, that plant produces the toxin internally in grain and other plant tissue. Bt plants are shown to have a thousand times more of the Cry protein expressed in kernels than would ever result from use of topical Bt toxin as a pesticide. Now, study after study by independent scientists, especially in the last two years, is producing evidence of the toxicity of the Bt transgene and the transgenic Bt plants.

Judy Carman and Gilles-Eric Séralini are two eminently qualified independent scientists, who have critiqued the feeding studies of the Mahyco bio-safety dossier of Bt brinjal. Their appraisals represent the first independent scientific scrutiny of any crop developer's safety dossier in India and the first of its kind for a "near commercialised food crop." They have stated that Bt brinjal has not been properly and adequately tested by Mahyco, is unsafe, and must not be released.

In reply, Mahyco says that "all its studies followed norms prescribed by GEAC [India's apex GM regulator]. We are at advanced stages of field trial for GM brinjal and our results are extremely promising."

Mahyco is quite right in saying that they have followed norms prescribed by the GEAC. This is exactly the point of the public interest Writ Petition in the Supreme Court of India: that there are no proper bio-safety regulations for the environmental release of transgenic crops in India, with the apex regulator, the GEAC, essentially adopting US-style lack of regulation for GMOs.

It gets worse. The regulators have seriously misled successive Indian prime ministers about the truth of GM crops and, in particular, the inadvisability of introducing these crops in a center of megadiversity like India. Thus, having received the political mandate they need, they now function under this mandate openly to promote GMOs without safety testing.

The lack of safety testing is very clear from four years of evidence submitted to the court on a whole range of issues including information under the RTI (Right to Information). The Mahyco Bt brinjal dossier of safety studies along with the critical issue of contamination of rice fields as a result of criminally negligent field trials of Bt rice in Jharkhand in July 2008 (in the corridor of the center of origin for rice), are the litmus test of the culpability of the Indian Regulator, who is now being asked to stand down.

India is one of the world's most biodiverse countries and a "center of origin" of many plants, the wild species of which have important traits for drought or insect resistance etc., (the same plant traits that biotechnology companies must rely on to produce their GMOs). India is also a center of domestication for many of these plants, and existing domesticated varieties that have been bred over hundreds or thousands of years are, properly, a part of farmers' capital. Transgenic crops, due to the inevitable threat of contamination, threaten biodiversity of both wild and domesticated varieties. Extreme caution should be required before India is exposed to GM crops—and especially a food crop like brinjal, for which India is a center of origin and diversity, and for rice, for which India is the center of origin.

No GM crop has been commercialized anywhere in the world in a country that is the center of origin for that crop. That the GEAC sees fit to pay scant attention to such a critical issue defines its approach and their culpability. I shall return to this point.

On the other hand, Mahyco-Monsanto has done exactly what was expected of them: to put their Bt brinjal in the best possible light by any available means. They have managed this thanks largely to the lack of obstacles in their way—particularly full, stringent, and scientifically rigorous regulation. Given the track record of Monsanto's performance in various countries, it can hardly be expected that the bio-safety dossier of Bt brinjal would include an admission of the inadequate design of Mahyco's safety studies.

The flaws and gaps in safety testing are significant and serious in the Bt brinjal dossier. With regard to environmental studies, the woefully inadequate gene flow studies and the lack of testing for non-target organisms, soil toxicity, and other routes to contamination is a disgrace. Furthermore, Mahyco should have been required by the regulators to undertake long-term, multi-generational feeding studies on a large number of animals (e.g. at least fifty rats per group), using species that are proper proxies for humans and to measure outcomes that are relevant to human health (such as full haematology, blood biochemistry, and histology on all rats). Such studies should also be designed well enough to stand a chance of determining whether GM brinjal causes any adverse effects on the animals.

Mahyco should also have been required to fully analyze the data and to properly report the findings of the study according to internationally accepted scientific standards (e.g. to at least report the full nature of each statistical test undertaken and the p-values resulting from the tests). It is of considerable concern that they did not do so. These studies were not subjected to any kind of independent scrutiny and oversight, nor did they have the benefit of public-funded safety-testing institutions that are internationally accredited—because we have none.

The inescapable conclusion of these feeding studies is that they have been "engineered" or designed to throw up "no significant differences." It is also clear that the Indian regulator either 1) did not understand that the information it was given by Mahyco was woefully inadequate, which suggests serious incompetence on the regulator's part, or 2) did understand that the information was inadequate but still passed it as adequate, which invites a charge of criminal negligence.

The GEAC is on record as wanting to "trust" the crop's developers because it would be wrong not to do so without reason—despite Monsanto's history of corporate criminality, including court indictments for some shocking violations.

This history includes the production of Agent Orange, dioxin, and PCBs—all of which they declared as safe.

The urgent question is this: Is India as a nation prepared to risk our entire future for all time, in terms of contamination of our biodiversity, health, farmers, farming environment, and food security because of an inappropriate investment of "trust" by our government and its regulator, in Mahyco-Monsanto and other GM crop developers?

Thus, India is at great threat from its own regulators. The result is that field trials in India have been conducted on every conceivable food crop—based mainly on the Bt gene, which is undeniably toxic—over a period of about a decade without proper biosafety tests being done.

The Bt brinjal dossier clearly shows what things have come to. Bt brinjal has not been properly and adequately tested, and is now declared to be unsafe by experts, yet it is on the verge of commercialization and would have been commercialized by now except for the courageous opposition to it by farmers and civil society groups plus legal opposition.

Dr. Pushpa Bhargava (the Supreme Court's nominee to the GEAC to provide some balance to this committee) has advocated a core list of tests that must be done before any GMO is approved for release. This list has the unqualified agreement of leading international scientists who state that they conform to world class scientific standards for safety assessment. These experts have supported the stand that Dr. Bhargava has taken in the GEAC despite facing severe opposition from the regulator that is both unscientific and unprofessional.

In the ultimate analyses, the Bt brinjal tests quite astoundingly amount to this: In the best of the tests, one study of ten rats, which have been caged for ninety days, has been conducted. It has been subjected to independent scientific analyses by Séralini and Carman, and even with its severe deficiencies, it shows worrying results both clinically and statistically on various parameters which Mahyco dismisses as not being significant.

Both scientists say the release of Bt brinjal must be forbidden until full and proper safety assessments are done to a proper standard, preferably by researchers who are independent of vested interests, and the results are published in a peer-reviewed scientific journal for other scientists to read and comment upon.

On the best construction of "intent" of the GEAC, on the basis of their "trust" in Mahyco-Monsanto, our government and its apex regulator are prepared to risk the health of one billion Indians—and in perpetuity, because once introduced into the environment, GMOs can never truly be recalled and thus can have irreversible impacts.

As Dr. Carman has pointed out, if only one in a thousand of exposed people later gets ill, or has an underlying illness made worse, more one million Indians would require treatment. This would result in a huge cost to the Indian government and community.

This risks a social cost and a health scam of almost unimaginable magnitude that will make " chicken-feed" of every other scam in the country. Clearly, the government of India must be made to see reason in its policy on GM crops. We must announce a moratorium of at least five years, while we get GM regulation on track in the manner required.

Hearts of Darkness: The Biotech Industry's Exploration of Southern African Famine

By Doreen Stabinsky

Doreen Stabinsky, PhD, *is a Professor of Environmental Politics at College of the Atlantic, Science Advisor for Greenpeace USA and Greenpeace International, and a former member if the CRG Board of Directors. She actively researches and writes about the impacts of climate change on agriculture and food security. This article originally appeared in* GeneWatch, *volume 15, number 6, November–December 2002.*

A famine currently threatens southern Africa. More than fourteen million people in six countries—Zambia, Zimbabwe, Malawi, Mozambique, Lesotho and Swaziland—are facing starvation due to years of domestic turmoil, a serious drought and subsequent failed harvests. International response has been swift: money, trucks, and food supplies are being donated to the region, and private and UN aid agencies are mobilizing to address the problem.

However, a large portion of world media's coverage of the famine hasn't actually been about the famine at all, but about the controversy surrounding genetically engineered (GE) food aid. At the center of the controversy is the US insistence on sending GE maize—as corn is called in the rest of the world—to southern Africa, although most of the countries there have stated their preference for non-GE maize.

The US has long used its food aid programs as a means of developing and expanding export markets for the nation's agricultural commodities, and in other ways promoting US foreign policy. In the words of the US Agency for International Development (USAID):

> The principal beneficiary of America's foreign assistance programs has always been the United States. Close to 80 percent of the US

> Agency for International Development's (USAID's) contracts and grants go directly to American firms. Foreign assistance programs have helped create major markets for agricultural goods, created new markets for American industrial exports and meant hundreds of thousands of jobs for Americans.

Recent years have seen no exception to this, as GE-contaminated food from the US has turned up in countries around the world, distributed by USAID and through the U.N.- affiliated World Food Program (WFP), the major recipient of US food aid. Southern Africa is just the latest location where this political game is being played, as the US tries to get rid of its surplus crops.

In contrast to all other countries providing aid to the region, only the US is providing aid-in-kind (actual commodities). The US is actually selling surplus wheat on the world market and using the proceeds to buy GE maize to ship to southern Africa, even though a number of the countries have requested wheat. But that's a longer story. Aid organizations, including the World Food Program, prefer monetary donations to in-kind donations because of the greater flexibility cash affords. World Food Program spokesman Richard Lee notes:

> All US aid to southern Africa has been in-kind while all other donations have been in the form of financial aid. We prefer cash donations as they offer us greater flexibility and speed things up. Financial aid also brings much needed cash into the region.

Initially Zambia, Zimbabwe, Malawi, and Mozambique all refused GE maize. But under US diplomatic pressure, three of the countries eventually acquiesced to accepting GE maize that was first milled (ground into meal). The principal concern of those governments was the potential biosafety hazard of transgenic maize seed being inadvertently planted and then contaminating local varieties and local production through cross-pollination or mixing of seed. They are especially concerned about maize intended for export, or used for feeding cattle that would then be exported to Europe. The European Union, which provides a crucial market for southern African agricultural exports, has strict policies against the importation of genetically engineered food.

Zambia has held fast in its rejection of GE food aid and has come under severe diplomatic pressure, as well as public pressure from mainstream

US media, including *The New York Times* and *Wall Street Journal*. Those newspapers have sought to portray the Zambian president as an autocrat who would rather starve his people than feed them GE foods. Right wing think-tanks are also getting involved in the act.

The public relations nightmare of an African country refusing GE food in the midst of a famine is clearly too much for the US to handle, and the US government has increased pressure on the Zambian government to accept GE maize. Numerous reports have come out of private voluntary aid agencies—Oxfam, Intervision, Africare, and so on—that USAID is pressuring them to sign a statement in support of GE food aid. While the statement has not yet materialized, the fact that reports are coming from a number of quarters lends credibility to the story.

All this is happening because the US refuses to acknowledge that GE foods are being segregated (separated from non-GE foods) domestically, and because it needs markets for its GE crops. The government, through USAID, has claimed that they do not have the ability to segregate GE crops and provide GE-free maize to the World Food Program. However, huge amounts of US grain are segregated for export. And quantities of GE components, regularly found in testing of foodstuffs in the United States, are well below the amounts found in bulk commerce. It is clear that segregation is happening in the US, both for domestic human consumption and for export.

Meanwhile, European markets for US corn and soy have almost evaporated with the introduction of GE varieties and lack of segregation. Consequently, there is a need to redirect hundreds of millions of dollars' worth of foodstuffs to less discriminating markets. The oversupply of US corn may explain why the US is so intent on sending genetically modified corn to southern Africa.

Do Zambians have a good case for refusing GE maize? Yes. Despite US regulatory claims that these are the most rigidly tested foods to date, they are not even tested as carefully as pesticides.

The maize being exported to Zambia is Bt maize, corn that has been engineered to contain the endotoxin protein of *Bacillus thuringiensis*. To test the toxicity of Bt maize, Monsanto produced large quantities of the toxin in bacteria, purified it, and administered it to mice by oral gavage—stuffing it directly into their stomachs. They tested ten mice. None of the mice died, and only superficial examinations of organs were conducted, after which regulators concluded that Bt maize is perfectly safe for human consumption.

But many scientists, including the scientific steering committee of the Health and Consumer Protection Directorate of the European Commission, note that in cases where there are uncertainties about equivalence with a traditional counterpart—as is the case with the Bt endotoxin, which has no equivalent counterpart in the human food supply—the whole food should also be tested. It is suggested that the testing program include at least 90-day feeding studies in rodents, though additional toxicology studies may be necessary. The only feeding studies done with Bt maize for US clearance were in chicken and quail. Chicken were fed Bt maize for 38 days and they gained weight at the same rate as chicken being fed non-Bt maize. Similar results were obtained with quail.

In evaluating the safety of a food product, it is also important to consider dietary uses. Though genetically engineered corn or its derivatives are present in a number of products consumed in the US, the amount of modified substances they contain is relatively low due to the segregation of the nation's food supply. Contrast that with the diet of the average Zambian, for whom maize is a dietary staple. Many Zambians don't even say that they've eaten unless they've had a serving of maize with their meal; they might eat maize three times daily. According to Charles Benbrook, former director of the National Academy of Sciences Board on Agriculture:

> If regulatory authorities had felt that a sizeable portion of the populations of people consuming this corn would eat it directly (largely unprocessed) and that moreover, the corn might make up as much as half or two-thirds of daily caloric intake, they would *never* have approved it based on the human safety data presented at the time. Anyone who claims that US and European regulatory reviews "prove" safety in the context of food aid to Africa is either ignorant of the factual basis of US and European regulatory reviews, or is willing to make some rather major assumptions.[1]

In addition to health concerns, there is the larger question of choice in situations of food shortage. The countries of Africa have made their positions on GE crops quite clear to the world community over the last several years. The African Group was quite instrumental in securing a strong protocol on biosafety under the UN Convention on Biological Diversity. Under the protocol, exporters are required to gain advance agreement from importing nations for engineered organisms that are to be deliberately introduced into the environment. It was only due to the intransigence of the United States and its allies that a similar provision was not adopted for engineered organisms entering a country for food, feed or processing. But the sentiment of the African countries on this issue was quite clear.

The principal goal of the international community in times of famine should be to make sure people are fed. But, sadly, food aid has too often been used for political purposes. The US government has been using food aid, and the World Food Program, to systematically introduce genetically modified organisms into countries around the world without notifying the importing governments. It has been up to citizen groups to test the imported products and publicize the results of those tests. In Ecuador, the government was forced to remove a controversial product from aid distribution; its importation was a violation of national law as the product was not approved for consumption in the country. In Nicaragua and Bolivia, food aid has been found to be contaminated with Starlink, a variety of GE maize that is not approved for human consumption in the United States.

Non-governmental organizations around the world are challenging the World Food Program in their role as distributor of surplus US GE products—products rejected by consumers all over the globe. Agricultural commodity traders insist there is adequate non-GE maize in commercial channels to address the food shortages in southern Africa. Numerous countries have promised to sell non-GE maize to Zambia; in fact, the only countries in the world to export GE maize are the United States and Argentina.

There are many unresolved questions about the safety of genetically engineered food. Furthermore, it is not right to disregard the social and cultural preferences of other people, even during times of crisis. The media spotlight should be directed first to the tragedy occurring in southern Africa, and then to the US government's outrageous attempt to exploit it.

Rooted Resistance: Indian Farmers Stand against Monsanto

By Mira Shiva

Dr. Mira Shiva, MD, *is a medical doctor and public health activist in India. She has tackled issues of health care access, misuse of medicines, and medical technology for the past thirty years. Dr. Shiva is director of the Initiative for Health, Equity and Society and is affiliated with Doctors for Food and Biosafety, Task Force on Safety of Food and Medicine, and numerous other public health organizations.* *This article originally appeared in* GeneWatch, *volume 24, number 5, August–September 2011.*

In 2002, Bt Cotton became India's first genetically modified crop when the country's Genetic Engineering Approval Committee approved three varieties developed by Maharashtra Hybrid Seed Company Limited (Mahyco) in collaboration with Monsanto. Genes from *Bacillus thuringiensis*, a naturally occurring bacterium, were introduced along with an antibiotic resistant marker gene and cauliflower mosaic virus gene to enhance expression of the Bt gene.

Monsanto and Mahyco (of which Monsanto owns a 26 percent stake) made tremendous profits while hiking up the price of cotton seeds to more than five hundred times what farmers used to pay, from rupees 7/kg to rupees 3600/kg ($0.14/kg to $74/kg). Nearly half of this came from royalty payments. The companies were collecting around ten billion rupees (more than two hundred million dollars) per year in royalty payments from Indian farmers before the government of Andhra Pradesh, a state in southeast India, sued Monsanto, leading to a cap on the price of cotton seeds.

Andhra Pradesh saw problems beyond seed prices. Farmers who had commonly grazed their animals on cotton fields after harvest reported losing 25 percent of the sheep that grazed on leftover Bt cotton plants. In 2006, shepherds in the village of Ippagudem lost 651 of their 2,601 sheep; in the village of Valeru, they lost 549 of 2,168.[1] The corporations and authorities denied any connection to the animal deaths.

They also denied any connection to the rash of farmer suicides in India. Since the introduction of Bt cotton, tens of thousands of farmers have committed suicide—17,368 in 2009 alone.[2] A disproportionate number of those farmers were cultivators of Bt cotton who had incurred enormous debt linked to high costs of seeds, as well as the fertilizers and pesticides promoted by Monsanto and Mahyco as a necessity in order to grow the new cotton varieties. Vastly increased costs of production, high interest rates for credit, and low cotton prices have created unprecedented levels of debt for Indian cotton farmers. With the indebtedness came humiliation for proud farmers who have for generations managed life and work with dignity; driving farmers to find any way out. For some, this has meant selling a kidney; for many others, suicide.

Cotton farmers have little choice but to grow Bt varieties. As non-Bt seeds have been systematically made unavailable, 95 percent of the cotton being cultivated in India now comes from Bt seeds. Through licensing arrangements with seed companies across India's cotton belt, Mahyco has ensured that seed dealers sell only Bt cotton seeds.

Next Up: Food Crops

While Bt cotton is now entrenched in India, Monsanto and Mahyco have set their sights on what would be the first genetically modified food crop in India, Bt brinjal. Brinjal (known elsewhere as eggplant or aubergine) was first cultivated in India, and today there are 4,000 different varieties in the country, each linked with different regional recipes. The crop is not in short supply, so Monsanto and Mahyco's introduction of Bt brinjal was seen with much concern. It was thanks to the outcry at public hearings that the Indian government placed a hold on Bt brinjal approval.

Serious concerns have been raised about Mahyco's biosafety studies on its Bt brinjal. The trials centered around rat feeding studies lasting a mere ninety days. The study stopped at one generation, neglecting to assess effects on fertility and progeny. Nevertheless, the Genetic Engineering Approval Committee quickly decided to approve the crop. The GEAC's enthusiasm for genetically modified crops was not limited to Bt brinjal; in one meeting, it cleared ten different food crops for 91 field trials.[3]

Other genetically modified food crops lined up for trials in India include papaya, cauliflower, potatoes, tomatoes, corn, groundnuts, mustard, cabbage and pigeon peas. GM rice trials were planned in Chattisgarh, home of the country's richest biodiversity of rice varieties, but were stalled by the regional government following protests.

Federal Push for GMOs

Protests and actions against the unhindered commercialization of genetically modified crops have come from public outcry and local governments. In the federal government, India's biotechnology regulators are, to say the least, corporate-friendly. A bill currently awaiting passage in the parliament, the Biotechnology Regulatory Authority of India Bill, not only creates a new agency (the Biotechnology Regulatory Authority of India) with more leeway to speedily approve GMOs, but also imposes fines and even jail sentences for those who mislead the public about the safety of GMOs. This provision was targeted not at biotech companies—as one might think—but at the opponents of genetically modified crops.

In 2004, India announced a new Seed Bill making "unregistered seeds" illegal. While ostensibly protecting farmers against unscrupulous seed dealers, the act does not provide any new protections or compensations for farmers, aside from punishments for those selling unregistered seeds. Rather, it threatens small farmers' way of life, making it illegal for them to sell their own seeds to each other, although they have saved and shared their seeds for generations.

The Act sells farmers out under the guise of protecting them. Its main beneficiaries are private seed companies, transnational corporations in particular. In other words, it is exactly what one has come to expect of the biotechnology interest in Indian government.

20

Why GM Crops Will Not Feed the World

By Bill Freese

Bill Freese *is science policy analyst at the Center for Food Safety, a nonprofit group that supports sustainable agriculture. This article originally appeared in* GeneWatch, *volume 22, number 1, January–February 2009.*

Last spring marked a tipping point for rising global food prices. Haiti's prime minister was ousted amid rice riots; Mexican tortillas have quadrupled in price. African countries were hit especially hard.[1] According to the World Bank, global food prices have risen a shocking 83 percent from 2005 to 2008.[2] And for the world's poor, high prices mean hunger. In fact, the food crisis recently prompted University of Minnesota food experts to double their projection of the number of the world's hungry by the year 2025—from 625 million to 1.2 billion.[3]

Many in the biotechnology industry seem to believe there's a simple solution to the global food crisis: genetically modified (GM or biotech) crops.[4] Biotech multinationals have been in media blitz mode ever since the food crisis first made headlines, touting miracle crops that will purportedly increase yields, tolerate drought, and cure all manner of ills.

Not everyone is convinced. The UN and World Bank recently completed an unprecedentedly broad scientific assessment of world agriculture, the International Assessment of Agricultural Knowledge, Science and Technology for Development, which concluded that biotech crops have very little potential to alleviate poverty and hunger.[5] This four-year effort, which engaged some 400 experts from multiple disciplines, originally included industry representatives. Just three months before the final report was released, however, agrichemical/seed giants Monsanto, Syngenta and BASF pulled out of the process, miffed by the poor marks given their favorite technology. This withdrawal upset even the industry-friendly journal *Nature*, which chided the companies in an editorial entitled "Deserting the Hungry?"[6]

GM Crops: The Facts on the Ground

GM crops are heavily concentrated in a handful of countries with industrialized, export-oriented agricultural sectors. Nearly 90 percent of the world's biotech acres in 2007 were found in just six countries of North and South America, with the US, Argentina and Brazil accounting for 80 percent.[7] GM soybeans rule in South America, and Argentina and Brazil are known for some of the largest soybean plantations in the world. In most other countries, including India and China, biotech crops (mainly GM cotton) account for 3 percent or less of total harvested crop area.[8]

GM soybeans, corn, cotton and canola, the same four GM crops that were grown a decade ago, comprise virtually 100 percent of world biotech crop acreage.[9] Soybeans and corn predominate and are used mainly to feed animals or fuel cars in rich nations. Argentina, Brazil and Paraguay export the great majority of their soybeans as livestock feed, while more than three-fourths of the US corn crop is either fed to animals or used to generate ethanol for automobiles. Expanding GM soybean monocultures in South America are displacing small farmers who grow food crops for local consumption, and thus contribute to food insecurity. In Argentina, production of potatoes, beans, beef, poultry, pork and milk have all fallen with rising GM soybean production, while hunger and poverty have increased.[10] In Paraguay, the poverty rate increased from 33 percent to 39 percent of the population from 2000 to 2005, the years in which huge soybean plantations (about 90 percent of them now GM soybeans) expanded to cover over half of Paraguay's total cropland.[11] The only other commercial GM crops are papaya, squash and beets, all grown on miniscule acreage, and only in the US

Most revealing, however, is what the biotech companies have engineered these crops for. Hype notwithstanding, there is not a single GM crop on the market engineered for increased yield, drought-tolerance, salt-tolerance, enhanced nutrition or other attractive-sounding traits touted by the industry. Disease-resistant GM crops are practically non-existent.

In fact, commercialized GM crops incorporate just two "traits"—herbicide tolerance and/or insect resistance. Insect-resistant or Bt cotton and corn produce their own built-in insecticide(s) derived from a soil bacterium, *Bacillus thuringiensis* (Bt), to protect against certain insect pests. Herbicide-tolerant crops are engineered to withstand direct application of an herbicide to more conveniently kill nearby weeds. Crops with herbicide tolerance predominate, occupying 82 percent of global biotech crop acreage in 2007.[12]

Herbicide-tolerant crops (mainly soybeans) are popular with larger growers because they simplify and reduce labor needs for weed control. They have thus facilitated the worldwide trend to concentration of farmland in fewer, ever bigger, farms. Gustavo Grobocopatel, who farms 200,000 acres of soybeans in Argentina (an area the size of New York City), prefers to plant Monsanto's GM herbicide-tolerant variety (Roundup Ready) for the sake of simplified weed control, even though he obtains consistently higher yields with conventional soybeans. According to the Argentine Sub-Secretary of Agriculture, this labor-saving effect means that only one new job is created for every 1,235 acres of land converted to GM soybeans. This same amount of land, devoted to conventional food crops on moderate-size family farms, supports four to five families and employs at least half a dozen.[13] Small wonder that family farmers are disappearing and food security declining. The rapid expansion of "labor-saving" GM soybeans in South America has led to "*agricultura sin agricultores*" ("farming without farmers").

Increased Pesticide Use, Resistant Weeds, Lower Yields

According to the most authoritative independent study to date, adoption of herbicide-tolerant GM crops in the US increased the overall amount of weed-killers applied by 138 million lbs. in the nine years from 1996 to 2004, while Bt corn and cotton reduced insecticide use by just 16 million lbs. Thus, GM crops have increased overall use of pesticides (herbicides and insecticides) in the US by 122 million lbs. in less than a decade.[14]

The vast majority of herbicide-tolerant crop acres are planted to Monsanto's "Roundup Ready" varieties, tolerant to the herbicide glyphosate (aka Roundup). The excessive use of glyphosate associated with continuous planting of Roundup Ready crops is responsible for a growing worldwide epidemic of weeds that have evolved resistance to this chemical, alarming the world's agronomists.[15] Millions of acres of cropland have become infested with glyphosate-resistant weeds in the US, Argentina and Brazil, precisely those countries that rely most heavily on Roundup Ready crops, leading to a vicious cycle of increasing pesticide use and evolution of still greater levels of weed resistance.[16] Hence a technology often fraudulently promoted as moving agriculture beyond the era of chemicals has in fact increased chemical dependency. And of course, expensive inputs like herbicides (the price of glyphosate has more than doubled over the past two years) are beyond the means of most poor farmers, especially in combination with more expensive GM seeds.

What about yield? The most widely cultivated biotech crop, Roundup Ready soybeans, suffers from a 5 to 10 percent "yield drag" versus conventional varieties, due to both adverse effects of glyphosate on plant health as well as unintended effects of the genetic engineering process used to create the plant.[17] Unintended, yield-lowering effects are a serious though little-acknowledged technical obstacle of genetic engineering, and are one of several factors foiling efforts to develop viable GM crops with drought tolerance.[18] While insect-resistant crops can reduce yield losses under conditions of heavy pest infestation, such conditions are relatively infrequent with corn. And because cotton is afflicted with so many pests not killed by the built-in insecticide, biotech cotton farmers in India, China and elsewhere often apply as much chemical insecticide as growers of conventional cotton. Only because they have paid up to four times as much for the biotech seed, they end up falling into debt. Each year, hundreds of Indian cotton farmers commit suicide from despair over insurmountable debts.[19]

Biotechnology = Patented Seeds + Chemicals

If biotech crops are not about feeding the world, what is the point? The agricultural biotechnology industry represents an historic merger of two distinct sectors—agrichemicals and seeds. In the 1990s, the world's largest pesticide makers—companies like Monsanto, DuPont, Syngenta and Bayer—began buying up the world's seed firms. These four biotech giants now control a substantial 41 percent of the world's commercial seed supply.[20] The motivations for this buying spree were two-fold: the new technology of genetic engineering, and the issuance of the first patents on seeds in the 1980s. As we have seen, biotech firms employ genetic engineering chiefly to develop herbicide-tolerant crops to exploit "synergies" between their seed and pesticide divisions. Seed patents ensure greater control of and higher profits from seeds, in part by allowing biotech firms to outlaw seed-saving.

While patents on biotech seeds normally apply to inserted genes (or methods for introducing the gene), courts have perversely interpreted these "gene patents" as granting biotech/seed firms comprehensive rights to the seeds that contain them. One consequence is that a farmer can be held liable for patent infringement even if the patented gene/plant appears in his fields through no fault of his own (e.g. cross-pollination or seed dispersal), as happened most famously to Canadian canola farmer Percy Schmeiser. Another consequence is that farmers can be sued for patent infringement if they engage in the millenia-old practice of seed-saving—that is, replanting seeds saved from their harvest.

In the US, industry leader Monsanto has pursued thousands of farmers for allegedly saving and replanting its patented Roundup Ready soybean seeds. An analysis by Center for Food Safety has documented court-imposed payments of more than $21 million from farmers to Monsanto for alleged patent infringement. However, when one includes the much greater number of pre-trial settlements, the total jumps to over $85 million dollars, collected from several thousand farmers.[21]

Spurred on by the biotech multinationals, the US and European governments are pressuring developing nations to adopt similar gene and seed patenting laws. This is being pursued through the World Trade Organization, which requires member nations to establish intellectual property regimes for plants, as well as through bilateral trade agreements. Since an estimated 80 to 90 percent of seeds planted in poorer nations are produced on-farm (i.e. saved seed), the revenue to be gained from elimination of seed-saving is considerable—conservatively estimated at $7 billion dollars.[22] If biotech/seed firms have their way, the "seed servitude" of US farmers could soon become a global reality.

Biotech firms also have Terminator technology waiting in the wings. Terminator is a genetic manipulation that renders harvested seed sterile, and represents a biological means to achieve the same end as patents: elimination of seed-saving. While international protests have thus far blocked deployment of Terminator, Monsanto recently purchased the seed company (Delta and Pine Land) that holds several major patents on the technology (together with USDA). And while Monsanto has "pledged" not to deploy Terminator, the company has clearly stated that this "pledge" is revocable at any time.

Private Profit Replaces Public Interest

The rise of GM crops has been accompanied by a massive shift in plant breeding from the public to the private sector. Breeders at universities and non-profit agricultural research institutes once played a major role in delivering useful new crop varieties, guided at least in part by the interests of farmers. Today, public sector breeding is fast dying, the victim of dramatic cutbacks in funding from rich nations and the World Bank. Organizations like the International Rice Research Institute and Center for Improvement of Maize and Wheat lack funds to even distribute useful new crop varieties they have already developed to farmers who need them—including conventionally-bred wheat and rice with high yield, disease- and/or insect-resistance. In contrast, GM crop development is overwhelmingly dominated by profit-seeking biotech firms. In the US, 96 percent of approved GM crop varieties were developed by private firms, 88 percent by the "big five" biotech companies.[23] Monsanto alone is responsible for the traits in at least 87 percent of GM crops worldwide.[24] Public relations aside, biotech firms continue to devote the bulk of their research efforts to develop new herbicide-tolerant crops for use with their proprietary chemicals, labor-saving crops best-suited to larger farmers.[25]

Fewer Seed Choices, Higher Seed Prices

To make matters worse, high-quality conventional seeds are rapidly disappearing, thanks to the biotech multinationals' tightening stranglehold on the world's seed supply. Biotech seeds presently cost two to more than four times as much as conventional varieties. The price ratchets up with each new "trait" that is introduced. Seeds with one trait were once the norm, but are rapidly being replaced with two- and three-trait versions. As Monsanto put it in a presentation to investors, its overriding goals are "acceleration of biotech trait penetration" and "to invest in "penetration of higher-[profit-]margin traits . . . "[26] Monsanto and Dow recently announced plans to introduce GM corn with 8 different traits (6 insecticides and tolerance to 2 different herbicides). Farmers who want more affordable conventional seed, or even biotech seed with just one or two traits, may soon be out of luck. As University of Kentucky agronomist Chad Lee put it: "The cost of corn seed keeps getting higher and there doesn't appear to be a stopping point in sight." The biotech industry's growing control of the world's seed supply ensures that farmers in developing countries that accept GM crops will face dramatically rising seed prices from "trait penetration."

True Solutions

The authors of the UN-World Bank-sponsored IAASTD report mentioned above recommend agroecological farming techniques as the most promising path forward for the world's small farmers. Ever since the Green Revolution, the agricultural development establishment has focused primarily on crop breeding and expensive inputs (e.g. fertilizers, pesticides and "improved seeds"), not least because input-centered schemes offer potential market opportunities to multinational agribusinesses. In contrast, agroecology minimizes inputs, and relies instead on innovative cultivation and pest control practices to increase food production. A 2001 review of 200 developing country agricultural projects involving a switch to agroecological techniques conducted by University of Essex researchers found an average yield gain of 93 percent.[27]

One strikingly successful example is the push-pull system, practiced by 10,000 farmers in East Africa. Push-pull involves intercropping maize with plants that naturally exude chemicals to control insect and weed pests, which increases yields while also enhancing soil fertility and providing a new source of fodder for livestock.[28] A new dryland rice farming technique called the System of Rice Intensification substantially increases yield, and is spreading rapidly in rice-growing nations despite dismissal by the agricultural development establishment. Small farmers like agroecological techniques because they foster independence and reduce expenditures on inputs.

Conclusion

The tremendous hype surrounding biotechnology has obscured some basic facts. Most GM crops feed animals or fuel cars in rich nations, are engineered for use with expensive weed killers to save labor, often have reduced yields, and are grown by larger farmers in industrial monocultures for export. The technology is dominated by multinational firms intent on controlling the world's seed supply, raising seed prices, and eliminating farmer seed-saving.

Real solutions will require radical changes. Rich nations must stop dumping their agricultural surpluses in the global South, respect the right of developing countries to support their farmers, and fund agroecological techniques to enhance small farmers' ability to feed their families and their nations' citizens.

PART 4

Corporate Control of Agriculture

A small number of corporations are taking legal and ownership control over the world's food supply. The result has been a decrease in seed and therefore crop biodiversity. The issues of patents on living organisms, ag-biotech monopolies, and the creation of monocultures all raise serious questions about the soundness of genetically engineering the world's most important crops.

Corporations like Monsanto are able to genetically engineer a particular seed with a foreign trait and then patent that seed. The Monsanto Corporation can then dictate the terms of use of its patented product. Some corporations holding patents on seeds and crops have required farmers to sign legal documents compelling them to grow only that company's seed, use only that company's chemicals, and pay "technology fees" for the genetically engineered seeds in addition to the cost of the seeds themselves. The availability of patent protection for these products increases the interest of investors, as patents help to ensure profits as long as farmers agree to plant the genetically engineered crops and consumers agree to buy the food.

Food security is further threatened by the fact that a smaller number of multinational corporations are taking control of the ownership of the food supply. These seed oligopolies threaten to squeeze out the voices of farmers and consumers in the debate about genetically engineered food and are prepared to use all their power to protect their financial interests to advance their technology. Relying on a handful of self-interested corporations to make important and far-reaching decisions about agriculture has resulted in inequitable policies. As an example, crop genetic engineering has threatened small organic farmers who bear the risks of genetic drift and genetic pollution.

GMO agriculture is an important piece of the current corporate focus on monocultures, which leaves no room for natural biodiversity. If a genetically engineered crop is susceptible to a new virus, for example, the whole crop will be susceptible since there will be very little chance that some of the crop may have a particular mutation to protect it from the virus.

The essays in this section explore the current trends in agricultural biotechnology and their impact on food security.

—Jeremy Gruber

Patented Seeds vs. Free Inquiry

By Martha L. Crouch

Martha L. Crouch, PhD, *was a graduate student at Yale University studying the development of seeds and flowers when genes were first cloned. By the time she headed her own plant molecular biology lab at Indiana University, plant genes were being patented. Now she consults from her home in Bloomington, Indiana, on issues of biotechnology and agriculture for non-profit groups and law firms. This article originally appeared in* GeneWatch, *volume 26, number 1, January–March 2013.*

Several years ago I was approached by a colleague who wanted to do preliminary experiments on translocation of the herbicide glyphosate into flowers of Roundup Ready soybeans. He needed a few handfuls of a named variety of seed. Since I live in soybean country he figured I might know a farmer willing to part with such a small quantity. "Sure," I said, "no problem."

It was, I discovered, a big problem. As part of the technology agreement that farmers sign when they purchase Roundup Ready soybeans, they . . . may *not* plant and may not transfer to others for planting any seed for crop breeding, research, or generation of herbicide registration data."[1]

This was a surprise to me. I was quite familiar with prohibitions on farmers replanting seeds of genetically engineered crops, a controversial innovation of Monsanto's[2] that they vigorously enforce and which has been adopted by other companies as well.[3] But as a scientist, I bridled at being prohibited from simply walking into a store or farmer's shed and leaving with material to study one of the most common plants in my environment. After all, we are swimming in these Roundup Ready soybeans here in Indiana. Every fourth acre of the state is planted in them. Seeds fall out of trucks, volunteer in subsequent crops, and are piled high in grain elevators along our county roads. Soybeans, soybeans everywhere, and not a seed to study?

Legally, the only way to study Roundup Ready soybeans or any other genetically engineered crop, commercialized or not, is to go through the company that owns the patents. If the company is willing, it will offer researchers—or more likely today, their institutions—confidential agreements with the terms under which research can be conducted.

In other words, the companies that stand to profit or lose from the results are ultimately in control of who gets to do research and who doesn't. Some scientists who have labored under such contracts think that this restricted access gives the seed industry too much say in the kinds of research scientists do and the way their data are reported to the public, stating that "no truly independent research can be legally conducted on many critical questions" involving these crops.[4,5,6]

What, then, are tangible consequences of biotech companies holding the keys to research on the genetically engineered crops that surround us? From looking at the literature, it is inherently difficult to say that a specific study is missing or skewed because of such a policy. However, since learning about the "no research" clause, I've had opportunities to ask agricultural scientists about how this particular dependence on agribusiness has affected them.[7]

Some admitted to being put off of particular projects even before asking the company for a contract. They weren't affiliated with an institution that normally conducts agricultural research and thus weren't covered by a blanket agreement, or they belonged to an institution that could not accept the terms offered. One scientist didn't want to disclose her methods or theories for fear that her ideas would be pilfered at an early stage by company scientists who had more resources and experience with the system she wanted to study.

In fact, it seems that the superior resources of corporations compared with many academic labs have discouraged some graduate students from doing projects with genetically engineered crops. Students asking for research materials reported being told by company scientists that "we know everything there is to know about [whatever]," leaving them with the impression that it would be difficult to carve out a niche in competition with the "big guys." I've also heard of students being strongly encouraged to change direction when it appeared that their research was heading for a conclusion that was not in the company's interest. In one case, a student said he was offered a grant if he dropped his current project in favor of another one after his preliminary results pointed to an issue with crop performance.

Researchers who did persevere and in the end reported results that might damage the company's bottom line were sometimes refused further access to seeds or other materials. They also faced coordinated attacks on their published work, well beyond what most academics experience during the normal give and take of scientific critique. Often their work was discounted for deficiencies in methodology, such as lack of the most appropriate control plants or reagents, at the same time that they were denied access to these materials.

Other factors weigh against independent research in agriculture, of course. Public research in agriculture is influenced by private money and guidance at every level. For years, as public money has dwindled, grants and contracts from corporations have increased. So have industry-sponsored endowed chairs, graduate student fellowships, undergraduate teaching grants, internships, and other partnerships that give corporations more say, including having seats on university boards.[8] I have argued that even public money for basic research is steered towards projects that will support agribusiness.[9]

It is a wonder that any truly independent studies of genetically engineered crops get done against the backdrop of all of these corporate influences. Some scientists rise to the challenge, but my sense is that many more find it too much of a hassle and decide to work on other issues.

Certainly, when I comb the scientific literature for impacts of particular genetically engineered crop systems I am dismayed at how few independent studies I find. This is especially true for impacts on non-target organisms, meaning all of us except weeds and pest insects. For example, recently I was searching for information about the levels of glyphosate in pollen and nectar of Roundup Ready crops in light of the crises honeybees and other pollinators are experiencing. I didn't find any relevant studies in the public domain.

Maybe if I had been able to send my colleague some Roundup Ready soybeans when he asked, we would know about glyphosate levels in pollen and nectar by now, and thus be better equipped to assess environmental impacts. And I wonder how many other studies are missing at least partly because of the impediments to free inquiry from patent-driven research restrictions.

I believe it is in the public's interest to bring seeds of genetically engineered crops back into the common sphere where they can be used freely in research. Although doing so won't remove all corporate influence from scientific studies, it is a concrete step in the right direction towards transparent, reliable information about these new and impactful technologies.

BGH and Beyond: Consolidating Rural America

By Jack Doyle

Jack Doyle *is the director of J.D. Associates, an investigative research firm specializing in business and environmental issues. He is the former director of the Agricultural Resources Project at the Environmental Policy Institute. He is the author of* Altered Harvest *published by Viking Press.* *This article originally appeared in* GeneWatch, *volume 4, number 2, March–April 1987.*

"You can convert your Ford into a Lincoln Continental," says Monsanto's Howard Schneiderman, using an automobile metaphor to hype his company's new genetically engineered bovine growth hormone (BGH) for dairy cows. "By giving your cow bovine somatotropin," he says, "all of a sudden it becomes a high performance cow." Indeed, some highly-pampered cows might produce up to 40 percent more milk with BGH, sending a great wash of new milk onto an already glutted dairy market, driving prices down, and forcing as many as one-third of all dairy farms—mostly the smaller ones—out of business. But BGH is only the tip of the iceberg.

In a recent FDA-commissioned study entitled, "Emerging Developments in Biotechnology," some 92 firms were identified pursuing 171 separate veterinary biotechnology projects in the US. W.R. Grace, Genentech, Biogen, Upjohn, Syntex, Amgen, California Biotechnology, TechAmerica are among the more prominently noted firms. In addition, corporations as diverse as Kodak, Burroughs-Welcome, Unilever, SmithKline-Beckman, A Akzo, and Diamond Shamrock are also involved. Experts polled in the FDA study predict that more than 200 distinct veterinary products and processes can be expected within the near future, with more than half achievable within the next eighteen months.

Concerns

Meanwhile, a number of social, ethical, and economic questions have emerged with the proposed use of the livestock biotechnologies. For example, will the use of BGH in dairy cows facilitate their "burn out?" Does BGH have any effect on disease susceptibility or fertility in dairy cows? If porcine growth hormone is widely used in the hog industry, what will a 25 percent reduction in corn feed demand mean to grain farmers in the Corn Belt? Will some agricultural regions benefit at the expense of others with the widespread application of livestock biotechnologies? Will consumers react differently to animal products derived from livestock engineered with human genes, such as human growth hormone?

In some cases, these concerns and others have resulted in farmer and public reaction to the planned introduction and use of products such as BGH, including the following:

- farmer protests in Wisconsin, with picketing of publicly and privately-sponsored university research on the hormone;
- a proposed national consumer boycott of BGH-produced milk by the Foundation on Economic Trends;
- a June 1986 Congressional hearing on BGH concerned with the USDA's role in sponsoring BGH research, the product's impact on family farmers, and its impact on the dairy price support program;
- two bills introduced in the US House of Representatives last year by Rep. Gunderson (R-WI), one prohibiting the FDA from approving the hormone until Congress is provided with an analysis of its effect on the environment, milk production and price supports; and another requiring the USDA to report to Congress on the impact of BGH on milk production and price supports;
- statements or resolutions from a number of farm and dairy groups, including the National Grange, Vermont's Cabot Co-op Creamery Co., the Empire State Grange of New York, and the National Milk Producers Federations, calling for a postponement or flat out prohibitions of the use of BGH-produced milk until all questions are put to rest.

Industry, meanwhile, is countering with a public relations effort aimed at minimizing consumer fears of foods made with genetically altered hormones, while emphasizing health-related benefits, such as the production of leaner pork and beef. Research at Pennsylvania State University has found that porcine somatropin (PST) reduces back fat in hogs by some 70 percent, and may reduce the amount of feed needed by 25 percent. In the pork industry, for example, there is some fear that consumers may confuse PST with steroids, such as DES, which was taken off the US market in 1979. Industry officials are encouraging their people to use the term "porcine somatotropin" rather than "porcine growth hormone."

A genetically engineered PST, estimated to cost $4 per treated pig at the farm level, may reach the market by 1988. Four corporate/biotech "teams" will compete for the estimated $300 million-a-year PST market in the US: International Minerals & Chemicals Corp./Biogen, Monsanto/Genentech; SmithKline-Beckman/Amgen; and American Cyanamid/BioTechnology General.

Other Developments

Several companies and universities are researching ways to genetically engineer growth and other traits directly into farm animals via embryo engineering. Work is being conducted on embryos at the one-cell stage, where cloned foreign genes governing various traits can be micro-injected into the pronuclei. The genes for growth hormone, prolactins (lactation stimulation), digestive enzymes, and interferons are some of the likely candidates.

Amgen, Inc., a California bio-technology company, is working on interferon to treat cattle shipping fever and a poultry somatotropin for improving breeding stock. Emberx Inc. of North Carolina, is working on ways to administer vaccines and antibiotics to turkeys and chickens while they are still embryos in the shell.

Others are researching the twinning gene in cattle, the use of human growth hormone gene in cattle and hogs, and the gene in sheep that makes certain breeds produce multiple births routinely. Still others are considering the possibility of modifying the digestion-aiding bacteria that live in the stomach of ruminant livestock—a few with the idea of giving these microbes the ability to digest lignin, the protein that is a major structural component of wood.

Vaccines

Worldwide, there are more than twenty billion farm animals susceptible to disease that vaccines could be used to treat or prevent. In the US alone, annual losses from animal diseases are estimated at $17 billion. Not surprisingly, there is a long list of companies conducting research on vaccine development. Hoechst and International Minerals & Chemicals Corp are nearing market with a vaccine for hoof and mouth disease; Molecular Genetics has plans to develop and market as many as thirty different livestock vaccines for treating pig and calf scourges.

Feeds & Feed Improvement

H.J. Heinz, Kodak and Monsanto have each made forays into biotechnology from the standpoint of animal nutrition and/or improving feeds. Kodak, for example, announced plans to build a multimillion-dollar animal nutrition research center in Tennessee. Two years later in October 1986, Kodak entered into a four-year $20 million cooperative research agreement with Molecular Genetics, Inc. of Minnesota to focus on the development of products for the "health, nutrition and productivity of animals."

Product "Synergy"

The feed and animal drug industries have an obvious stake in the genetic make-up of farm animals that need to be fed and/or treated. Some major companies—such as W.R. Grace, Cargill, Upjohn, Merck, Pfizer, Perdue and others—have fully or partially entered the livestock genetics business, and are quite aware that changes in the genetic make-up of farm animals can change feeding and treatment strategies. Others see possibilities for "product synergy" with particular breeds, or among and between products such as vaccines, hormones, feeds, and antibiotics. For example, Unilever's Philip Porter explains that "in modern agriculture, with its intensive rearing practices, there is a synergy between antibiotics and vaccines that has not yet been exploited." In addition, the recent decision of the US patent office allowing the patenting of man-made, recombinant animals will heighten corporate interest in the livestock genetics industry resulting in further acquisitions and consolidation.

In fact, economic consolidation throughout agriculture at every level will continue in the near future, pushed by an agricultural biotechnology that results in ever-mounting farm surpluses, depressed farm prices, and more farm failures. More public policy emphasis needs to be turned toward an agricultural research orientation that will diversify American agriculture, reduce production costs for farmers, and revitalize the farming economies of small town America. For if agricultural biotechnology succeeds only in producing more, in a simple volume sense, that will certainly mean less in real economic return to rural America and the nation.

23

Changing Seeds or Seeds of Change?

By Natalie DeGraaf

Natalie DeGraaf *is program specialist for Global Science Engagement, International Affairs at the American Society for Microbiology. This article originally appeared in* GeneWatch, *volume 24, number 6, October–November 2011.*

Recently, technological solutions to public health issues took on the form of a seed, bringing small farmers to a new crossroads between traditional forms of agriculture and industrial biotech agriculture. There is an extensive and sordid history of the impact that GM crops inflict upon nations, including the US, Argentina and India. Three predominant concerns arise with the integration of GM seeds into African agriculture, all feeding into larger public health priorities. The first concern identifies GM seed integration and corresponding impacts as problematic to individuals and environments. The second highlights GM crops' ineffectiveness at helping farmers or improving food security in the developing world. Lastly, the triumvirate of private industry, international corporations and governments generate squalid attempts to promote political agendas through GM seed introduction into previously untapped markets and communities.

The latter two concerns are particularly crucial to developing nations due to lack of political transparency and accurate assessments of new biotech solutions. As a continent afflicted with severe drought and pest resilience, Africa provides an environment that continues to be exploited by international corporations claiming "massive potential for crop yields." Amidst the hype, there is a growing opposition to the implementation of GM seeds and the corporations willing to donate them. While many private companies and foundations finance initiatives and research to address the global health issue of food security and hunger, using GM crops to meet these ends will ultimately create a system of dependence on foreign corporations and further deepen social, economic and environmental disparities in African farming communities.

Concern 1: Problematic individual and environmental impacts from GMOs

There is increasing evidence from the United Nations and World Health Organization that a strong causation exists between the adoption of GM seeds and environmental degradation, including deforestation. Most research shows a decrease in biodiversity with the introduction of GMOs. This means using GM seeds may actually make agricultural conditions worse than they are presently, not to mention the added threat to the health of humans, insects, and animals.

Current international trade policies have heavily regulated importing GM products into the EU as a political response to social outcry of lack of evidence on safety. The safety of GM seeds on the public health and environment is still highly unknown due to a severe lack of unbiased research being conducted external to the reports issued by GM company laboratories. In 2002, the nations of Zimbabwe, Zambia and Mozambique actually refused requested food aid that contained GMOs for fear of health and safety. Farmers in rural India have noted instances of animals dying from grazing on GM crops and new reports are investigating the relationship between increased allergy prevalence and GM foods as well as transference of antibiotic resistance to consumers. Most citizens in developing nations fail to consider these potential health effects because of the perception that government regulation would address such issues.

Concern 2: High economic costs associated with using industrial agricultural methods and the ineffectiveness of GM seeds at addressing food security and hunger.

Effectiveness of GM seeds to increase crop yield has been repeatedly refuted, along with the economic feasibility of their use. In 2010 GM seed giant Monsanto discovered their seed Bollgard 1 was no longer effective at eradicating the pest 'bollworm' that threatens the crops of cotton farmers in India. The bollworm pest developed a resistance to the technology that only a year earlier was deemed a significant technological success by the Union Science Minister. Monsanto responded by creating a new seed, Bollgard 2, and recommending the increased use of pesticides at a higher price to the consumer. This situation proved two points: that GM seed modifications are unreliable and can cause further issues in the long term; and that the cyclical trend of pest resistance necessitates the ongoing development of new, costlier seeds which can trap farmers in the GM web. It's easy to see how this cycle leaves farmers deeply in debt after spending so much money on the seeds and necessary additives.

The cost to mitigate and sustain GM seeds has historically led numerous farmers into deeper poverty, as they require expensive fertilizers and pesticides. Furthermore companies that produce GM seeds prohibit seed saving, a process that small farmers have relied upon for centuries to generate income and ensure livelihoods. Recent advancements have also allowed Monsanto to now genetically modify seeds to self-destruct after one season, preventing farmers from saving seeds from their crop to plant next year and instead requiring them to return to Monsanto for new seeds every year. Those hardest hit economically by GM use are the farmers most willing to support the corporations that advertise the benefits of their use. This occurs when farmers enter into deals with GM seed corporations without knowledge, understanding and awareness of the plethora of social, economic and environmental costs. While companies like Monsanto and DuPont may be willing to donate their seeds to small African farmers initially, once the farmers have utilized the seed, they are locked into purchasing them from these companies. This ethically questionable trend puts money only in the pocketbooks of corporations as uneducated farmers make uninformed decisions.

In a recent study on the impact of biotech in West Africa, researchers found "little evidence of practical application of biotechnology and benefit to farmers and the wider community."[1] Additionally, a recent Worldwatch Report noted the lack of correlation between GM seed food production and rural development; in fact the opposite occurred, with more farmers moving to the city as their lands are taken over by large industrial farmers. Collaborative studies done in Africa also found that GM seeds magnified the gap between socio-economic classes. With a 90 percent share of food production in Africa attributed to small farmers, about 20 percent more than the global average, the introduction of GM seeds threaten to push smallholders out of business in favor of large scale, high-input agriculture.

What does all this mean for Africa's future? It means that GM seeds will place Africa in a poorly situated position to address imperative food issues and promote sustainable economic growth. What is most unfortunate is that Africa already has the tools to combat its food security issues through traditional agro-ecological farming practices. Insufficient evidence exists to show that GM crops produce higher yields and better pest protection than traditional farming practices. A United Nations Rapporteur on food rights concluded that natural farming methods actually fared substantially better than chemical fertilizers in terms of yield and protection against pests. Agro-ecological methods encourage natural seed breeding, including organic, and utilize an integrated soil-plant-animal cropping system. Many critics of GM seeds argue for a more grassroots approach, acknowledging the importance that lay expertise provides in African farming and allowing certain communities to preserve local seeds that have been enhanced through natural, localized selection processes to be pest resistant and drought resistant. This is supported by a 2011 UN press release that noted "scant attention has been paid to agro-ecological methods that have been shown to improve food production and farmers' incomes, while at the same time protecting the soil, water, and climate."[2] This transition to eco-agriculture has been noted in India and Malawi, with notable positive outcomes for approximately 100,000 small farmers in Malawi. Both these nations are instigating a national shift to agro-ecological practices that have since produced significantly higher crop yields without the risk of environmental effects. What's more, 25 to 50 percent[3] of the yearly harvest is wasted in Africa due to infrastructure constraints such as lack of storage, transportation to get the crop to market, and a lack of post-harvest technologies; this is a hefty sum to ignore in nations stereotyped as having no food. The promotion of GM seeds in developing nations is not the most effective manner to address hunger and food security.

Concern 3: Attempts to promote political agendas through GM seed introduction into previously untapped markets and communities

If the traditional agricultural systems are capable of producing better results, why are organizations funding GM seeds? Some speculate it is a joint effort to usher private corporations into developing nations to manipulate the global agenda. This is accomplished through large international corporations taking advantage of weak national biosafety regulations and laws that are designed to protect citizens but instead protect biotech companies' investments by enforcing laws regarding patent protection of GM seeds. Africa is full of poor farming communities that are more than willing to accept free GM seeds that corporations such as Monsanto, in connection with the Bill and Melinda Gates Foundation, provide to NGOs and governments for distribution. These seemingly mutually beneficial relationships allow corporations to use public health platforms to infiltrate new markets and take hold of the local industry. Groups such as the African Center for Biosafety assert that the relations between large corporations and private philanthropic foundations like the Gates Foundation, which donated $1.7 billion to kick-start the second "Green Revolution," harbor too much power and foster hidden agendas, as represented by the Gates Foundation purchase of $27.6 million in shares of Monsanto stock over three months in 2010. In South Africa, the Department of Agriculture is purported to use "attractive offers to provide farming equipment, water piping and seeds"[4] as a means to promote GMO's to small farmers, often uneducated and illiterate with no understanding of patents and property rights. The current legal system in Africa is fragmented with different levels of patent law and IP rights between nations. It comes as no surprise that multinational GMO corporations are promoting common legislation for African biosafety assessment and GM seed utilization, making it easier for companies to commercialize their biotech products.

If a farmer's GM seed cross-pollinates into a neighboring farm's non-GM crop, the neighbor's farm would then fall under patent violation and could therefore no longer store seeds for the following year. The entwined system of agriculture and patent law has resulted in numerous multi-million dollar lawsuits against poor farmers and countless farms being forced to form cooperatives with neighboring farms to offset the costs of GM seeds. The decrease in the number of workers and creation of a centralized food production system has detrimental effects on national sovereignty, food security, and individual rights to choose agricultural methods. These legal actions create a system of foreign exploitation of natural resources and a monopoly on global food production. International Property Rights (IPR) protect the laws that heavily favor the corporations, but some nations are refusing to adopt these IPRs and acknowledge the patent rights of GMO corporations on crops. Some nations have demonstrated the inequitable nature of this process and are pursuing legal action against these corporations. For instance, Argentina chose a national plan to subsidize Monsanto's product Roundup Ready Soya but lacked the patent laws to enforce royalty payments to Monsanto while simultaneously prohibiting seed saving. This resulted in enormous GMO cross-pollination between fields, with patented strains found in more than 95 percent of the market for soya, all of which would have to pay royalties and purchase seed yearly. The ramifications were massive unemployment and emergence of large monopolistic farms that engulfed the smaller bankrupted farms. In retaliation for not receiving their royalties, Monsanto has since blocked shipments of soybeans from Argentina to other countries.

Western efforts to aid developing nations have at times been shown to cause more harm than good. The Bill and Melinda Gates Foundation maintains a narrow scope of solutions to broad public health issues, resulting in implanting westernized plans to aid developing nations without having localized knowledge of the situation. This "outsiders" approach contributes to the problem rather than the solution. Biotech expert Philip Bereano has noted, "Big donors are undermining huge numbers of local initiatives to increase food security and protect biodiversity when they exclude small-scale projects in favor of industrial ones that actually have consequences counter to such goals."[5] The Gates Foundation is known for their technologically sophisticated solutions, which are appealing to decision makers and donors looking to invest dollars in fast outcomes that look great on paper and provide clearly defined results, such as vaccines. Unfortunately, the reality is that the unsubstantiated impact projections and bar charts that sold the promise of biotech to investors often dissipate quickly once solutions are implemented and shown to fail to meet expectations. Agricultural technologies cannot provide complete solutions to hunger and food sovereignty. Social, political and economic factors must first be addressed in order to ensure food access and appropriate development.

A more proactive regulatory approach toward biotech solutions could help to buffer developing nations against the harmful impacts of the implementation of new biotechnologies. The failures of policy and decision makers to generate a buffer are illustrated globally where the utilization of GM seeds required higher crop prices at the market to offset the investment costs of expensive seeds and fertilizers. This generates difficulties with selling crops and contributes to issues of food waste. In fact, a significant challenge for farmers with GM produce is the lack of partnership with EU nations for the sale of GM products, leaving farmers with a surplus of crops without a market. These conditions are not orientated toward the goal of achieving food sovereignty and addressing hunger issues in Africa. Instead, they are the product of top-down interventions that accumulate profit for their shareholders. The International Institute for Environment and Development, a leading proponent of revised food sovereignty policies, has outlined principles to define food sovereignty as empowering citizens to define their own agricultural management system unrestricted by intellectual property rights and GM patents.

Civil rights author Maya Angelou has a saying: "I did the best that I knew how to do. When I knew better, I did better." Non-profits and foundations working to address global health issues are presented with the arduous task of creating infrastructure for healthy and sustainable development to enhance food security in nations where very little economic or political structure exists to address these needs. While many mistakes have been made due to the masquerading of GM seeds as the best solution to food security and hunger, now is the time to know better.

Food, Made from Scratch

By Eric Hoffman

Eric Hoffman *is the former food and technology campaigner for Friends of the Earth. This article originally appeared in* GeneWatch, *volume 26, number 1, January–March 2013.*

Agriculture as we know it needs to disappear. . . . We can design better and healthier proteins than we get from nature. —J. Craig Venter[1]

A (Very) Short History of Agriculture

For ten thousand years humans have been manipulating plants for food production. This began at a very basic level, saving slips or seeds from the fastest growing, highest yielding, best tasting and most nutritious plants for the following season. This form of conventional breeding eventually led to the development of hybrid crops which involved cross-breeding two genetically different lines in the same genus and usually the same species. These changes in the plants were limited to the genes already present within the plants.

This all changed dramatically with the advent of genetic engineering in the 1970s and 1980s. Genetic engineering allowed the transfer of genes between species, even between species of different kingdoms, as when bacteria DNA were inserted into plants—and court decisions allowed, for the first time, patents on life. Since then, genetically engineered organisms, often called genetically modified organisms (GMOs), have become a ubiquitous feature of industrial agriculture in the US, comprising roughly 88 percent of the corn, 94 percent of the soybeans, 90 percent of the canola, 90 percent of the cotton, and 95 percent of the sugar beets grown in the country.[2] These crops have been engineered and patented by chemical companies, including Monsanto and Bayer, to either withstand increasingly heavy doses of herbicides or to produce their own systemic pesticide.

Synthetic Biology—Extreme Genetic Engineering

In the second decade of the twenty-first century, we are likely to see even more radical changes on the horizon, this time via a rapidly growing field known as synthetic biology. Synthetic biology is a broad term used to describe a collection of new biotechnologies that push the limits of what was previously possible with "conventional" genetic engineering. Rather than moving one or two genes between different organisms, synthetic biology enables the writing and re-writing of genetic code on a computer, working with hundreds and thousands of DNA sequences at a time and even trying to reengineer entire biological systems. Synthetic biology's techniques, scale, and its use of novel and synthetic genetic sequences make it, in essence, an extreme form of genetic engineering.

Synthetic biology is a nascent but rapidly growing field, worth more than $1.6 billion in annual sales today and expected to grow to 10.8 billion by 2016.[3] Many of the largest energy, chemical, forestry, pharmaceutical, food and agribusiness corporations are investing in synthetic biology research and development or establishing joint ventures, and a handful of products have already reached the cosmetic, food, and medical sectors with many others not far behind. Much of this focus is being placed on agriculture applications to become the next wave of GMOs—let's call them synthetically modified organisms (SMOs).

Synthetically Modified Organisms

Monsanto, the biotech and chemical giant, recently announced a joint venture with Sapphire Energy, a synthetic biology algae company. Monsanto is interested in algae because most types of algae reproduce daily, compared to traditional agriculture crops which only reproduce once or twice a year. Monsanto hopes to isolate traits in algae at a much faster rate than can be done in plants, which could then be engineered and inserted into crops.[4] Such technologies will potentially allow increasing numbers of (and more extreme) genetically engineered crops on our fields.

J. Craig Venter, a leading synthetic biologist who built the world's first synthetic genome by copying a rather simple goat pathogen's genome, created a new company, Agradis, to focus on applying synthetic biology to agriculture. Agradis aims to create "superior" crops and improved methods for crop growth and crop protection. The company plans to create higher-yielding castor and sweet sorghum for biofuels through undisclosed "genomic technologies."[5]

There are even plans to "improve" photosynthesis in plants through synthetic biology. Researchers at the Department of Energy's National Renewable Energy Laboratory in Colorado believe that "the efficiency of photosynthesis could be improved by re-engineering the structure of plants through modern synthetic biology and genetic manipulation. Using synthetic biology, these engineers hope plants can be built from scratch, starting with amino acid building blocks, allowing the formation of optimum biological band gaps,"[6] meaning plants could turn a broader spectrum of light into energy than done naturally through photosynthesis.

Other food and agriculture applications of synthetic biology in the works include food flavorings, stevia, coconut oil, animal feed additives, and even genetically engineered animals with synthesized genes. Food flavorings may sound benign, but actually pose another set of risks: economic risks to farmers. These natural botanical markets are worth an estimated $65 billion annually and currently provide livelihoods for small farmers, particularly in the global South.[7] Replacing the natural production of these products by farmers with synthetic biology in biotech vats in the US and Europe will have major socio-economic impacts and may drive smallholder farmers further into poverty.

The Perils of Synthetic Biology

While some of these developments sound promising, synthetic biology also has a dark side. If an SMO were to be released into the environment, either intentionally (say, as an agricultural crop) or unintentionally from a lab, it could have serious and irreversible impacts on the ecosystem. Synthetic organisms may become our next invasive organisms, finding an ecological niche, displacing wild populations and disrupting entire ecosystems.[8] SMOs will lead to genetic pollution—as happens commonly with GMOs—and create synthetic genetic pollution which will be impossible to clean up or recall. Using genes synthesized on a computer instead of those originally found in nature also raises questions about human safety and the possibility that SMOs could become a new source of food allergens or toxins.

What's different and possibly more hazardous about synthetic biology is that the DNA sequences and genes being used are increasingly different than those found in nature. Our ability to synthesize new genes has far outpaced our understanding of how these genes, and the biological systems they are being inserted into, actually work. It is already difficult to assess the safety of a single genetically engineered organism, and synthetic biology raises this level of complexity enormously. To date, there has been no scientific effort to thoroughly assess the environmental or health risk of any synthetic organism, which can have tens or hundreds of entirely novel genetic sequences.

Biotechnology is already regulated poorly in the US, and SMOs will only push the boundaries of this antiquated regulatory system. For example, the US Department of Agriculture regulates GMOs through plant pest laws, since most have been engineered through a plant virus. Synthetic biology opens up the possibility for SMOs to be created without plant viruses, meaning those crops may be completely unregulated by the USDA—or any agency.

Our risk assessment models for biotechnology are quickly becoming outdated as well. Safety of GMOs is typically determined if it is "substantially equivalent" to its natural counterpart. This idea of "substantial equivalence" quickly breaks down when looking at the risk of an SMO, which has genes that have never existed before in nature and whose "parent is a computer."[9]

An End to Industrial Agriculture As We Know It

Synthetic biology may hold some promises, but is a dangerous path to follow if we don't know better where it leads. The past few decades of agricultural biotechnology have produced a multitude of problems, many of which will be exacerbated by synthetic biology, including genetic contamination, superweeds, an increasing dependence on ever more toxic industrial chemicals, larger areas of unsustainable monocultures, fights over intellectual property and the suing of farmers, and the further concentration of corporate control over our food supply.

Far from making "agriculture as we know it disappear," as Craig Venter hopes to do, we should work to make industrial agriculture as we know it (and its dependence on biotechnology and toxic chemicals) disappear, refocusing our energies on agricultural systems we know to work, such as agro-ecology and organic farming. For example, a recent USDA study found that simple sustainable changes in farming, such as crop rotation, produced better yields, significantly reduced the need for nitrogen fertilizer and herbicides, and reduced the amounts of toxins in groundwater, all without having any impact on farm profit.[10] Such systems have shown to be equally if not more productive than industrial agriculture systems, and are also better for the planet and our climate[11] and produce food that is healthier and more nutritious without a dependence on hazardous, expensive and unproven technologies.

A moratorium on the environmental release and commercial use of synthetic biology is necessary to ensure that our ability to assess its risks and regulate it to protect human health and the environment keep pace with the technology's rapid developments, and to provide time to explore and support alternatives.[12] Instead of continuing down the road of GMOs to SMOs, let's look to solutions that already exist to create a vibrant, healthy, sustainable and just food system.

Future Imperfect: Discussing the Industrialization of Agriculture with Deborah Koons Garcia

By Evan Lerner

Evan Lerner *is Science News Officer at the University of Pennsylvania and the former director of Communications of the Council for Responsible Genetics and the editor of* GeneWatch. **Deborah Koons Garcia** *is the director of the film* The Future of Food. *Produced in 2004, it was re-released in theaters and on a special edition two-disc DVD in December 2005. The DVD can be purchased on the Web at www.thefutureoffood.com. This article originally appeared in* GeneWatch, *volume 19, number 3, May–June 2006.*

GeneWatch: Have you always been active in the field of food and agriculture?

Deborah Koons Garcia: The subject interested me since I was very young. When I was fifteen, I polyploided* plants as part of a science fair experiment, and saw the difference between the polyploided and the regular varieties. Just using colchicine and radiation in my bedroom, it became very clear to me that I could have a significant impact on these plants. And because the plants with polyploidy were bigger, thicker and deformed, I knew I didn't want to eat anything like that. So I decided to keep an eye on agriculture on a larger scale, thinking "if I could do that to a plant by myself, what else could be done to them?"

In college, I became very interested in the relationships between food and health, and food and politics. Making a film that could explore those ideas for people who weren't concerned about them yet had been on my mind for

* Polyploidy is when a cell or organism has more than two copies of its chromosomes. Colchicine is a chemical that prevents chromosomes from segregating during meiosis, resulting in a polyploid gamete and a gamete with no chromosomes.

decades. And as the stakes were getting higher, I thought "now is the time to make this film."

GeneWatch: Was there a particular event that acted as a catalyst for your decision to make this film?

Garcia: The film I thought I was going to make was on pesticide use and sustainable versus industrialized agriculture, until an organic farmer friend of mine told me about the genetic engineering and patenting that was making Roundup Ready crops possible, as well as Monsanto's consolidation of seed companies. This was in 2001, and I had thought of myself as being very sophisticated and informed about this topic, but I didn't know about any of these things. And if I, a food fanatic, didn't know anything about them, then I figured the general public certainly didn't know about them, either. I became determined to make the film to tell these people about the major shift in our agriculture systems, so that they could decide for themselves whether or not they wanted these shifts to happen.

GeneWatch: Your film looks at both the biological and socioeconomic repercussions of this transformation. Do you see one aspect as being more dangerous than the other? In your experience, does one aspect activate or outrage people more than the other?

Garcia: You can't separate the two anymore. With the ability to patent a gene, corporate ownership has invaded biology at the most microscopic level. And with the ability of these genes to invade other plants, it's like the takeover bully of the plant world is doing the work of the takeover bully in the corporate world.

The general population is getting fed up with the amount of control corporations are being given by our government. That aspect of the industrialization of agriculture, how corporate insiders are being allowed to make up the rules for their own benefit, is outraging more and more people. Concern over the biological aspects, i.e. the health effects of genetically engineered food, is less widespread in America, but it is starting to gain traction with people who have young children in public schools. When these people start trying to get healthy food served in their children's schools, they learn about genetically engineered food and go from there. However, as of now, Europeans and Japanese are certainly more concerned about the health aspects than most Americans.

GeneWatch: Do you have a sense of why this topic has become such a big issue in the rest of the world, but hasn't sparked the same kind of outrage in the US? Has the reaction to *The Future of Food* been different overseas?

Garcia: It depends on how connected you are to the food you eat. Americans are more disconnected from where their food is grown and how it gets to their tables, as smaller family farms continue to be replaced with factory farms. It is changing though; organics are the fastest growing section of the industry.

We've gotten orders from around the globe from people who want to see *The Future of Food*. There's been a kind of populist movement in screening the film privately, though we had it shown in theaters in thirty cities in the US and Canada, and are in negotiations to have it shown in France, Germany and the Netherlands. There has been a lot of interest in the film since the World Trade Organization decision [regarding the European Communities' implementation of rules and procedures for approving GMOs].

GeneWatch: Speaking of the WTO, how do you see these national differences in light of GM agriculture on the international scale?

Garcia: The fact that European countries and places like New Zealand enact the kind of moratoriums that the WTO strikes down is another thing that differentiates their responses from the largely indifferent one typical in the United States. This comes with its own set of problems, however. The screenings I went to in New Zealand were packed, and everyone there was very angry with their government for softening its stance toward Monsanto and GM agriculture in general, when the populace had made it very clear that it wanted nothing to do with it. That's why we had Portuguese and Spanish translation teams making copies to bring to Curitiba, Brazil for the MOP3 conference. Everyone in New Zealand was very concerned about the government not representing the populace's views there.

This is the same thing that's happening in Europe right now. The citizens said "no" when they were first asked if they wanted GM crops in their country, which led to the moratorium. Because of that, however, the citizens felt like it was a done deal. They didn't realize that the companies that own the patents on these plants would just keep pushing until they got their foot in the door. It starts with test fields and grows from there. For example, genetically engineered potatoes are now being planted in Ireland, so I was glad that activists there are using the film to help organize local resistance.

GeneWatch: The decision to strike down the moratorium stems from the fact that the EU's safety approval procedures don't match up with industry-based ones the WTO has mandated. This is not so dissimilar from how things work in America, where regulatory agencies ask for voluntary assurances from industry members, or cite industry-based testing as sufficient. How do we fight this information war, when GMO advocates seem to be stacking the deck?

Garcia: The public needs to be made aware of these relationships, and of the fact that governmental regulatory bodies have passed the responsibility of assuring that a GM food product is safe to the product's manufacturers. But the most important thing that is missing from this equation is funding for large-scale neutral testing of these products. Alternative long-term health and safety studies are only just being started, but there is very little funding for them.

The bigger barrier to convincing people that this kind of consolidation of control presents a problem is that people implicitly want to trust that their food is safe. There are so many other important issues in the world that it's hard to weigh the consequences of your food choices every day. Food serves as a comfort against those other stresses, as it can be something that doesn't require that much thought. It will take a shock—a visceral reaction on the part of consumers—to get them to start looking at their food in a critical manner.

This also speaks to the difference between America and the rest of the world when it comes to genetically modified foods. Americans have become disassociated not only from who grows their food, but from the processes of cooking and eating as well. So much of what Americans buy is heavily processed and made for convenience; really thinking about the contents doesn't enter into the equation. But more Americans are opposed to genetically modified foods than you would think; they simply don't realize that they are eating them. The prevailing winds are changing, though, as many Americans are starting to demand healthier food. If cultural norms are indeed changing so that more thought goes into everyday food choices, people will be more wary of who is altering and controlling their food.

GeneWatch: Genetically modified food and sustainable agriculture are such huge topics, it's hard to encapsulate the entire argument in a two-hour film. What ended up on the cutting-room floor? What else would you have included that has occurred since the film's release?

Garcia: In a sense, not much has changed. The system of consolidating control over what used to be in the public domain, specifically, plants and ways of farming them, continues. If anything, the sides have become more polarized. Obviously, the WTO decision exemplifies this, with American corporations pushing harder than ever to make sure their way of doing business survives.

While the film does touch on it, I would have loved to include more information about farm subsidies. It's a very complex problem, but it boils down to taxpayer money being handed over to these large corporations so they can get their products on more American fields. That's an issue that cuts across political leanings; people want to know what they are paying for with their taxes. But there are many other facets, especially since these subsidies allow for GM food products to be sold for less than alternatives. That makes a difference domestically, for people trying to choose among products in the supermarket, but it is a far bigger problem at the level of international trade.

All of the topics contained under the umbrella of GM agriculture remain battlegrounds. The Center for Food Safety is behind a recent lawsuit over genetically modified alfalfa in South Dakota. While the plant in question has changed, it's the same story all over again: the GM strain will contaminate organic alfalfa, and will lead to much heavier pesticide use.

GeneWatch: If you had to distill the film's message to a single point, what would it be?

Garcia: In the end, I really want to emphasize that the issue of GM agriculture is a battle between public and private control of the commons. When people see the film, or hear the story of Percy Schmeiser, their response is usually something like, "How can that be?" This is a lesson that Americans, in particular, need to learn: that things that used to be publicly owned and controlled can, and are, being bought up. I deliberately made a populist advocacy film, one that would cut through political affiliations so that everyone can be confronted with this problem. I've had Republicans come up to me to say how much they loved the film, and order fifty copies for their friends. The same has happened with farmers who formerly grew genetically engineered crops.

It's so vital that this message be spread now, because in five or ten years, it will be too late. We won't be able to remove the contamination from the environment, or the patents from the private sector. That the biological or legal techniques are a means to the end of taking control away from the public is at the heart of this issue, and it goes beyond food and agriculture, into culture in general. We can't forget that freedom is a thing we need to constantly struggle for. Using the topic of food is a good place to start, though, as it isn't just a liberal issue or an activist issue. It's a human issue. When push comes to shove, the ability of food to bring all levels of society and all political beliefs together is going to be a real strength for defending the public interest.

Stealing Wisdom, Stealing Seeds: The Neem Tree of India Becomes a Symbol of Greed

By Vandana Shiva

Vandana Shiva *is an environmental activist and author. She is a leader and board member of the International Forum on Globalization and she directs the Research Foundation for Science, Technology and Natural Resource Policy.* *Reprinted from* Third World Resource #63. *This article originally appeared in* GeneWatch, *volume 10, number 2–3, October 1996.*

Of all the plants that have proved useful to humanity, a few are distinguished by astonishing versatility. The coconut palm is one, bamboo another. In the more arid areas of India, this distinction is held by a hard, fast-growing evergreen of up to 20 meters in height: *Azadirachta indica*, commonly known as the neem tree.

The neem's many virtues are to a large degree attributable to its chemical constituents. From its roots to its spreading crown, the tree contains a number of potent compounds, notably a chemical named *azadirachtin*, useful for medicine, fuel, and agriculture, that is found in its seeds.

These benefits, known to Indians for millennia, have led to the tree's being called in Sanskrit *Sarva Roga Nivarini*, "the curer of all ailments." Access to its various products has been free or cheap: There are some 14 million neem trees in India, and the age-old village techniques for extracting the seed oil and pesticidal emulsions do not require expensive equipment.

In the past seventy years, considerable research on the properties of neem has been carried out by groups ranging from the Indian Agricultural Research Institute and the Malaria Research Center to the Tata Energy Research Institute and the Khadi and Village Industries Commission (KVIC). Much of this research was fostered by Gandhian movements, such as the Boycott of Foreign

Goods movement, which encouraged the development and manufacture of local Indian products. A number of neem-based chemical products, including pesticides, medicines, and cosmetics, have come to the market in recent years, some of them produced in the small-scale sector under the banner of KVIC, others by medium-sized laboratories. There has been no attempt to acquire proprietary ownership of formulas since, under Indian law, agricultural and medicinal products are not patentable.

For centuries the Western world ignored the neem tree and its properties, but growing opposition to chemical products in the West has led to a sudden enthusiasm for the pharmaceutical properties of neem.

Since 1985, more than 30 US patents have been taken out by US and Japanese firms on formulas for stable neem-based solutions, emulsions, and even for a neem-based toothpaste. At least four of these patents are owned by the US-based multinational chemical company W.R. Grace; three are owned by another US company, the Native Plant Institute; and two are owned by the Japanese Terumo Corporation.

W.R. Grace's aggressive interest in Indian neem production has provoked a chorus of objections from Indian scientists, farmers, and political activists, who assert that multinational corporations have no right to expropriate the fruit of centuries of Indian scientific research. This has stimulated a bitter transcontinental debate about the ethics of international property and patent rights.

W.R. Grace claimed that its extraction processes constitute a genuine innovation: "Although traditional knowledge inspired the research and development that led to these patented compositions and processes, they were considered sufficiently novel and different from the original product of nature and the traditional method of use to be patentable. *Azadirachin*, which was being destroyed during conventional processing of Neem Oil/Neem Cake, is being additionally extracted in the form of WaterSoluble Neem Extract and hence it is an add-on rather than a substitute to the current neem industry in India."

In other words, the processes are supposedly novel and an advance on Indian techniques. But this apparent novelty is mainly a product of Western ignorance. The allegation that *azadirachtin* was being destroyed during traditional processing is inaccurate. The extracts were subject to degradation, but this was not a problem since farmers put them to use as they needed them. The problem of stabilization arose only with respect to packaging and commercial marketing.

Moreover, stabilization and other advances attributable to modern technology were developed by Indian scientists in the 1960s and 1970s, well before US and Japanese companies expressed interest in them.

Finally, the argument made by W.R. Grace that its neem project benefits the Indian economy is hard to justify. The company argues that its project has been beneficial by "providing employment opportunities at the local level and higher remuneration to the farmers as the price of neem seeds has gone up in recent times because value is being added to it during its processing. Over the last twenty years the price of neem seed has gone up from 300 rupees a ton to current levels of 3,000 to 4,000 rupees a ton."

In fact, the price has risen considerably more than this: In 1992 Grace was facing prices of up to $300 (over 8,000 rupees) per ton.

The increase in the price of neem seeds has turned an often free resource into an exorbitantly priced one, and the local user is now competing for the seed with an industry supplying consumers in the North. Since the local farmer cannot afford the price that the industry can pay, the diversion of the seed as raw material from the community to industry will ultimately establish a regime in which a handful of companies holding patents will control all access to neem as raw material and all production processes.

PART 5

Regulation, Policy, and Law

For the most part, it is not possible to predict the full implications of tinkering with the genetic make-up of living organisms. When transformative technologies are introduced into society, there is usually a lag time between their introduction and the proper control and regulation of their applications to support their benefits and minimize their harmful effects.

At this stage in the development of genetically engineered food and crops, the US government has given the assessment and monitoring role to industry. The US government fails to require independent safety testing and works under the assumption that there is no cause for concern. The result is that the American consumer is being offered a false level of confidence in his government's regulation and oversight of GMO safety.

Agencies that have the legal power to take more of an interest in the regulation of genetically engineered food and crops have made a conscious decision not to. For example, the FDA under the Food, Drug, and Cosmetic Act (FDCA) could determine that foods with foreign genes added to them qualify as regulated food additives. This would place the burden on industry to provide scientific evidence of the safety of the substances added to genetically engineered foods. This standard would place genetically engineered foods in the same category as other foods with additives, instead of giving them a wholesale exemption, which is the current FDA practice.

This essays in this section explore the current state of regulatory and legal governance of GMOs and its limitations.

—Jeremy Gruber

AG Biotech Policy: 2012 in Review

By Colin O'Neil

Colin O'Neil *is the Director of Government Affairs at the Center for Food Safety. This article originally appeared in* GeneWatch, *volume 26, number 1, January–March 2013.*

In a year filled with attention-grabbing headlines about reelection campaigns and congressional roadblocks, the narrowly defeated California Proposition 37 may be the story about genetically engineered (GE) foods that people most recall. However, for those directly affected by biotech policy, 2012 was shaped by more than just Prop. 37. Its issues were driven by chemical companies, farmers and consumers.

Farmer and Environmental Opposition Slows the Chemical Arms Race

The year began like most years in biotech policy, with a major announcement by the Administration during the holidays. In this case, 2012 was ushered in with a statement by the US Department of Agriculture (USDA) revealing that it was moving forward with Dow AgroSciences' application for its controversial Enlist GE corn that is engineered to withstand exposure to the herbicide 2,4-D, a component in the Vietnam era defoliant Agent Orange that has been linked to a number of human health and environmental harms.[1] This announcement brought sharp criticism from farmers, environmental groups and consumers, and in January 2013, after a year of strong opposition, Dow announced that it would be delaying the release of its 2,4-D corn until at least the 2014 planting season.[2] Monsanto's dicamba-tolerant soybean also inched further toward approval in 2012, yet by year's end no decision had been made by the agency likely as a result of sustained opposition by farmers, environmental groups and consumers.

GE Labeling Movement Marches Forward

In the fall of 2011 the Center for Food Safety filed a groundbreaking legal petition with the US Food and Drug Administration (FDA) demanding that the agency utilize its existing authorities to require the labeling of all food produced using genetic engineering. On March 12, 2012, fifty-five Members of Congress joined a bicameral letter led by Senator Barbara Boxer (D-CA) and Representative Peter DeFazio (D-OR) that was sent to the FDA Commissioner in support of the labeling petition. By March 27, more than one million public comments had been submitted to the FDA in support of the petition-the largest public response the FDA has ever received.

The labeling movement did not stop at the FDA. More than a dozen labeling bills were introduced at the state level in 2012 that would have required the labeling of GE fish, GE wholefoods or all GE foods. Ultimately, the bills failed under industry pressure; but already in 2013, thirty-four more bills have been introduced in twenty-one states including Hawaii, Washington, Indiana, Missouri and Vermont, with many more expected by year's end.

Biotech Riders Emerge in House Bills

Despite increased calls for labeling and better oversight of GE crops and foods, Republican members in the US House of Representatives attempted multiple times to roll back the clock on GE crop regulations.

On the heels of federal court decisions that found approvals of several GE crops to be unlawful, a dangerous policy rider (Sec. 733) was inserted into the FY 2013 House Agriculture Appropriations bill. The rider was intended to strip federal courts of their authority to halt the sale and planting of GE crops and compel the USDA to allow continued planting of those crops upon request by industry. The rider drew sharp criticism from groups like the American Civil Liberties Union, Earthjustice, and the National Family Farm Coalition, who viewed it as an assault on the fundamental safeguards of our judicial system and one that would negatively impact the environment, public health and farmers across America.

Following the appropriations rider, a suite of policy riders (Sec. 10011, 10013, and 10014) were buried in the House Agriculture Committee's draft 2012 Farm Bill. These riders sought to: dramatically weaken the oversight and regulation of GE crops and specifically eliminate the critical roles of our most important environmental laws; dramatically shrink the time the USDA has to analyze biotech crops, while withholding funds for the USDA to conduct environmental reviews; limit the regulatory authority of the EPA and other agencies; establish multiple backdoor approval mechanisms for GE crop applications; and force the USDA to adopt a national policy of allowable levels of GE contamination in crops and foods. The riders were widely opposed by industry including the Grocery Manufacturers' Association, the National Grain and Feed Association, the Snack Food Association and the Corn Refiners Association, as well as environmental, consumer and farm groups.

Neither the appropriations rider nor the Farm Bill riders were included in any final legislation.

A Fish with a Drug Problem

In keeping with holiday tradition, on December 21, 2012, FDA officials released their Draft Environmental Assessment (EA) and opened a public comment period concerning the AquAdvantage Salmon produced by AquaBounty Technologies. The GE Atlantic salmon being considered for approval under the FDA's new animal drug law was developed by artificially combining growth hormone genes from an unrelated Pacific salmon with DNA from the anti-freeze genes of an arctic eelpout. This modification causes production of growth hormone year-round, creating a fish the company claims grows at twice the rate of conventional farmed salmon, allowing factory fish farms to further confine fish and still get high production rates.

Since the FDA first announced its approval process for GE salmon in 2010, numerous environmental, health, economic and animal safety concerns have been raised by advocacy groups and the scientific community. A 2011 study published by Canadian scientists concluded that if GE Atlantic salmon males, like those used in the company's facility, were to escape from captivity they could succeed in breeding and passing their genes into the wild.[3] More recently, previously hidden documents surfaced during a Canadian investigation which found that AquaBounty's Prince Edward Island facility was contaminated in 2009 with a new strain of Infectious Salmon Anemia (ISA),[4] the deadly fish flu that is devastating fish stocks around the world. This information was hidden from the public and potentially other Federal agencies and the FDA's own Veterinary Medicine Advisory Committee (VMAC).

Looming Battles in 2013

Much remains to be seen about the impact that Prop. 37 will have on other state labeling initiates and where continued opposition to the next generation of GE crops can be maintained. However, it is clear that members of Congress are now ready to intervene on the chemical industry's behalf and only with the strong will of farmers, advocacy groups and members of industry working together will we be able to halt the march toward the further industrialization of agriculture.

28

EPA and Regulations

By Sheldon Krimsky

Sheldon Krimsky *is the board chair of the Council for Responsible Genetics. He is a former member of the Recombinant DNA Advisory Committee to the National Institutes of Health, and the author of many books including* Genetic Alchemy, A Social History of the Recombinant DNA Controversy. *He is the Lenore Stern Professor of Humanities and Social Sciences at Tufts University and teaches in the Urban and Environmental Policy & Planning Department as well as serving pro bono as the Board Chair of the Council for Responsible Genetics. This article originally appeared in* GeneWatch, *November–December 1983.*

In August 1983, the Environmental Protection Agency (EPA) announced it was dispatching a new unit in its Division of Pesticides and Toxic Substances to develop policies that would define the agency's role in regulating biotechnology. The action came after the EPA Administrator's Toxic Substances Advisory Committee (ATSAC) recommended that the Toxic Substances Control Act (TSCA) be applied to regulate two applications of biotechnology. TSCA is a major federal law passed in 1976 that regulates the manufacture and industrial use of toxic chemicals. First, ATSAC cited large-scale uses of nonliving biotechnology products in the environment. Second, it targeted the intentional release of genetically engineered living organisms.

The new EPA unit is expected to develop policies that will serve as the basis for future rulemaking in biotechnology. Although it is too early to predict what the agency will come up with, its investigation is already limited by several important conditions. The most significant of these is TSCA itself. At this stage EPA is exploring what it needs to know to apply TSCA to new biological agents that are designed to be released into the environment.

The EPA has used TSCA to regulate chemical agents for several years. Its extension to biotechnology is expected to cover microorganisms exclusively (bacteria, viruses, and fungi). If the agency does develop rules in the next several years, they will not apply to novel plants, genetically engineered seeds, insect species used for purposes other than pest control, or genetically modified higher organisms that are released into the environment.

Through its current initiative, the EPA is trying to respond to criticisms that the federal oversight of biotechnology is inadequate. Thus far, the National Institutes of Health (NIH) have taken the leading role in assessing the environmental and human health risks of releasing modified organisms and plants into the environment. However, NIH is not a regulatory agency and has no legal authority to oversee private industry. Moreover, since the agency supports and promotes scientific research in genetic engineering, its oversight of gene splicing represents a conflict of purpose.

Recently, the NIH Recombinant DNA Advisory Committee (RAC) approved the release of genetically engineered strains of the bacteria Pseudomonas and Erwinia in field tests so that a team of scientists at Berkeley could investigate how well these strains reduce frost damage in plants. Jeremy Rifkin, the Foundation on Economic Trends, and several environmental groups responded with a legal challenge because they believe that the federal oversight for this unprecedented release was insufficient from an ecological perspective. The lawsuit cites inadequate risk assessment and NIH's failure to comply with the National Environmental Policy Act (NEPA), which requires an environmental impact review for major federal actions affecting the environment. The scientists have postponed the experiment in response to the suit.

The only area in biotechnology for which EPA is reasonably well prepared is the oversight of biological pesticides. In 1980, the agency drafted a 433-page set of working guidelines which established requirements for the registration of biological control agents, toxicological data, and performance standards. This extensive document also established criteria for assessing the environmental survival of released agents, their host range, and their potential effects on non-target organisms.

Since 1948, ten biological agents have been registered in the US for insect control. At present no novel organisms have been submitted for pesticide registration. The data requirements for genetically engineered organisms will be determined by EPA on a case-by-case review.

Over the past several years a number of research programs have gotten underway which could result in industrial releases of novel microorganisms into the environment. In addition to the bacteria being developed to reduce frost damage to plants, other projects include oil-eating microbes and bacteria engineered to degrade toxic chemicals. Except for NIH's limited oversight of university research and its voluntary compliance program for industry, no federal regulatory body is prepared, at this time, to issue guidelines on intentional release of novel biotypes. This is the primary regulatory gap that EPA intends to fill.

It is likely that EPA's biotechnology project will result in a framework for formulating regulations modeled somewhat on the guidelines for biological pesticides. But a number of important questions have to be resolved before EPA begins the rulemaking process. The agency must determine which biological agents would be covered by TSCA. Tentatively, the agency is limiting its focus to microorganisms. But which microorganisms should be considered? The EPA will have to develop criteria that defines a novel organism. It will have to review the techniques currently used to alter a genotype and determine whether the creation of novelty is related to them.

The comparison of genetically engineered organisms with inert chemicals is not a helpful one. Chemicals can be uniquely identified by their well-defined structures. A change in a single chemical bond constitutes novelty. The issue is far more complex for microorganisms, which are not uniquely characterized by a chemical structure. The EPA will have to decide, from the point of view of applying TSCA, what conditions determine a novel genotype.

Over the past several decades, scientists have paid considerable attention to testing the toxicity and mutagenicity of chemical agents. But these tests are not transferable to biological agents. EPA officials have emphasized that there are almost no accepted methodologies for evaluating the safety of genetically engineered products. New sets of protocols will have to be developed for testing the toxicity of biological agents.

Survivability for biologicals will also be handled differently from chemicals. If the EPA plans to regulate the intentional release of genetically modified organisms, it will have to determine relevant criteria for survival, the exchange of genetic information with other organisms, and the effect of novel organisms on the environment. The EPA currently uses TSCA to review new chemicals prior to manufacture and before the substances enter commerce and the environment. The agency pays relatively little attention to chemicals already in use. In biotechnology the distinction between novel and natural organisms will be a critical factor in determining the scope of regulations.

The EPA will not provide oversight of industrial fermentation and large-scale cultures of genetically modified organisms. Regulations for worker health and safety in the biotechnology industry will have to be developed independently by the Occupational Safety and Health Administration (OSHA). Currently, OSHA has not initiated any programs in this area.

What can you do? EPA is at an early stage in the development of its policies on biotechnology. If you are a microbial ecologist, plant ecologist, or population biologist, or if you understand what it takes to make TSCA effective for biotechnology, then let EPA know about your interest in this initiative. Get on its mailing list. Ask to see policy documents as they are released. Make yourself available as a consultant or commentator.

GM Food Legislation: Modified Foods in the Halls of Power

By Lara Freeman

Lara Freeman *is assistant chaplain and instructor in religion at St. George's School.* *This article originally appeared in* GeneWatch, *volume 18, number 1, January–February 2005.*

Trying to follow the burrowing of US public policy into genetically modified (GM) agriculture is like chasing earthworms through a cornfield. At first glance, the paths of GM legislation don't seem to be a long any certain trajectory, just a dizzying course of competing interests. Legislation to regulate the use of GM crops in one part of the country is met with its full encouragement in another.

What is the stimulus channeling this legislation between federal regulation agencies, government houses, town hall meetings, and conversations at the seed store? Legislation promoting GM agriculture is being introduced by agribusiness interests and, more often than not, kept alive long enough to be signed into law. However, legislation to regulate GM agriculture, under agribusiness influence, withers or gets watered down long before it leaves state agricultural committees. Concern for environmental precaution and human health is rarely at the heart of legislation on GM agriculture, and the rhetoric of economic benefits underlies the influence of industry money.

According to the Pew Initiative on Biotechnology, roughly forty percent of the bills introduced at the state level in 2002 to 2003 supported the use of genetically modified agriculture, specifically through research and development, tax exemptions for biotechnology corporations, and the promotion of agribusiness development. In the present recession, many states face fiscal deficits, and thus exploring the promotion of biotechnology, looking for ways to boost their economies.

The increasing use of GM crops worldwide brings new attention to liability and agricultural contracts. A battle is now being waged between farmers, seed distributors, and agribusiness owners over who should be responsible for genetic contamination of neighboring fields and who will cover the costs of unexpected repercussions. Task forces to study GM technology, as well as bills to regulate GM organisms in the environment or place a moratorium on their use, made up a significant portion of the GM legislation introduced last year-eighteen percent and fourteen percent respectively. The latter group, however, has fallen in frequency from previous years. Only 8 percent of the legislation introduced in 2002 to 2003 dealt with labeling, as compared with 16 percent in 2001 to 2002. Moratorium bills fell from 9 to 6 percent.

While some GM legislation is introduced with the idea of bolstering dilapidated state economies through greater agricultural production and business development, not all states see GM agriculture as a moneymaker, and perhaps no state leader should see it as economically beneficial. For many states, GM agriculture will reduce the marketability of their crops, further enforce trends of unsustainable monocrop agriculture, and continue to disenfranchise farmers from their trade. The organic farming communities in California and Vermont have been especially vocal in the promotion of GM regulation, as have conventional and organic farmers in the mid-west. North Dakota, for instance, produces 48 percent of all US wheat, and risks the loss of its international wheat markets where 60 percent of its crop is sold—if it moves towards the use of GM wheat. As Congressman Earl Pomeroy pointed out in his statement to the federal government's Interim Agricultural Committee, important consumers of US wheat such as Japan, Mexico, Algeria and the Philippines, don't want GM wheat inside their borders or on their tables. Pomeroy questioned the wisdom of growing something without consumer demand. The North Dakota Wheat Commission, on the other hand, has stated that its goal in the coming year is to create such demand, by encouraging market acceptance of GM products and reassure consumers of the safety and wholesomeness of American wheat.

The recent legislation in North Dakota reflects significant agribusiness influence on the legislators in that state, the dampening of other views, and the tenuous market outlook for GM wheat. According to Janet Jacobson of the Northern Plains Sustainable Agriculture Alliance, "Legislators don't listen to the science. The party line is 'We won't stand in the way of technology.' Not many will buck that line. Senator Bowman tried to introduce a liability bill and the Senate made gestures to take this seriously, but in the end they rewrote it to be meaningless and defeated it anyway." The testimony and hearings in the Agricultural Committee on GM wheat have been dominated by representatives from Monsanto. At one hearing, Monsanto was given six hours of presentation time, while citizens and those opposed to biotechnology were given just one hour. According to Jacobson, "even [the legislators'] body language says they aren't listening. They sat with their backs to the citizens, some were sleeping, and some were playing on their laptops."

Over the past year, all attempts at legislation to regulate GM in North Dakota have been defeated, including the establishment of a wheat board (HB1026), certificates of approval for the sale of GM wheat (SB2408), and a bill concerning cross-contamination (SB2304), among others. While there are some state and regional groups trying to organize and educate farmers, few people attend the meetings and those who do tend to be polarized. The discussion seems to be dominated by agribusiness and the state legislators whose hands they have tied.

New England states and pockets of mid-sized farming communities across the country share concerns with farmers in the heartland. States with vigorous organic agricultural markets, specifically Vermont, seem to be distancing themselves from GM agriculture. With 24 percent of its vegetable acreage under organic management, Vermont has the greatest proportion of organic vegetable acreage of any state. It is not surprising that organic and conventional farmers in Vermont are concerned about GM contamination hurting the marketability of their produce and dairy products. And, unlike North Dakota, Vermont's organic farmers have found a voice in the state houses through effective organization.

The combination of a system of local government that requires extensive constituent involvement, a growing number of well-informed farmers and consumers, and effective grassroots organizing creates an atmosphere in Vermont where citizens can develop an informed opinion on GM agriculture. The outcome of these efforts has been significant. According to Brian Tokar of the Institute for Social Ecology, seventy-nine towns in Vermont, as well as eleven in Massachusetts and a few in New Hampshire, have passed resolutions at town meetings stating that they are opposed to GM agriculture. While these are advisory only, the Vermont state legislators are paying attention. They have responded with three groundbreaking bills. On April 12, the Vermont House of Representatives voted 125 to 10 on final passage to endorse the Farmers-Right-to-Know Seed Labeling Bill (H.0777). This bill defines genetically modified seeds as different from conventional seeds and mandates the labeling of all genetically modified seeds sold in the state of Vermont. This follows on the heels of a March 10 vote where Vermont Senators voted unanimously (28 to 0) in support of the Farmer Protection Act (S.0164), a bill that will hold biotechnology corporations liable for unintended contamination of conventional or organic crops by genetically modified plant materials. The Time Out on GMOs Bill (S.0162), calling for a two-year moratorium on the planting of genetically modified crops in Vermont, is awaiting action in the Senate.

If the Vermont legislature passes these bills into law, they will join Mendocino County, California, as the leaders in domestic policy to ban genetically modified crops. Measure H, Mendocino County's initiative, was passed on March 2, 2004. The success of this legislation can be attributed to the hard work of local business owners, farmers, and concerned residents organized by Els and Allen Cooperriders of the Ukiah Brewery Company and Restaurant, who led the locally inspired campaign against GM. As it stands, this measure is a symbolic act, since genetically modified crops are not grown in the area; however, some voters felt that it could be used as a marketing tool for their organic goods, especially in the European market, where consumers strongly oppose GM foods. Others felt that Measure H set a precedent for neighboring counties in Northern California and other communities nationwide that are becoming more aware of the implications of GM crops and foods. Six other California counties—Butte, Marin, San Luis Obispo, Sonoma, Santa Barbara, and Humboldt—had similar measures on the November 2 ballots, though only the Marin County measure passed. At the state level, however, there is concern that agribusiness lobbyists will introduce legislation that will remove the authority of counties to pass agricultural legislation, thus nullifying the Marin and Mendocino measures.

Legislators in coastal states have responded much more decisively to regulate GM aquaculture. National scientific advisory groups, supported by international studies such as the North Atlantic Salmon Conservation Organization's Guidelines for Action on Transgenic Salmon, and vocal aquaculture communities have aided this legislative activity. Additionally, the corporations' genetically engineering fish are generally smaller and less politically influential than others in agribusiness. Because the distinction between "wild" and "farmed" ocean areas are blurry, often separated only by an easily broken net, and because fish have much greater mobility than land crops for spreading their genes, the potential for the decline or complete destruction of competing wild fish stocks is imminent if precautions are not taken. Genetically modified fish are engineered to grow quickly and have a larger adult body size than their wild counterparts; GM males tend to out-compete the wild males, and their size is more attractive to females.

For these reasons, if GM fish were accidentally released into the wild, they might be more successful reproducers than their non-engineered competitors. However, GM offspring would be less fit and less likely to survive. As a result of this combination, scientists predict that GM fish could cause some species to become extinct if they escape from open ocean farms. This potential impact of GM fish is a significant threat to coastal fishing communities whose economies depend on the continuation of wild fish populations.

Tracie Letterman of the Center for Food Safety's Campaign on Genetically Engineered Fish suggests that the states, responding to the demands of fishermen and scientists alike, are taking up this issue because the federal government has not. Eight states passed laws between 2002 and 2004 regulating either the release or use of GM fish in state waters. These include California, Maryland, Michigan, Minnesota, Mississippi, Oregon, Washington, and Wisconsin. California's legislation (Cal. Fish and Game Code 15007) bans the raising of GM fish and the entry of salmon farms in California ocean waters. While GM fish are not currently being raised in California's oceans—or any ocean—at this time, there is concern that, if the USDA approves GM salmon for human consumption this year, the industry will soon take root.

Thirty-one states proposed legislation on the genetic modification of agriculture or aquaculture in 2002 to 2003. The intended goal of much of this legislation is economic growth, but it should be asked whose economic growth will truly be realized: that of states and their constituents, or that of distant corporations? Tax breaks for national or transnational corporations and state-funded construction of buildings for agribusiness research do not help a government prosper. Liability for cross-contamination that falls on citizens of the state rather than the corporations that produced the contaminating seed does not improve the economic situation of the residents of the state in the long run.

As the legislation is sorted out, it becomes clear that much of the pro-GM legislation is funded and promoted by agribusiness lobbyists, while legislation to regulate GM is painstakingly proposed through collective efforts of concerned and informed constituents. While those who seek to regulate GM also have economic interests in mind, these tend to represent the broader interest of sustainable livelihoods, markets, agriculture, and health. According to Brian Tokar, the success of legislation to regulate GM agriculture in his state rests with a democratic system where the voices of the people have been heard over private corporate interest. The efforts there began in town hall meetings where only citizens of each township were allowed to speak. When seventy-nine towns spoke in favor of regulating genetically modified agriculture, state legislators had to listen. May the people of Vermont continue to be heard the world over.

Goliath vs. Schmeiser: Canadian Court Decision May Leave Multinationals Vulnerable

By Phil Bereano and Martin Phillipson

Philip L. Bereano, JD, PhD, *is professor emeritus in the field of Technology and Public Policy at the University of Washington a cofounder of AGRA Watch, the Washington Biotechnology Action Council and the Council for Responsible Genetics.* **Martin Phillipson** *is Vice Provost, College of Medicine Organizational Restructuring and professor of law at the University of Saskatchewan.* *The authors would like to acknowledge the tireless efforts of Mr. Schmeiser, whose campaigning against GMOs has significantly raised the visibility of issues of monopolies over life forms, farmers' rights, GMO contamination, and corporate control of agriculture. This article originally appeared in* GeneWatch, *volume 17, number 4, July–August 2004.*

Are genes patentable? Are transgenic plants and animals patentable? In the United States the answers are affirmative, and over the past two decades the US has pressured other countries to adopt the same sort of patent rules. Yet, two years ago, in *President & Fellows of Harvard College v. Canada*, the so-called "Harvard Mouse" case, the highest Canadian court held that "higher life forms" could not be subjected to patent monopolies.

This spring, the Supreme Court of Canada rendered judgment in another closely followed case: Monsanto's suit against Percy Schmeiser, which alleged that the Saskatchewan farmer had infringed their patent on Roundup Ready canola. The result was mixed. The Court affirmed the Harvard ruling that plants are not patentable in Canada, but said that genes are. Schmeiser, though he had infringed, was not held responsible for monetary damages.

In 1993 Canadian Patent No. 1,313,830 was issued to Monsanto Canada for "Glyphosate Resistant Plants." However, the patent did not cover the plants themselves, but only the process by which genes resistant to herbicides (in this case, Monsanto's own Roundup) were developed, as well as the modified genes and cells. By the year 2000, 40 percent of all canola grown in Canada was "Roundup Ready."

In order to use Roundup Ready canola, farmers must sign a Technology User Agreement (TUA), paying a royalty fee of $15 per acre to Monsanto Canada, agreeing not to save and replant seed, promising to use Roundup herbicide, and allowing Monsanto to inspect their crops in order to verify compliance with the terms of the TUA.

On March 29, 2001, a trial judge found Schmeiser to have committed multiple infringements of Monsanto's patent and fined him $20,000, asserting that the levels of Roundup Ready Canola on Schmeiser's property were such that he "knew or ought to have known" that his crop was planted with Roundup Ready seeds. Since Schmeiser had no agreement with Monsanto, he was guilty of using their patented product without a license.

The findings of fact of the trial judge are crucial to the overall outcome of the legal battle between Monsanto and Schmeiser. Generally, once a trial judge has made findings of fact, appellate courts will overturn them only in exceptional circumstances. Appellate courts only have the original transcripts of the trial before them and there are no new witnesses present or new evidence accepted.

Although discussions of *Monsanto v. Schmeiser* have been based on wildly diverging versions of "what actually happened," the only version of events that matters legally is the one accepted by the trial judge. The Supreme Court highlighted the most significant aspects of this factual history in paragraphs 59 through 68 of its judgment:

> In 1996 Mr. Schmeiser grew canola on his property on Field Number One, the seed which was the subject matter of Monsanto's allegations could be traced to this 370-acre field on Mr. Schmeiser's property . . . . [I]n the spring of 1997, Mr. Schmeiser planted the seeds saved from Field Number One. He sprayed a 3-acre patch of this field with Roundup and found that 60 percent of the plants survived, a clear indication that these plants contained Monsanto's patented gene and cell . . . . [I]n the fall of 1997 Mr. Schmeiser harvested the Roundup Ready Canola from the 3-acre patch he had

> sprayed with Roundup. He did not sell it. He instead kept it separate, and stored it over the winter in the back of a pick-up truck. A Monsanto investigator took samples of canola from the public road allowance bordering two of Mr. Schmeiser's fields in 1997, and all samples contained Roundup Ready Canola. In March 1998, Monsanto put Mr. Schmeiser on notice of their belief that he had grown Roundup Ready Canola without a license. Mr. Schmeiser nevertheless took the harvest he had saved in the pick-up truck and had it treated for use as seed. Once treated, it could be put to no other use. Mr. Schmeiser planted the treated seed in nine fields, covering approximately 1,000 acres in all. Samples were taken from the canola plants grown from this seed . . . and a series of independent tests by different experts confirmed that the canola Mr. Schmeiser planted and grew in 1998 was 95 to 98 percent Roundup resistant.

The trial judge found that there was no other "reasonable explanation" for the concentration or extent of Roundup Ready canola of commercial quality evident from the results of tests on Schmeiser's crop. Given these uncontested (according to the court) findings of fact, the only legal issue to be decided by the Supreme Court was whether these actions amounted to "use" of Monsanto's patented genes and cells, and whether (in the wake of the Harvard Mouse case) Monsanto's patent was invalid as constituting a patent over a "higher life form."

The Court was at pains to point out that its decision was based on the facts as found at trial and that in different factual circumstances, a different legal outcome might result. "The issue is not the perhaps adventitious arrival of Roundup Ready Canola on Mr. Schmeiser's land in 1998. What is at stake in this case is the sowing and cultivation which necessarily involves deliberate and careful activity on the part of the farmer" (*Paragraph 92*). Schmeiser was, however, spared the insult of having to pay damages to the multinational corporation, since the majority found that he had not profited additionally from the sale of the patented genes in his canola.

The monopoly granted by a country's patent extends only within the boundaries of that nation. So, literally, the Monsanto v. Schmeiser case only governs the nature of patent law in Canada. Yet some cases (particularly the

1980 US *Chakrabarty* decision, the first in the world to find a living organism patentable) have had impacts far beyond the country's borders. Abetted by cajoling and pressures from all recent US administrations, patent doctrines favoring the biotech industry have spread rapidly, consistent with the growth of corporate globalization, international trade harmonization agreements, and the desire of multinationals to operate under uniform rules. Monsanto and its governmental allies may try to extend aspects of the Schmeiser case to more lands. Thus, it is important to dig beneath the corporate spin and understand exactly what the Canadian court did, and did not, decide.

The following are the major elements of this decision:

- In Canada, plants are not patentable. In this regard, one should also note that the subject of the litigation was Monsanto's patent on the altered gene and the process for making it, which did not even claim the resulting plant.
- Although the general rule of patent infringement is that any unauthorized use, even unknowing or minimal, is infringement (although the damages would depend on such factors), this decision says that for gene patents the basis for a successful suit depends on the intention of the defendant and the nature and extent of the defendant's use.
- Thus, the Schmeiser case centers on the nature of his use; any liability is highly fact-dependent. The judges split 5-4 over whether the "use" of protected genes in unpatentable crop plants could amount to infringement; the minority said no, since the plants cannot be monopolized. However, the majority held that, because the factual use of the crop containing Monsanto's patented genes was extensive, was in a commercial context, and was found to be done "knowingly," it did legally constitute "use" of Monsanto's invention and therefore amounted to infringement (*Paragraph 87*).
- Contamination—the "accidental and unwelcome" presence of the transgenes—by itself is not automatically patent infringement in Canada (*Paragraph 86*). The subsequent conduct of farmers upon discovering the existence of Roundup Ready Canola in their fields will be more determinative of their legal liability than the mere factual existence of the crop on their property (*Paragraph 95*).
- Also, this case says nothing about whether contamination is actionable against a patent holder like Monsanto (for example, under the common

law doctrines of nuisance, trespass, or—like a pending Saskatchewan case—violation of environmental protection statutes).

- Farmers' rights are not inherently jeopardized by this decision, no matter what the industry says. Canada has a Plant Breeders Rights Act, which allows for a form of intellectual property protection over novel plant varieties. The rights granted under the Plant Breeders Rights Act are not as extensive as those granted under the Patent Act, but of significance in light of *Monsanto v. Schmeiser* is the fact that the Act contains a specific "farmers' privilege." Farmers are allowed to save and replant seeds from a protected variety subject to certain conditions. In Canada, therefore, a traditional feature of intellectual property law remains intact—i.e., that if something is protected under one piece of intellectual property legislation, it cannot be simultaneously protected under another. This is contrary to the position in the United States, where in 2001 the Supreme Court held in the *Pioneer* case that regular patent protection was available for plant varieties in spite of the existence of two separate legislative schemes to give other protection to them.

In conclusion, we must understand that the results of this case were heavily dependent upon the facts found by the trial court. It is a confusing decision. Monsanto was able to exert legal control over crop plants even though the law does not allow plants to be patented. This is why the minority dissented. They stated the old adage of patent law, that "what is not claimed is automatically disclaimed." Monsanto claimed only the gene and the process; ergo they disclaimed the plant (which in Canada is non-patentable in any event), and Schmeiser could not be guilty of patent infringement by "using" the canola plants. The majority found this view of "use" to be unrealistic and disagreed, stating that by cultivating a plant containing the patented gene and composed of the patented cells Mr. Schmeiser of necessity "used" the patented material. In many respects, this finding is the most significant (and most troubling) outcome of the *Monsanto v. Schmeiser* battle, because it gives Monsanto control over something which it cannot patent—the Roundup Ready Canola plants themselves. Although in many ways the Schmeiser case is rightly seen as a setback for GMO critics, it also sets a useful precedent for arguing that such contamination is not an infringing use of patented biological materials if a corporation were to try to raise an infringement argument in defending against a contamination lawsuit. In the future, opponents of genetically modified organisms will be able to argue that the contamination by GMOs that is already occurring—and which governmental regulations have not yet been effective in preventing—can be the basis for litigation; the possibility of the award of damages will pressure corporations to avoid further contamination.

Legal Challenge to Genetically Engineered Bt Crops Marches On

By Joseph Mendelson

Joseph Mendelson *is the Chief Climate Counsel at the U.S. Senate Committee on Environment and Public Works and former legal director at the Center for Food Safety.*
This article originally appeared in GeneWatch, *volume 13, number 1, February 2000.*

On February 18, 1999, the Council for Responsible Genetics (CRG), Greenpeace International, the International Federation of Organic Agricultural Movements (IFOAM), the Center for Food Safety (CFS), and more than seventy other plaintiffs, including individual organic farmers, environmental organizations and impacted businesses filed a lawsuit against the Environmental Protection Agency (EPA) in the US District Court for the District of Columbia.

The law suit charged the EPA with the wanton destruction of the world's most important natural biological pesticide. The toxins produced by a bacterium, called *Bacillus thuringiensis* (Bt), are essential to a twenty-first-century agriculture based on biological controls and not on the use of synthetic insecticides. Specifically, the lawsuit alleges that, by allowing genetically engineered Bt plants onto the market, the EPA violated the Federal Fungicide, Insecticide and Rodenticide Act (FIFRA), Endangered Species Act (ESA), Regulatory Flexibility Act, National Environmental Policy Act (NEPA), and the common law Public Trust Doctrine.

As the case weaves its way through the courts, scientific findings continue to validate the plaintiffs' claims. In May of 1999, entomologists at Cornell University disclosed that they found significant adverse effects in monarch butterfly caterpillars fed pollen from Bt corn. Almost one half of the monarch caterpillars eating Bt corn pollen died after four days, a substantially higher number than was observed among those which ate normal pollen. The importance of this finding gains in its impact because of the popularity of the monarch butterfly and because the effects of Bt pollen on other butterflies and moths, some of which are endangered species, are unknown.

Similarly, in August, University of Arizona entomologists released a study that cast doubt upon the effectiveness of "insect refuge strategies." Under the "refuge" theory, by planting buffer zones of non-Bt crops near Bt crops, farmers will be able to ensure that there are enough conventional pests available to mate with Bt resistant insects, to slow the ability of pests to develop Bt resistance.

At the same time as these studies were released, the EPA filed an answer to the plaintiffs' claim that the agency unreasonably delayed responding to their original petition (filed in September of 1997). The agency also filed a motion asking the court to dismiss the remaining claims, in which plaintiffs asked the court to cancel the registration of all Bt plants, suspend further Bt registration approvals, consult with the Department of Interior regarding the effect of Bt plants on beneficial and non-target organisms, and immediately perform assessments of the impacts of Bt crops on the environment and on small businesses.

This summer, four biotechnology industry groups filed a motion seeking to participate in the Bt lawsuit. The four industry groups—the American Crop Protection Association, the Biotechnology Industry Organization, the National Cotton Council and the Bt Registrants' Task Force—claim that they should be allowed to intervene in the lawsuit because they have a substantial interest in the case. CFS filed an opposition to the request to intervene. However, the court recently allowed the industry groups to participate in the lawsuit.

The case continued to pick up steam this winter. At a January 18, 2000 oral hearing, federal district Judge Louis Oberdorfer gave the plaintiffs their first victory in the case by ordering the EPA to answer the environmental concerns initially brought to the agency's attention over two years ago via legal petition. Coinciding with the 'court's ruling, the EPA released new restrictions on the use of Bt crops during the 2000 growing season, and recently admitted that it needs to require stricter testing of the Bt crops' impacts on wildlife such as mallard ducks, bobwhite quail, rainbow trout, channel catfish, honey bees, and earthworms.

32

A Primer on GMOs and International Law

By Phil Bereano

Phil Bereano, JD, PhD, *is Professor of Technology and Public Policy at the University of Washington, Seattle. He is on the roster of experts for the Cartagena Protocol, co-founder of the Council for Responsible Genetics, and currently represents the Washington Biotechnology Action Council and the 49th Parallel Biotechnology Consortium at international meetings.* *An earlier version of this article appeared in the April 2004 issue of* Seedling *magazine, published by Genetic Resources Action International (GRAIN). This article originally appeared in* GeneWatch, *volume 25, number 3, April–May 2012.*

Two international instruments changed the playing field in the past decade regarding the international regulation of genetically engineered organisms. One is the Cartagena Protocol on Biosafety, which is intended to regulate the international transfer of "living modified organisms" (LMOs). The second is a set of guidelines, the Risk Analysis Principles for Foods Derived from Biotechnology, established by a little-known United Nations body called the Codex Alimentarius Commission.

These two instruments signal attempts by the world community to establish rules governing the production, trade and use of genetically modified foodstuffs. Both agreements emphasize the rights of consumers and farmers, and the protection of ecosystems. However, it is still not completely clear how their provisions will work alongside the free-trade rules of the World Trade Organization (WTO).

The Cartagena Protocol: A Greener Way

By joining the WTO, countries agree to limit their freedom to impose restrictions on foreign trade. The Cartagena Protocol, however, stresses that trade considerations need not always be given precedence over other national objectives. It recognizes that the need to protect biodiversity, the environment and human health are valid priorities in decision-making. As of today, some 163 countries (minus several of the most important agricultural exporters, including the United States, Canada, Argentina and Australia) have ratified the Protocol, which came into force on September 11, 2003.

The Protocol establishes a procedure called Advanced Informed Agreement. Under an AIA, those planning to export LMOs for introduction into the environment must notify the country to which they are being sent. That country is then entitled to authorize or refuse permission for the shipment, based on a risk assessment. Furthermore, the Protocol allows the recipient nation to invoke precautionary regulation if, in its judgment, there is not enough scientific information to make a proper assessment: "Lack of scientific certainty due to insufficient relevant scientific information and knowledge regarding the extent of the potential adverse effects of a living modified organism on the conservation and sustainable use of biological diversity in the Party of import, taking also into account risks to human health, shall not prevent that Party from taking a decision, as appropriate, with regard to the import of that living modified organism . . ."

The Protocol does not specify how to resolve any conflict between its own rules allowing an importing country to control trade in LMOs and that country's obligations not to impede trade if it is also a member of the WTO.

The state of international law regarding LMOs is intentionally fuzzy in some respects; diplomatic concerns for the WTO resulted in having a Protocol Preamble containing three intentionally conflicting provisions: that trade and the environment should be "mutually supportive"; that the agreement does not change any Party's international rights and obligations; and that the Protocol should not be interpreted as being "subordinate" to any other treaty. In particular, the Protocol's adoption of the precautionary principle—the idea that an action should not be carried out if the consequences of it are unknown but highly likely to be negative—is claimed by trade interests to run counter to the WTO mandate.

Those involved in drafting the Protocol, along with other observers, also acknowledge that there are a number of outstanding issues relating to the oversight of genetic manipulation technologies even after adoption of the Protocol text. These include:

- "Living modified organisms" (LMOs) is a more restricted category than "genetically modified organisms" (GMOs), since it excludes those no longer alive, and their products.
- "Intentional introduction into the environment" may not address situations where the exporter knows that some shipped modified grain, for instance, will be planted within the importing country, but does not necessarily "intend" this to happen.
- Many important countries are not members of the Protocol, including the largest growers and exporters of LMOs: the United States, Canada, Argentina and Australia.
- The Protocol's provisions on trade in LMOs between a party and a non-party state do not require that its procedures be followed.
- The Protocol says nothing about any regulatory oversight within a country.

In the fall of 2010, a Supplemental Protocol on issues of liability and redress for damages caused by LMOs was adopted after seven years of intense negotiations, and is in the process of being ratified by the requisite forty countries.

The Codex Alimentarius: Focus on Food Safety

Two months before the Protocol entered into force, a separate breakthrough took place. In July 2003, with the backing of all its 168 member nations, the Codex Alimentarius Commission produced the first set of international guidelines for assessing and managing any health risks posed by GM foods.

A relatively obscure United Nations agency, the commission is charged with the key global task of setting international guidelines for food quality and safety. It was established in 1963 by the Food and Agriculture Organization (FAO) and the World Health Organization (WHO), and given the mandate of "protecting the health of the consumers and ensuring fair practices in the food trade." The commission draws up voluntary international food guidelines through negotiations in approximately thirty committees and task forces.

The most significant element of the 2003 guidelines is that they call for safety assessments of all GM foods prior to their approval for commercial sale. This has important implications for WTO members. In 1995, the WTO had agreed that Codex norms should be the reference point for evaluating the legitimacy of food regulatory measures that are challenged as restrictions on trade. Thus, although the Codex guidelines are strictly voluntary, they have legal significance for WTO members as a defense to charges of "unfair trade." Also significant is that all of the major countries growing GMOs—the US, Canada, Argentina, and Australia—are Codex members and agreed to these risk assessment guidelines.

The Codex risk assessment guidelines contain much language about the need for a "scientific" evaluation of the actual hazards presented by the new foods. But they also recommend that "risk managers should take into account the uncertainties identified in the risk assessment and implement appropriate measures to manage these uncertainties." This wording appears to acknowledge the validity of a precautionary regulatory regime, similar to that allowed for international shipments under the Cartagena Protocol.

The Codex also recognizes that "Other Legitimate Factors"—non-scientific in nature—can form a valid basis for regulations, such as using halal or kosher standards. Other provisions within the guidelines call for a "transparent" safety assessment, which should be communicated to "all interested parties" that have opportunities to participate in "interactive" and "responsive consultative processes" where their views are "sought" by the regulators.

These non-scientific aspects are consistent with the second prong of the Codex mandate, namely its role in deterring deceptive practices. Such practices might, for example, include selling or distributing GM foods to consumers without labeling them as such. As a top world food exporter, the United States has vigorously advocated that only "objective" and "scientific" health claims be used as the basis for regulating GM foods, but consumer groups have vigorously contested this position. In the summer of 2011, after eighteen years of struggle, Codex finally adopted a guidance document recognizing that countries can adopt laws and regulations covering the labeling of GE foods, including mandatory labeling.

Too Rich a Mix?

It is not obvious how the Protocol, the Codex guidelines and WTO rules mesh together. Seeking a simple answer to this question assumes that the negotiation of these agreements was guided by a logical process. In fact, they were produced at different times, by delegations from different national ministries with various missions (trade, environment, food, agriculture, health, etc.), and without any reference to the bigger picture. These agreements also reflect the different configurations of industry and public interest groups that helped shape them.

Environmentalists argue that the new Codex guidelines on GM foods simply underscore how easy it has been for industry to bring GM foods to market without regulatory supervision, for example in the US. This practice has been criticized by many activist organizations and a growing number of scientists, as well as several international authorities on food safety matters.

Many of these critics point out that there is virtually no peer-reviewed, published scientific research on the risks or benefits of GM food that would allow for safety claims to be tested. They argue that the lack of evidence of risk is not the same as evidence of no risk. Many civil society organizations have insisted that precautionary steps should be taken to avert potential risks. Even the WTO Appellate Body, which settles its disputes, has recognized that divergent scientific views may be considered in making assessments, such as those evaluating food risks.

Using the precautionary principle to manage risks also puts the burden of proof on those seeking to introduce the new technology. The United States and other exporters of GM foods have blocked efforts to incorporate the principle explicitly into the Codex guidelines. But some commentators and activists believe that, despite no actual mention of it in those guidelines, the precautionary principle is implicit in the document's suggestions for risk analysis because these call for the safety of a GM food to be analyzed before it is produced and sold.

The governments blocking the inclusion of the precautionary principle into the Codex guidelines have argued that if it were to be applied to regulating GM foods, it could be used to justify regulations intended primarily to protect domestic industries from foreign competitors—in violation of the WTO agreements. Others point out, however, that it is not the purpose of the Codex guidelines to stimulate trade, but rather, to protect consumers. The WTO is supposed to follow Codex norms, not vice versa.

Whither GMO Politics?

The political storm raging round GM foods continues to grow in intensity, largely because the economic stakes rise steadily while scientific debate remains unresolved. Given the frameworks described above, what conclusion can one draw about the prospects for adequate regulatory supervision of the technology, and for proper protection of human health and the environment?

The four countries keen to export GM crops—the United States, Canada, Argentina and Australia—are all Codex members, but none of them is a party to the Cartagena Protocol. Therefore, one could argue that it would be inappropriate for such countries to object about others that choose to use the Codex risk assessments, since they all voted in Codex to adopt them.

On the other hand, as the countries that signed the Protocol meet to work out the details for carrying out risk assessments under its aegis, and to set rules on traceability and liability, none of these four nations will be legally able to block action taken under the Protocol. In reality, however, several nations which are Parties to the Protocol seem to be acting to protect the interests of these exporters.

As a result, the Protocol is likely to lead to rules that focus on protecting biodiversity and health more than any rules devised by the WTO. On that basis, there are grounds for believing that the future will see better environmental and health protection than exists at present.

A different situation, however, is likely to unfold behind the scenes as GM food exporters—particularly the United States—put pressure on countries, one by one, to waive their rights under international law. This already happened before the Protocol was enacted, where weak nations such as Croatia and Thailand had been subjected to pressure by the United States. And last year, Kenya—under enormous pressures from the US, Monsanto, the Gates Foundation and GE interests in South Africa—adopted a very weak "biosafety" law that will likely lead to the large-scale introduction of GE crops being grown in that country. Thus the responses of civil society will be crucial to ensure democratic and transparent oversight of this technology.

33

GMOs Stalled in Europe: The Strength of Citizens' Involvement

By Arnaud Apoteker

Arnaud Apoteker *is a GMO campaigner with The Greens/EFA Group in the European Parliament.*

Within a global review of GM crops, it is interesting to reflect on how Europe and the US have diverged in their adoption of GM crops since the 1980s and 1990s. Indeed, when GM crops were still in the research phase before their commercialization, the EU and the US (the two world's largest agricultural markets) were both rushing enthusiastically towards the agro-bio-technology revolution. It was arguably in Europe that the first transgenic plant was created in a lab in Flanders (Belgium) by European scientist Van Montagu and his team. France was second after the US for the number of experimental field trials of GMO crops and the EU had adopted a GMO regulation that was to pave the way for authorizing GMO crops in the EU territory in 1990 already. Very few would have predicted at that time that GMO expansion would be so different in the US and the EU. It is at the stage of commercializing GM crops that the two continents took very different paths.

The first marketing authorizations for GMOs happened almost at the same time in the US and in Europe, with marketing approvals for GM soybean and GM maize granted in the US and in Europe in 1995 and 1996. The first GM crop to be grown commercially on a large scale was soybean. It was authorized and grown for the first time in the US in 1995, and authorized for marketing in the EU in 1996. There had been no request for a growing authorization of this GM crop in the EU, because the EU grows very little soybean.

At the time the marketing authorization for GM soybean was granted by the EU, followed very soon after by GM maize, EU citizens were still under the shockwave created by "mad cow disease." It had indeed opened people's eyes to the complexity of modern food production that was increasingly getting extremely technical, unnatural, and profit-driven, and leading to completely new potential health risks. Environmental NGOs in the EU soon realized that a formidable GMO steamroller was coming with unpredictable and irreversible ecological consequences, and that the brakes needed to be put on now, because later it would be impossible to stop.

Greenpeace began to stage highly visible activities in European ports interfering with the unloading of the first shipments of GMO by painting ships' hulls with the X symbol from the television series *The X-Files* in order to symbolize the unpredictable and the invisible threat that releasing GMOs into the environment carries. In France, the French daily newspaper *Libération* headlined its cover with, "Alert, mad soybean is coming in," making an allusion to mad cow disease, which had been revealed a few months previously. Although the comparison was not based on scientific facts, it highlighted the risks of adopting widely a new technology without proper risk assessments and echoed a real concern in the readers' minds, that the way you feed animals may have an impact on animal and human health.

Greenpeace activities, supported by family farmers' organizations, other environmental groups, and some consumers groups have quickly generated a lot of public attention towards these new GMOs and their role in industrial and massive food production. The issue was very "popular" in the sense that ordinary citizens from all sectors of society showed a huge interest in GMOs and food issues. GMOs were quickly linked to industrial farming, the imposition of unwanted and potentially dangerous products to unwilling citizens through international trade (which was highly contested in EU civil society during the GATT agreements and the beginning of the WTO negotiations), patents on life, and the control of the food chain by a handful of corporate interests that threatened world food security. These connections have helped to create a tsunami of resistance to GMOs stemming from all sectors of society and many civil society organizations. GMOs became a symbol of citizens' struggles in a wide array of issues, including political and economic, fair trade, family and organic farming, consumers' choices, health, hunger relief, and development.

Despite different goals, NGOs have been able to rally around a specific demand: a moratorium of GMO releases in Europe. This has created a common message that was carried by many NGOs and all sectors of society and for which all citizens could get active at their desired level. Indeed, activities for all levels of citizens' involvement have been developed by a growing network of NGOs, from letter writing to field actions with the network *Faucheurs Volontaires* (a French anti-GMO activist group), from hunger strikes to labelling activities in supermarkets, etc. This diversity of choice for individual activism against GMOs has been fundamental in maintaining public pressure over all these years.

In Europe, these actions have forced EU governments and the EU bureaucracy to lead a number of public initiatives to discuss and try to reach consensus on the issue. These initiatives include national and regional debates, citizens' conferences, and parliamentary hearings. Citizens, although not all opposed to GMOs on principle, have consistently asked for a moratorium on commercial GM crops in Europe, to give some time for more research on their long-term health, environmental, agrarian, and socio-economic impacts. Formally though, this demand has rarely been followed by an EU state or by the EU that remained largely in favor of GM crops. The governmental neglect of public opinion in turn strengthened public rejection of GMOs after citizens realized their voices had not been heard at all.

The intense public attention on GM crops from citizens of all political inclinations had political outcomes. At the regional and local level, decision makers felt that they needed to take initiatives to protect their constituencies from GM crops and began to declare "GMO-free regions." This movement would spread throughout Europe from formal EU administrative regions to smaller localities to cities, fields, and even individual households. The concept proved very useful in involving the public, as citizens often feel they have more leverage on the local level. Although the legal power varies greatly from one member state to the next, civil action has led to regional prohibitions. For example, GMO-free regions currently cover about half the EU area.

Given the wide debate on GMOs and local initiatives, national and EU politicians have been pushed to express a position on GMO crops. This showed that opposition to GMOs was spread among all political groups, most of which publicly supported this technology apart from the Greens who have always been unanimously and unambiguously opposed to GM crops.

Similarly, the position of each government within the EU on GMO authorizations was not defined along typical political lines. For example, the

UK and Spanish governments have always been staunch supporters of the biotech industry, regardless of whether they are led by right- or left-leaning parties. Other countries, like Greece or Austria, have consistently opposed the cultivation of GMOs in the EU and in their territories. The entry into the EU of ten new member states did not change the balance, as these new members were as divided on the issue as the "old" members were.

Citizens and political pressure led to new EU regulations of GMOs that are considered the strictest in the world, although they have in practice failed to enforce precautionary principles or implement long-term risk assessments as required. The European Food Safety Authority (EFSA), the legislative body in charge of risk assessment, has been plagued from its creation in 2002 by conflicts of interest and undue industry influence and as a result lacks credibility.

The EU legislation on GMOs is very complex, as it involves twenty-eight national member states, the EU Parliament, and the Commission. The regulation of GMOs in the EU is basically covered by two main pieces of legislation: Directive 2001/18/EC, which regulates deliberate releases of GMOs into the environment (growing authorizations), and Regulation 1829/2003/EC, which regulates the authorization of imports and the use of GMOs in food and feed.

Both laws involve all member states through the so-called "Comitology procedure," meaning that a Commission's proposal to allow a GMO must be accepted or rejected by a qualified majority of the member states in a regulatory committee made up with representatives of the member states. A qualified majority is a majority of 62 percent of the votes, with each member state allowed a number of votes according to its size. When no qualified majority is reached, which has been systematically the case for all proposals to authorize GMOs, the Commission must act; in other words, it must authorize the GMO. The result is that the EU Commission, the executive and pro-GMO body of the EU, is systematically authorizing GMOs without the qualified majority of the member states. In doing so, it is making politically sensitive decisions which are way above its duty of implementing EU law, thereby exceeding the implementing powers that the EU treaties have conferred to it.

The GMO EU regulations provide member states with safeguard clauses or emergency measures. To apply such measures, a member state must prove that it has justifiable reasons to consider that any GMO that has received written consent for being placed on the market constitutes a risk to human health or the environment. These measures have been applied on many occasions by many member states. The scientific evidence provided by these member states as justification for their measures has been submitted to the Scientific Committee(s) of the European Union for opinion. In all of the cases that have been reviewed so far, the Committee(s) unsurprisingly considered that there was no new evidence that would justify overturning the original authorization decision. These bans are still in place, though.

GMOs Authorized for Cultivation in the EU in 2013

As GMO crops have been a very controversial issue among European citizens, scientists, and politicians, and are only authorized through a complex regulatory procedure, the EU has been largely free from GMO cultivation its its territories. Almost twenty years after the first authorization to grow a GM crop in the EU, only one crop is permitted to be grown in the EU: GM maize MON810 from Monsanto.

The second authorization for a GMO crop after MON810 was granted in 2010 for the GM potato Amflora from BASF. It was planted on a few hundred hectares in Sweden and Germany in 2011, but was a commercial failure and has since been prohibited by an EU Court of Justice decision in December 2013.

Today, only MON810 is planted in Europe. Authorized since 1998, it has been banned through safeguard clauses in nine European countries and is grown in only four. It covers a bit more than 100,000 hectares (about 0.1 percent of the entire EU agricultural area), mainly in Spain. Portugal grows a few thousand hectares, while the Czech Republic and Slovakia grow minimal amunts. This is in sharp contrast with the euphoria following the first growing authorizations, when projections from the biotech industry were that half of the maize growing area would be converted to GM maize in ten years.

GMOs Authorized for Imports, Food, and Feed Uses in the EU in 2013

Authorizations for GMO imports have been much more difficult to oppose, and numerous authorization requests have been processed since 1996. More than fifty GMO products have been authorized for food and feed uses in the EU, and the list is constantly growing.

Although only four GM crops have been authorized (soybean, maize, rapeseed oil and cotton), GMOs are potentially in almost all industrial food products, as soy lecithin, maize starch and maize glucose syrup are ubiquitous in processed food. This explains why the EU regulation on labelling has been so important for the anti-GMO movement.

EU Regulation on Labelling

EU Regulation 1829/2003, implemented in April 2004, lays down specific labelling requirements for food that contains GMOs. Genetically modified foods must be labelled, regardless of whether DNA or proteins derived from genetic modification are contained in the final product or not. The labelling requirement also includes highly refined products, such as oil obtained from GM maize.

The same rules apply to animal feed, including any compound feed that contains GM soy. Corn gluten feed produced from GM maize must also be labelled, so as to provide livestock farmers with accurate information on the composition and properties of feed.

Since governments are under pressure from the food and farming industries, the labelling of animal products such as meat, milk or eggs obtained from animals fed with genetically modified feed was not required in Reg. 1829/2003 despite consumers' demands.

The labelling regulation has had far-reaching consequences. Faced with the new labelling requirements and the general rejection of GM products by EU consumers, the food industry has shifted to non-GMO ingredients and all the major food brands have clearly indicated this to their suppliers. As a consequence, there are very few GMO-labelled food products on the EU shelves.

But millions of tons of GMOs enter the EU food chain through animal feed because animal products are not labelled. The food industry does not have the same requirements for feed as for food, and consumers are kept in the dark about it due to this lack of labelling. As more countries grow increasing amounts of GM soybean and maize, it is getting more difficult and costly for smaller producers to be supplied with GMO-free feed. As a consequence, more than 80 percent of animal products consumed in the EU are coming from animals that have been fed with GMOs. For consumers to have a real choice about avoiding GMOs in their food, there has to be sufficient availability of GMO-free supplies.

Some member states (Austria, France, Germany, and Luxembourg) have taken legislative steps to allow and set the conditions for GMO-free labelling in the last few years. As a result, some companies have begun labelling animal products as coming from GMO-free fed animals, and some brands and supermarket chains have begun to supply "GMO-free fed" labelled animal products. The success of these initiatives is fundamental for the future of a GMO-free Europe as it is consumers' demand for GMO-free food products (all along the food production chain) that will ensure supplies of GMO-free animal feed. Labelling has proven to be one of the most powerful instruments to prevent the EU from being covered with GMO fields and flooded with GMO food products.

Challenges and Dangers Ahead

While EU fields and vegetarian plates are still reasonably spared from GMOs, the situation remains very fragile. It has been the result of a fifteen-year struggle from all corners of society against the most powerful biotech multinationals, including Bayer and Novartis, which have enormous financial resources and political networks. Paradoxically, having prevented the cultivation of GMOs in most parts of the EU, citizens seems less interested in the issue, even though public polls show a steady level of GMO rejection. Biotech companies have decided to step up their communication strategy after having "patiently" waited for more years of citizens' resistance. Pro-GMO developments have occurred at the end of 2013 that include:

- new authorizations for multiple genetic modifications (examples include Smartstax or Powercore);
- requests for GM insect releases (fruit flies);
- a current proposal for a new seed regulation that will benefit the companies that are the main suppliers of GMO seeds;
- a legal challenge won by Pioneer Seed Company for the Commission's failure to act expeditiously enough in granting authorization for growing GMO maize 1507 (brand name Herculex). As a consequence, the Commission proposed to the Council to authorize GMO maize cultivation. This may be the first GMO maize authorized in fifteen years.
- Lastly, the recently launched negotiations for a EU–US free trade agreement (TTIP) may ring the death knell for EU precautionary measures on GMOs.

Some key dates in European Union GMO history:

1990	First EU regulation on deliberate releases of GMOs 1990/220
1996	Approval of imports and use of Roundup Ready herbicide
1997	Approval of the marketing and growing of the first GM maize (Bt 176)
1997	First labelling regulation which covered only products with measureable = GMO content. It excluded, for example, soybean oil made with 100 percent GMO soybean, as the GMO content could not be detected in the oil
1998	Destruction of GM maize seeds from Syngenta by a French farmers' union
1998	Authorization of GM maize MON810 from Monsanto
1999–2004	Informal moratorium on new releases of GMOs, imposed by a blocking minority of member states within the EU Council
1999	World Trade Organization fails in Seattle, partly because of GMO and patents issues
2000	Biosafety Protocol is adopted
2001	EU Directive 2001/18 on deliberate releases of GMOs is passed, repelling directive 90/220
2002	Creation of EFSA (European Food Safety Agency)
2003	Faucheurs Volontaires in France for field actions
2003	Novel Food Regulation imposes labelling and traceability of GMOs
2008	Environment Council requires the Commission to improve risk assessments of GMOs
2010	Approval of transgenic potato, first growing approval since 1998
2010	First labels for "fed without GMOs" animal products appear in Austria, Germany, France
	The EU Commission, frustrated that no GMO authorization proposal meets the required qualified majority, proposes to give member states the ability to reject an EU-wide authorization of GMOs.
2012	The European Court of Justice declares that GMO contaminated honey must be labelled
2013	Court of Justice against Commission on Pioneer 1507's case
2013	Court of Justice against Commission on Amflora potato authorization

PART 6

Ecology and Sustainability

In the laboratory, scientists can control the conditions under which genetically engineered crops are grown by regulating what comes into contact with the plants, in addition to all aspects of their environment. In the field, however, genetically engineered plants come into contact with all sorts of other living organisms, including weeds, other plants, insects, people, birds, and various other wildlife. In addition, there can be strong winds, heavy rains, excessive sunlight, and a whole range of other environmental conditions that affect and are affected by the genetically modified plant. Once genetically modified plants are released into the environment, scientists no longer have any control over them, and as they reproduce, migrate, and mutate, they raise several concerns.

First, genetically modified plants produce pollen that may also contain the foreign genetic material that was inserted into them. The pollen can be picked up by insects, birds, wind, or rain and carried into neighboring fields or wild areas. If the neighboring farmer happens to be farming organically, the genetically modified pollen could do catastrophic damage to the farmer's entire crop.

The DNA from genetically modified plants can also transfer to wild relatives, creating hybrid populations over which scientists have no control. If the genetically engineered traits of herbicide tolerance and pest resistance spread into wild populations, for example, they could result in the creation of super-tolerant plants and pests. These super-bugs and super-plants will require stronger and more toxic chemicals to control and eliminate them. The genetically engineered pollen that has the potential to create the genetic pollution described above has also been shown to kill beneficial insects, not to mention affecting the biodiversity of native plant populations.

GM crops have also supported the applications of greater amounts of harmful herbicides resulting in the spread of herbicide-resistant weeds. This has two effects: 1) it results in farmers having to rely on older, more toxic herbicides that destroy food sources for beneficial insects such as the Monarch butterfly, and 2) it threatens biological diversity by drifting beyond field boundaries and damaging both neighboring crops and wild plants.

GMOs have serious implications for the environment. The essays in this section focus on those concerns.

—Jeremy Gruber

Environmental Release of Genetically Engineered Organisms: Recasting the Debate

BY THE GENEWATCH EDITORS

This article originally appeared in GeneWatch, *volume 5, numbers 2–3, March–June 1988.*

The introduction of genetically engineered organisms (GEOs) into the environment raises crucial questions for society about their potential public health and ecological impacts. Discussion of these impacts to date has been largely limited to a technical debate between ecologists and molecular biologists about how to assess the release of novel, self-replicating organisms. Until now the debate has also been limited to consideration of the immediate risks surrounding the first few small-scale field tests.

Within this technical discussion, molecular geneticists emphasize the continuity between classical genetics and recombinant DNA techniques. They identify potential hazards primarily through study of the inserted genes and the host organism, holding that GEOs possess no unique hazards and are less adaptive to the environment beyond a selected niche. If the inserted genes and their products are not known to pose any problems, and if the host is not pathogenic, then, most molecular biologists believe, the modified organism is safe for environmental release.

In contrast, ecologists emphasize the inherent uncertainties of introductions. They cite a body of relevant literature in the introduction of exotic species into new environments. They stress the importance of in situ tests, of pretesting in microcosms, and of modeling entire ecosystems in preparation for environmental releases. No one can predict the results, ecologists assert, merely by knowing the genes and their hosts. While biologists play an indispensible role in shaping the concerns surrounding the environmental release of newly altered organisms, the decisions society must make about the possible consequences of these releases are not simply technical decisions. They require broad public participation by those members of society who will bear the risks in order to include the full range of social, economic, and environmental issues raised. Furthermore, a realistic appraisal of the risks associated with environmental release must deal not only with the increasing number and scale of field tests, but must also weigh the impacts of commercial-scale manufacture and dispersion of GEOs.

This special report from the *GeneWatch* editorial committee will address the social, economic, and regulatory dilemmas posed by environmental release of GEOs. We will review:

- the lessons for biotechnology from the chemical revolution;
- the inadequacies of current environmental release regulation;
- the role of scientific expertise;
- accidental and unsanctioned releases;
- the assessment of socio-economic impacts.

Our analysis will end with an outline of specific proposals to regulate biotechnology that addresses the concerns currently left out of the debate. In doing so, we hope to see valuable research conducted safely and in the public interest.

What Can Biotech Learn from the Chemical Revolution?

The early regulatory history of the chemical revolution provides an important backdrop for thinking about how to regulate the relatively new biotechnology industry. Prior to 1970, the uncoordinated regulation of chemical substances functioned primarily as crisis management. The science of chemical risk assessment barely existed. The first laws that officials used to regulate chemicals were public health statutes dating from the turn of the century and modified thereafter. These laws originally dealt with infectious diseases caused by naturally occurring organisms that entered food and water supplies.

In 1970 a new generation of laws, specific to the chemical industry, made important changes in how government defined the problem of chemical hazards. These laws included the Federal Insecticide, Fungicide, and Rodenticide Act (FIFRA) and the Toxic Substances Control Act (TSCA). While the new regulatory philosophy for the chemical industry emphasized risk assessment and pre-market evaluation, tens of thousands of untested chemicals were grandfathered in under the new laws. Some companies either performed inadequate tests on scores of chemicals or withheld test results from those at risk. And the dangerous effects of still other chemicals probably could not be predicted given the state of risk assessment at the time. During the 1970s and 1980s regulators played frantic catchup after decades of neglect. The EPA is still decades away from a full assessment of the pesticides currently in use.

The primary lesson from the long and complex history of the revolution in synthetic chemicals is that public awareness of, and government response to, chemical hazards came only after substantial damage to humans and the environment had occurred. By this time, the nation's economy was dependent on the products and procedures of the firmly established chemical industry. Options for reviewing new developments were thus limited to cost benefit calculations based on already existing industry practices.

While little debate or concern for potential consequences accompanied the initial commercialization of synthetic chemicals, it is still possible for the public to examine the risks of planned environmental releases. The US biotechnology industry invests from 1.5 to 2 billion dollars annually, largely in efforts to develop commercial applications for GEOs. In many cases, these applications will require the introduction of newly engineered organisms into the environment. The biotech entrepreneurs envision a future where genetically altered microbes digest oil spills, and toxic waste; soil bacteria are engineered to poison crop-damaging insects; and rabies vaccines made with modified viruses immunize both domestic and wild animals.

Biotechnology's advocates often represent its promise as virtually limitless, and their claims recall those of the early chemical industry's most extravagant proponents. These chemical pioneers continued to tout the advantages of chemical pesticides, for instance, while denying mounting environmental problems. Now, a generation after the revolution in synthetic chemicals, society is faced with serious public health problems and the yearly production of some 60 million tons of hazardous chemical wastes. Unlike the revolution in synthetic chemicals, the revolution based on biotechnology involves products that are designed to reproduce in our environment. We must learn from the past and begin to draft regulation tailored specifically to biotechnology.

Table 1. Existing Regulatory Structure For Oversight Of Environmental Release

On June 26, 1986 the federal government outlined a "Coordinated Framework" for the regulation of environmental release of genetically engineered organisms. The framework's focus is the adaptation of existing federal statutes to this new field. It maintains oversight authority in existing agencies, including the Environmental Protection Agency (EPA), the Food and Drug Administration (FDA), the National Institutes of Health (NIH), the Occupational Safety and Health Administration (OSHA) and the US Department of Agriculture (USDA). An interagency Biotechnology Science Coordinating Committee (BSCC) is responsible for coordination and consistency of regulations and policy for environmental releases. The responsibilities of these agencies are delineated in the table below.

Agency	**Statute**	**Product Regulated**
EPA	Federal Insecticide, Fungicide, and Rodenticide Act (FIFRA) Toxic Substances Control Act (TSCA)	Pesticides New Microbes not falling in other categories
FDA	Public Health Service Act	Drugs, food additives, medical devices
NIH	NIH Guidelines	NIH-funded research
OSHA	Occupational Safety and Health Act	Workplace hazards
USDA	USDA Guidelines Virus-Serum-Toxic Act Federal Plant Pest Act	USDA-funded research Veterinary biologics Plant pests

Source: *Regulating Environmental Release of Genetically Engineered Organisms: The State Perspective*, published by the National Center for Policy Alternatives

What is the Present Role of Regulatory Agencies in Biotechnology?

Just as in the case of the nascent chemical industry, the current federal system for regulating environmental release is based on a reinterpretation of existing statutes, all of which predate genetic engineering. The federal policies are a loose patchwork, pieced together over five different agencies and at least ten different laws (see Table 1.) The EPA is attempting to regulate releases of GEOs into the environment under pre-existing statutes, namely FIFRA and TSCA, intended for chemical substances, not self-replicating organisms. The regulations neither address the special environmental and public health risks raised by experimentation with new genetic techniques, nor establish a program to improve the identification of those risks. Most of the regulations have very weak public information requirements and current laws lack standards for protection of workers or environmental monitoring.

We now face a regulatory dilemma in biotechnology. The present jerry-built framework that relies on TSCA and FIFRA does not take into account the complexity of assessing the risks of biological organisms designed for use in the environment. No standard tests or even definitions for genetically novel strains exist. Researchers may take years to design and implement certain microcosm or field experiments. Because there exists scientific uncertainty, assessments from a wide range of the scientific community could assist in both identifying possible risks and averting low probability, but high consequence, events. This lack of consensus within the scientific community itself on how to evaluate the risks of GEOs places a greater burden on the risk assessment and management process. Once the number of applications to EPA increases, the agency will be forced to use a "triage" system to determine which GEO it should choose for more careful review. When the pace of submissions reaches thirty to fifty per year, the agency will be unable to sustain case-by-case review and will search for mechanisms to expedite its process. In fact, we are already seeing EPA proposals to ease its burden by establishing Environmental Biosafety Committees within institutions. One of the most glaring weaknesses of current regulation lies with the significant obstacles TSCA creates. Unlike FIFRA, TSCA is a notification statute; it does not require that a firm obtain a license to manufacture a substance. A notification system places the burden of proof on the regulator rather than the manufacturer. The manufacturer is not even obligated to provide data that attests to the safety of its products. The EPA has a ninety-day period to review a TSCA submission. If the agency determines that a particular substance might present an unacceptable risk, it may issue a set of rules applicable only to that product. This rulemaking process to regulate the biological in question is cumbersome, laborious, and injects considerable discretion into regulatory decisions.

The role of the USDA presents another serious problem. No agency whose mandate includes promoting biotechnology in industry, commerce, and agriculture should have a responsibility to regulate its use in those sectors. Yet this is exactly the position that the USDA is currently in. Moreover, the agency lacks a coherent policy on engineered plants and organisms. The USDA also has a poor record of encouraging citizen participation in its decision processes for biotechnology or of communicating to the scientific community how it plans to evaluate new products.

What is the Role of Scientific Expertise?

Many scientists participated in debates over the first few GEO releases because of the media attention, through litigation, or because a regulatory agency solicited their expertise. As the review process normalizes, what incentives will exist to encourage a strong level of involvement among scientists? Even with effective environmental statutes for regulating GEOs, the role of outside experts is critical, particularly since no standard tests exist to determine what is safe. As long as scientific uncertainty remains high, the experts within environmental agencies must be aided by a broad network of scientists in related disciplines. Experts are needed to posit scenarios where the organism might be hazardous to some components of the ecosystem. If researchers discover such a scenario, agency support should be granted for the necessary empirical studies.

Scientists do not benefit professionally from troubleshooting new products. Therefore, it is in the public interest to provide financial incentives to attract this type of analysis. Annual investment in US biotechnology innovation from private, state, and federal sources totals nearly $5 billion; investment in expanding our knowledge of how these innovations affect the environment and public health are barely one thousandth of that figure. If scientists in the future know more about predictive ecology, and if they can develop a set of standardized risk assessment tests, like the tests for chemical mutagenicity, then the need for broad review may diminish.

Even with scientists available to critically evaluate new biotechnology products, modeling the introductions of novel organisms still has important limitations. If a product review reveals no hazards, then either it is safe or we do not know enough to prove otherwise. The sheer increase in the number of environmental releases raises the odds for an accident, with unpredictable consequences. The challenge we face is to spot any telltale signs of disaster before it happens. The more informed people we have thinking about possible adverse consequences, the better our chances in succeeding to identify and avoid them.

What about Accidental and Unsanctioned Releases?

Any discussion of the environmental release of genetically engineered organisms would be incomplete without consideration of accidental and unsanctioned releases. These types of releases of altered organisms become increasingly important as the biotechnology industry grows and matures. As the production of genetically altered organisms expands to a far greater scale there will be an increase in the potential for accidental releases during the entire "lifecycle" of the organisms—from the laboratory to the production/fermentation facility, greenhouse, field test, and waste stream.

A burgeoning and maturing biotechnology industry does not only mean the existence of a greater quantity of GEOs but a greater diversity as well. As a wider variety of organisms are created in the laboratory, the possibility that an accidental release could cause considerable and even irrevocable environmental damage grows accordingly. For instance, the sharp rise in Biological Defense Program research employing highly pathogenic microorganisms highlights the very real danger that an accidental release could pose, even despite a high level of physical containment in the laboratory.

And finally, as the routine use of genetic engineering techniques becomes more widespread, it raises a range of questions about the potential for and logistics of regulatory oversight. This issue is underlined by the notorious spate of unsanctioned environmental releases that have occurred in the biotechnology industry to date. A pattern of regulatory abuse has already been established with unsanctioned, illegal releases in Montana, Nebraska, South Dakota, Texas, and several different cases in California. In fact, there have been nearly as many unsanctioned as authorized releases of GEOs into the environment. The prevalence of unsanctioned releases, coupled with the possibility of greater numbers of researchers working with unquestionably dangerous organisms in a lax regulatory environment is cause for concern. The following bizarre example illustrates how strange unintentional releases may take place. In August 1987 a biochemist working in his home laboratory in Kingston, Massachusetts, was recombining the genes of sea organisms to create a new type of building material. His beachfront home collapsed. The scientist stated that none of the organisms or the menagerie of animals (mice, rats, parakeets, a parrot) that escaped his home were dangerous. The investigator in this incident violated no law; since private funds were involved he was not obligated to follow any guidelines.

In fact, aside from the releases that have taken place in violation of standing regulations, whole classes of release experiments deemed to be "non-commercial" are exempt from regulation. This includes research conducted by universities with funding sources other than the National Institutes of Health (NIH) or the USDA. As greater numbers of researchers begin to conduct experiments outside of the traditionally regulated regime, these types of regulatory loopholes will become all the more significant. Not only do they provide more "unseen," unregulated potential for accidental release, but laboratory workers can serve as pathways for the release of GEOs and can themselves face potential occupational health risks. Clearly, in terms of a potentially damaging accidental release, the risks posed are independent of what type or size of organization conducts the test, a fact which highlights the lack of a rational basis for such regulatory exemptions.

As for the possibility of a modified organism finding an environmental niche after "escaping" during a field test, again, while the chances of such an occurrence are slim, any regulatory structure must be designed to handle a vastly greater quantity and diversity of such field tests in the near future. When asked about the dangers of field tests, David Baltimore, director of the Whitehead Institute, has quipped, "Would corn planted at the edge of the forest take over the forest?" But in several instances introduced organisms—kudzu and the gypsy moth are well-known examples—have indeed taken over the forest, causing widespread damage.

But if the regulation of the above areas is inadequate and shortsighted, the regulatory framework for dealing with biogenetic waste is virtually nonexistent. The "new" biotechnology industry has brought with it a new form of waste which has the ability to live and multiply in the environment. As living organisms, biological waste has the potential to spread disease and/or undergo genetic exchange. Genetic material can be transferred between organisms of different species, genera, and even families. The rapid spread of antibiotic resistance among bacteria in clinical settings is an obvious example of the ease with which certain kinds of genetic exchange take place. In considering the possible effects of biogenetic waste it is important to remember that perhaps the major legacy of the chemical industry is the untold billions of metric tons of toxic waste that poison our environment today.

How Do We Assess Socio-Economic Impacts?

The Reagan Administration directed regulators to consider the adverse economic impacts on business that new regulations might have. With this policy the administration sought to counter the health and environmental laws of the 1970s without delegislating them. However, a product that will create distributional inequities within the industrial community or that threatens social transformation leaves the regulators silent for lack of authority. The Tulelake, California, farmers who opposed the field testing of ice minus (a variant of the common bacterium *Pseudomonas syringae*) in their community felt economically threatened by ice minus. They believed its commercial use might increase the land available for growing potatoes, intensifying competition in what was already a low profit margin enterprise. Or consider the case of the bovine growth hormone (BGH), which can now be produced using genetically engineered organisms. When injected into cattle the hormone can increase milk production up to 30 percent. But what economic impact will marketing BGH now have when milk surpluses are at record highs? The regulatory framework that can include such socio- economic assessments does not yet exist.

When an industry designs a new product for release into an environment where there are finite risks, we must ask: What do we gain? What could we lose? What are we displacing? Does the new product fill real social needs? Advanced Genetic Sciences exploited environmental symbols in their promotion of ice minus—under its trade name Frostban—as a way to heighten the importance of their product. They explicitly stated that chemical pesticides are unsafe and that they are developing biological pest control agents. The firm's promotional material implied that ice minus is a substitute for chemical pesticides, even though the widespread use of biocides has never been the treatment of choice for preventing frost damage. Meanwhile other agrichemical and biotechnology companies are genetically engineering crops resistant to harmful side effects of pesticides that will expand and prolong the use of such chemicals in agriculture. For some of these corporations pesticides are their most profitable products. Yet biotechnology could also be used to develop non-pesticide alternatives which would lessen farmers' dependence on a few big agrichemical companies.

Evaluation of the desirability of a new product whose development or use will pose risks to the environment should not be left only to the company that stands to profit from its marketing. In deciding what degree of risk is acceptable, industry interests must be balanced by those of the larger society. If the overall social and economic effects of a product are negative, then any environmental risk that accompanies its testing or use may be deemed unacceptable. At present, no regulatory mechanisms exist either for assessing environmental risks within the context of socio- economic impacts or for answering the questions about such impacts raised by affected individuals in a community.

Proposals for a Sound Regulatory Process

The present ambiguous and conflicting state of regulation satisfies no one. It creates difficulties and confusion for the biotechnology industry, it fails to adequately safeguard the public health and environment, and it does not include a framework for meaningful public involvement. The following proposals for crafting a sound regulatory process respond to the issues we have identified in recasting the debate over environmental release of genetically engineered organisms.

These proposals are:

- **Designation of the EPA as the lead agency.** The designation of one agency to oversee the entire field of deliberate environmental release of genetically engineered organisms, coupled with a single permitting process, would ensure careful review of risks prior to authorizing releases. The EPA could also be provided with the authority to prevent accidental releases. Due to the unpredictability and significance of potential risks of environmental release, companies proposing the releases should bear the burden of proving that releases are safe and deserve to be permitted.

The mission of the agency with respect to environmental release should be to protect the environment and public health. The agency should have no promotion mission, such as funding of research geared to developing biotechnology.

- **Amendment of the TSCA.** The Toxic Substances Control Act does not meet the needs of biotechnology. Organisms designated for release into the environment must fall under a licensing law that places a burden on the manufacturer to demonstrate safety and efficacy. The law should include provisions for risk reduction when a new technology is replacing an established one. The EPA should have sufficient latitude to request more information and additional tests. The model of drug regulation is appropriate in this case. Giving the EPA this authority will not eliminate the possibility of a hazardous product; it will simply reduce the probability when expertise is brought in at an early stage. We must remember that there is no public imperative to market most of the products of biotechnology. And if such an imperative could be demonstrated, as it has been in the development of certain drugs, then society might be willing to expedite the review process.

Any breach of the permit system should be considered a serious violation, as the threat posed by so-called minor infractions (i.e. a small release) could be severe. Yet some firms conducting genetic engineering research have shown a rather casual disregard for environmental safety. Regulators need to have the ability to penalize wrongdoers quickly and effectively; the penalties should be high enough to discourage violations. The agency should be provided with administrative penalties powers, rather than requiring court action to assess penalties for non compliance. Such an approach shifts the burden and risks of instigating litigation to the violators.

- **Provision of regulators with resources and in-house capabilities.** The massive expansion of industry's genetic experimentation will place tremendous economic pressure on regulators to act quickly, and without due care. One key to providing adequate oversight within a time constraint that industry can live with will be to fully fund the regulators. This funding will need to grow as the industry grows and could be secured by levying a tax on the firms conducting research.

These resources would be put to good use in developing staff expertise within government. Oversight of environmental release of genetically engineered organisms demands a range of professionals, such as soil ecologists and microbiologists. The complexity of the risk to be analyzed requires staff to be kept up to date on the state-of-the-art. In this way, regulators can be alert for omissions and errors in documentation submitted by applicants.

- **Establishment of ongoing advisory boards.** In addition to professional staff, outside advisory boards can make a substantial contribution to environmental release permit programs. Such advisory boards can bring a fresh perspective to difficult technical problems and advise on social and economic issues outside of the expertise of the regulators.

Appropriate conflict of interest rules should govern appointments to advisory boards. Such a board would typically draw heavily upon university scientists. However, in recent years, many molecular biologists in academia have become affiliated with, and/or taken equity in, biotechnology firms. All such relationships should be viewed as potentially prejudicing the views of these experts, and they should be disclosed before appointments are made. Advisory boards should be balanced to ensure that public interest and community groups are fully represented.

- **Establishment of case-by-case independent external review.** Funding must be available to bring the expertise of ecologists to bear on assessing the risks of new products. Publishing the name of an organism in the Federal Register and expecting scientists to divert themselves from busy schedules to examine the ecological consequences of a large-scale release is unrealistic. Scientific research and the assessment of genetically modified organisms must be coordinated. Regulatory review must include outreach efforts to relevant scientific groups and incentives to obtain the necessary critical review.
- **Public involvement.** Regulation of environmental release should provide for extensive public involvement, given the magnitude of risks, so that control over these experiments is not completely vested in the hands of a small group of technical experts. Although consideration of technical issues is essential, expert assessment of "acceptable" risks may differ dramatically from views of the public which bears those risks. The social, ethical, and economic implications of decisions to permit such experimentation must be considered. Public involvement in the regulatory process is vital. Citizens of communities facing potential releases may have information otherwise unavailable to regulators who do not live close to the location of a release. The alternative to early and thorough public involvement may often result

in polarized public opposition resulting in local battles and ultimately the obstruction of local releases.

Good precedents for informing community residents exist under other laws. For instance, the provisions of the community's right to know established for chemical products in the 1986 Superfund Amendments could be extended to the storage and release The ability of local residents to understand and participate in the regulatory process could be enhanced by technical assistance grants to local organizations.

Regulators must use care in identifying and protecting legitimate trade secrets. But the burden of proof for trade secrecy must be on the industry, so that spurious claims are disallowed, and the public is enabled to make a thorough evaluation of risks. As a general rule, the biological entity and its function is the only information that can be justified as proprietary across the board.

Public involvement need not be merely reactive. Representation of diverse points of view can be integrated into the policy-making process. For example, the Cambridge Experimentation Review Board played a responsible and constructive role in the development of the city's regulation of experiments involving recombinant DNA.

- **Risk assessments.** Regulators will need to create an acceptable basis for analyzing risks. Certainly, all proposed environmental releases should have a review, much like an environmental impact statement, conducted prior to permitting. Such studies should review worst case scenarios that may occur as a result of the releases, and should specify all nonstandard assumptions used in developing such assessments.
- **Post-release monitoring and mitigation.** After an approved release takes place, the survival success of the modified organisms, and the extent and conditions of their dispersion, must be carefully monitored. This monitoring is both a precaution, in case negative impacts occur and corrective measures are needed, and a research tool to strengthen future risk assessments. As a prerequisite to obtaining a permit, the applicant should also demonstrate some consideration to mitigate the worst impacts of a planned release.
- **Prevention of accidental release.** The potential for a small release of genetically engineered organisms to have a big effect makes prevention of accidental releases critical. Preventing environmental releases from indoor

testing has not been adequately addressed. One avenue for such releases involves effluents from fermentation tanks, laboratories, and greenhouses. Facilities' waste streams should be carefully regulated and monitored to ensure that no unintended releases occur. Physical barriers could also be incorporated, such as secondary containment chambers. Workers in the facilities potentially represent a pathway for engineered organisms to escape the lab since many organisms are designed to survive inside the human host. Current guidelines are inadequate to protect these workers.

- **Liability and insurance.** Regulation is one means provided by the legal system for the protection of public health and the environment. Tort law—the area of law allowing suits for recovery of damages by injured persons—is another. With or without adequate regulation, firms or institutions conducting releases take the risk of injuring people or natural resources. For releases of genetically engineered organisms, strict liability should apply regardless of whether the plaintiff is the government or anyone made ill by a release. Under the federal Superfund law for hazardous waste cases brought by the government against persons causing a toxic chemical release, the plaintiffs need only show a cause and effect relationship between release and injury. Complex proofs of negligence are not required. This would offer the strongest protection of victims, and provide the greatest liability incentives to corporations to consider more carefully a release that has a chance of causing harm.

Regulations can also require the releasers of genetically engineered organisms to carry adequate insurance to cover any claims. Such an approach could strengthen the development of risk assessment methodologies by adding insurance company interest and involvement in the development of such assessment methods.

- **Establishment of international coordination.** Communication and collaboration between nations must be established and encouraged to address the issues presented by the environmental release of GEOs. Environmental safety is a global, not a domestic concern, especially when an accidental release could conceivably have worldwide consequences on food crops or even human health. Whether through regular, international scientific meetings and exchanges, or through more formal, international channels, these issues must be addressed from a global perspective and on an ongoing basis.

In the United States, environmental provisions are needed to prevent American multinational corporations from conducting field tests or other procedures abroad that have been prohibited at home. Similarly, scientific data about environmental safety must be widely shared so that nations (especially in the Third World) will be apprised of the potential long-term environmental dangers presented when a country allows ill-tested procedures within its borders, even for the sake of a promised short-term economic gain. Only through such coordination can it be assured that international competition in biotechnology does not come at the expense of global environmental health and safety.

At present, regulators must make hard decisions about managing many new applications for environmental release with limited internal expertise, inadequate laws, and a science of ecology that raises many more questions than it can answer. Crafting a workable regulatory system, along the lines we have described, is a first step in the difficult task of insuring that environmental releases of GEOs are conducted safely. Deliberate and unintentional releases of genetically modified life forms must not be taken for granted as a *fait accomplis*.

While the probability of a disastrous outcome is small, the hazards posed by the environmental release of genetically altered organisms are unpredictable. Introductions of GEOs may threaten to create new human diseases, or to spawn new plant or animal pests that damage agriculture, or to otherwise disrupt delicate ecological balances, just as introductions of exotic species—like the gypsy moth or citrus canker—have done in the past.

As releases increase in number and variety, so will the potential for harm to the public or the environment. As ecologist Martin Alexander has cautioned, if an undesirable event has a probability of occurring once in one thousand "uses" of a given technology, the risk from a few uses of that technology would surely be low. Complacency should disappear, however, if six hundred or one thousand or more uses are envisioned. As we stand at the threshold of a burgeoning biotechnology industry, we must force ourselves to think not only of the dangers of the handful of field tests currently underway, but of the potential effects—accidental, unsanctioned, or planned—of a full-scale industry.

35

The Role of GMOs in Sustainable Agriculture

By Doug Gurian-Sherman

Doug Gurian-Sherman, PhD, *is a Senior Scientist with the Food and Environment Program at the Union of Concerned Scientists.*

Agricultural technologies must improve sustainability by addressing huge challenges to food distribution while reducing environmental harm. Genetic engineering (GE) is most often evaluated based on whether it may cause direct harm through consumption of engineered foods, or direct harm to the environment, such as through killing beneficial organisms. Also important, though less often discussed, is whether it can contribute to reversing the tremendous harm that industrial agriculture is now causing to the environment and public health, and whether the unsustainable use of natural resources, such as fresh water, can be reduced. We need to ask whether GE is addressing, and will address, these challenges. While some of the direct harms from GE remain uncertain, the harms already caused by industrial agriculture that are perpetuated by current genetically modified crops (GMOs), are well documented.

Scientists have recognized that environmental impacts, for which agriculture is a large contributor; including loss of biodiversity, climate change, and nitrogen and phosphorus pollution (major contributors to water pollution such as coastal "dead zones"), are at global tipping points[1] And agriculture is also the primary human use of both land and fresh water, at about 70 percent of fresh water withdrawal.

Also important is whether GMOs can contribute, and are needed, to produce enough food sustainably. We produce enough food now. India has the most food insecure people of any country, yet it still exports food. The US has many food insecure citizens, but produces more than enough food. So production is not currently the limiting factor for food security; poverty and disempowerment create the problem. Still, the increasing world population and the increasing demand for animal products, which are an inefficient means of supplying nutrition, and the use of food crops for biofuels, will probably increase demand for food.

It will also be critically important to conserve and rebuild soil fertility, empower women and smallholders, improve infrastructure such as water and food storage and roads, and to increase the resilience of food production to climate change.[2]

Pesticides have negative impacts on biodiversity, and therefore pesticide use is one measure of the environmental impact of GMOs. Genetic engineering has reduced insecticide use by small amounts, but has greatly increased herbicide use in the US, the largest producer of GE crops.[3] Some increases in soil-preserving conservation tillage are attributable to GE crops, but are threatened by herbicide-resistant weeds exacerbated by those crops. And it is clear that conservation tillage can be accomplished economically without GMOs. For example, in the US, most gains in the adoption of conservation tillage occurred prior to the introduction of GMO crops in the US,[4] mainly due to changes in farm policy in the 1980s.

Further, resilience to climate change and agricultural pollution has changed little due to the introduction of GMOs, and it has so far not contributed meaningfully to reducing nitrogen pollution. Often overlooked is the fact that GE crops are also much more expensive to develop and usually much less effective, than viable alternatives such as breeding and agroecology.

Perhaps most fundamentally, genetic engineering has so far been coupled to and has reinforced industrial monoculture farming systems, which inherently foster the need for pesticides,[5] are less resilient, cause more water pollution[6] and harm to soil fertility, and are less productive than agroecological methods like crop rotation and cover crops.[7] Given the realities of corporate control of the technology[8] through intellectual property and economic concentration in the seed industry, questions remain about whether the technology can be used in more sustainable and democratic ways.

Minimal Contributions to Productivity or Reduction of Environmental Impact

Engineered traits have produced modest productivity gains in corn, and little or none in soybeans in the US in recent years, with most productivity improvements instead coming from breeding and crop production methods.[9] Most yield gains in European countries in recent years have occurred without GMOs, and are as high or higher than in the US, where the same crops containing engineered genes are grown under similar conditions, suggesting that GMOs has not significantly increased productivity.[10]

Yield improvements from Bt traits in developing countries often are less than those from low-cost agroecologically-based methods like push-pull, a system which uses multiple crops to repel and trap insect pests.[11] For example, in South Africa, yield gains of between 17 and 33 percent have been attributed to Bt in maize.[12] But this is being threatened by the development of resistance in the target insect—the stem borer. By comparison, push-pull often more than doubles yields, controls the same insect (stem borer) as well as the parasitic weed striga, builds soil fertility, and costs poor farmers little (unlike GE seeds) and therefore does not contribute to crushing debt.

Molecular analyses show that crop breeders have used only a small fraction of genetic diversity in crop species and wild crop relatives.[13] Therefore substantial untapped potential remains for breeding to greatly improve yield and other important traits without using GMOs.

Genetic engineering has so far produced no successful nitrogen-use-efficiency (NUE) trait, while breeding has improved NUE in several crops by about 30 to 40 percent over several decades,[14] and ecologically based farming methods are most effective at reducing nitrogen pollution.[15]

Drought tolerance in corn has improved about 1 percent *per year* for several decades in the US through breeding and agronomy, while the single engineered drought-tolerance trait in corn functions well only under moderate drought, and provides *total* improvement in productivity of only about 1 percent in the US.[16] And there are many crops in addition to corn where multiple drought resistant varieties have been developed through breeding in recent years including rice, cassava, wheat, sorghum, and millet.[17]

Crop breeding is also far less costly than GE, and develops traits as rapidly.[18] It is often argued that the cost of regulations holds back the development of GMOs, but the large majority of costs are for research and development, marketing, and so on. Regulatory costs according to the industry are about 10 to 20 percent of the total, which averages about $140 million for a successful engineered trait.[19] By comparison, development of typical traits through conventional breeding costs around $1 million.[20]

Major challenges also lie ahead for GE technology. Successful GE traits are simple—single genes that code directly for insect or herbicide resistance. Traits like drought tolerance are genetically complex, often environment-specific, and controlled by many genes.[21] GE contributions to these traits are likely to be modest for the foreseeable future. This is in part because many traits have been substantially optimized by evolution and breeding already. Trying to improve their function by the addition of one or a few genes in a piecemeal way often leads to unexpected negative impacts on other important agronomic traits. Breeding, by selecting for a trait rather than single genes, can often be more effective, although complex traits remain a major challenge.

The Broader Context—Industrial Monoculture and Agroecological Alternatives

The technology and seed industries are dominated by transnational corporations interested in large commodity markets, which are driven in part by the very high costs of developing GE traits. Only large markets will support these costs. Gene patents in many countries, and the high cost of development, allow large companies to control most uses of the technology. This restricts most private sector applications of GE to a few major traits and commodity crops.

By contrast, patents on plant varieties developed through breeding are not allowed by most countries, or exemptions are provided for research and to allow farmers to save seed. This, along with lower cost, allows for farmer participation in local and regional development of crops with resilience to local conditions and for multiple crops used in rotations.

Current GE traits—mainly Bt insect resistance and glyphosate herbicide resistance—have reduced labor and thereby facilitated the simplification of agroecosystems and increasing farm size. Most GE acreage consists of large monocultures—production of the same crop at a site year after year—of corn and soybeans in the US, Brazil, and Argentina. Monocultures are contrary to agroecologically sound farming systems based on crop and ecosystem diversity. Farming based on agroecology addresses the whole farm and surrounding landscape in ways that reflect the farm's actual interaction with the environment. This reduces nutrient pollution[22] and improves resilience while remaining highly productive and profitable by rotating (alternating) crops, using cover crops and green manures to protect and enrich soil, growing perennials, and recycling nutrients through composts and manure. This greatly reduces the need for expensive and polluting fertilizers and pesticides.[23] Monocultures of herbicide-resistant crops have exacerbated the development and spread of herbicide-resistant weeds due to overreliance on these crops. This is leading to a new generation of crops engineered to use older, more toxic herbicides, which threaten to further increase herbicide use.[24]

Insects resistant to Bt, such as rootworm in the US and secondary insect pests not controlled by Bt, are leading to a rebound in the use of chemical insecticides in the US, India and China. The monoculture systems that GE crops are part of are vulnerable to pests because larger areas planted with the same crop encourage greater pest populations.

This perpetuates the need for large amounts of harmful pesticides. For example, it has often been noted that Bt crops in the US have reduced the amount of insecticide on those crops. But more acreage of corn is now treated with neonicotinoids insecticides in the form of seed coatings that have become almost ubiquitous, to control insects not susceptible to Bt. Fewer acres were subject to spraying or soil treatments previously for the targets of Bt, corn rootworm and corn borer. Systemic neonicotinoid insecticides are implicated in extensive harm to honeybees and other beneficial insects.

Some scientists envision GE crops produced by the public sector, designed for use at low cost in sustainable systems. It is unclear how this would be accomplished given the domination by corporate actors. The corporate sector lobbies for a continuation of farm policies that perpetuate large monocultures where they can market their products. Agroecology, by contrast, is knowledge and method-based, and reduces the need for purchased inputs like fertilizers and pesticides, and is largely neglected by the public and private sectors.

While it is not out of the question that GMOs may add some value to sustainable systems, it comes at a high cost compared to greatly underfunded systems approaches like agroecology or participatory breeding that are better suited and leave more profit with farmers. Most public effort should therefore be devoted to more effective and lower-cost approaches that enhance biodiversity and the sovereignty of small farmers in developing countries. There can be substantial opportunity costs from devoting resources to GMOs at the expense of better options.

Genetically Modified Crops and the Intensification of Agriculture

By Bill Freese

Bill Freese *is a science policy analyst with the Center for Food Safety. Bill played a key role in the discovery of unapproved StarLink corn in the food supply in 2000 and 2001. His comprehensive report on genetically engineered (GE) pharmaceutical crops in 2002 helped initiate public debate on "biopharming." In 2004, he teamed up with Salk Institute cell biologist David Schubert to write a comprehensive, peer-reviewed scientific critique of the regulation and safety testing of GE foods. Bill has given numerous public presentations on agricultural biotechnology to State Department officers, international regulatory officials, farm groups and the general public.*

Anyone who looks objectively at the world's food and agricultural situation today cannot help feeling pessimistic. Between 850 million and 1.5 billion people suffer from hunger,[1] while one in three children in developing countries is malnourished.[2] In North and South America, the industrialization of farming proceeds apace, spawning ever larger monoculture farms, rising agrichemical use, and massive emissions of greenhouse gases. Understandably, this bleak situation generates a hunger for good news.

The chemical-seed industry has long understood that feeding this particular hunger is critical to its success. Fourteen years ago, chemical firms launched a $50 million-per-year PR campaign promising genetically modified organisms (GMOs) that would reduce chemical use, enhance nutrition, and help developing country farmers.[3] Having failed to deliver, these same firms launched a renewed media blitz earlier this year.[4]

Naïve journalists enthralled by the hype churn out a steady stream of stories touting the ever-future "potential" of experimental GMOs to save the world, ignoring both the plethora of past failures and the hundreds of millions of acres of GM crops actually being grown in the world today. In this essay, I critically examine real-world biotechnology, and argue that GMOs intensify many long familiar and pernicious aspects of industrial agriculture. As detailed below, they tend to promote more of what's bad—agrichemical use, monocultures, and behemoth farms—and less of what's good—skillful farming, biological and agricultural diversity, small farmers, and viable rural communities.

Ground-truthing

Let's start with a few key facts. GM crops are grown on roughly 400 million acres worldwide, but are heavily concentrated in a handful of countries with industrialized, export-oriented agricultural sectors. Nearly 87 percent of the world's biotech acres in 2011 were found in just seven countries of North and South America, with the U.S., Canada, Argentina and Brazil accounting for 83 percent.[5] GM crop farming in these nations takes place on huge, monoculture farms, involves massive agrichemical use, and employs extremely few farmers and agricultural workers. In most other countries, including India and China, biotech crops (mainly GM cotton) account for a miniscule portion of total harvested cropland.[6]

GM soybeans, corn, cotton and canola—the same four GM crops that were grown a decade ago—comprise 99.5 percent of world biotech crop acreage.[7] Soybeans and corn predominate, and are used mainly to feed animals or fuel cars in rich nations. Argentina, Brazil and Paraguay export the majority of their soybeans for use as livestock feed, while more than three-fourths of the U.S. corn crop is either fed to animals or used to generate ethanol for automobiles.

Most revealing, however, is what the chemical companies have engineered these crops for. Virtually 100 percent of GMOs, by acreage planted, incorporate just two "traits": insect and/or herbicide resistance. Insect-resistant cotton and corn produce their own built-in insecticide(s) derived from a soil bacterium, *Bacillus thuringiensis* (Bt), to protect against certain insect pests. Herbicide-resistant crops are engineered to withstand heavy application of one or more herbicides to permit season-long eradication of weeds without damaging the crop. Crops with herbicide resistance are most prevalent, comprising more than 5 of every 6 acres of GMOs grown worldwide.[8]

Pesticide Use, Weed Resistance, and the Chemical Arms Race

Nearly all current herbicide-resistant crops—soybeans, corn, cotton, canola, alfalfa and sugar beets—are engineered for immunity to glyphosate, sold by Monsanto as Roundup. Monsanto's "Roundup Ready" (RR) corn, soybeans and cotton have triggered an overall increase in herbicide use of 527 million lbs. in the sixteen years from 1996 to 2012 in the U.S. alone.[9] Although Monsanto assured everyone it wouldn't happen,[10] the deluge of glyphosate has generated an epidemic of glyphosate-resistant weeds, on the same principle by which bacteria evolve resistance to antibiotics. Glyphosate-resistant weeds—virtually unknown prior to RR crops—now infest an estimated 61 million acres in the U.S. alone,[11] an area the size of the state of Wyoming.

Other chemical firms are poised to introduce a host of "next-generation" GM crops resistant to more toxic herbicides as a false "solution" to glyphosate-resistant weeds. These include Dow Chemical Company's 2,4-D-resistant corn and soybeans, and Monsanto's dicamba-resistant soybeans and cotton. Bayer, BASF and Syngenta have similar herbicide-resistant (HR) crops waiting in the wings, and they comprise the majority of GM crops in the industry's near-term pipeline.[12] The profit potential of HR crops is clear, given that herbicides comprise two-thirds of all pesticide use in the U.S.[13] If introduced, each will lead to a dramatic increase in the use of the associated herbicides. For instance, the USDA projects that annual 2,4-D use in agriculture would rise from 26 million lbs. at present to anywhere from 78 to 176 million lbs. with 2,4-D-resistant corn and soybeans.[14]

Increased spraying will foster more intractable weeds resistant to multiple herbicides,[15] triggering a chemical arms race between crops and weeds. One harbinger of the future may be glimpsed in a patent awarded to DuPont, which envisions crops resistant to seven or more classes of herbicide.[16]

Biological Diversity under Siege

The massive use of herbicides with HR crops has serious biological consequences. Fields wiped clean of all plant life contribute nothing to biological diversity. In agriculture-dominated landscapes like the Midwest, the results can be disastrous.

A prime example is the plight of the Monarch butterfly in North America, which has experienced a dramatic decline over the past twenty years.[17] Monarch larvae feed almost exclusively on common milkweed in their chief Midwestern breeding range. Milkweed—a common agricultural weed even fifteen years ago—has been virtually eradicated from Midwestern corn and soybean fields thanks to intensive use of glyphosate with Roundup Ready crops.[18] Because too little milkweed grows outside of agriculture to sustain Monarch populations, their very existence is threatened. Other less charismatic plants and animals are likely also at risk.

The dramatic increase in use of 2,4-D and dicamba herbicides projected to occur if crops resistant to them are introduced also threatens biological diversity. Both herbicides are extremely prone to drift beyond field boundaries, which damages not only neighboring crops but also wild plants growing near fields.[19] In an assessment based on EPA data, 2,4-D and dicamba are ranked as posing 400 and 75 times greater risk to seedlings of terrestrial plants, respectively, than glyphosate.[20] Even with relatively modest use today, 2,4-D is likely threatening several endangered species that depend for habitat on wild plants that the herbicides kill. These include the Alameda whipsnake, California red-legged frog,[21] and Pacific salmon.[22] These impacts would increase dramatically with the 200 percent to nearly 600 percent increase in use projected with introduction of 2,4-D-resistant crops. Because endangered species act as sentinels of ecosystem health, one can expect much more widespread impacts to scores of plants and animals.

The Conservation Tillage Myth

Chemical firms maintain that herbicide-resistant crops reduce soil erosion by decreasing the use of tillage to control weeds. On the contrary, USDA data shows clearly that the major reductions in tillage operations and soil erosion in the U.S. occurred in the 1980s and early 1990s,[23] before GM crops had been introduced. Reduced soil erosion in this period is attributable to farm policy enacted in 1985 that made farmer subsidies dependent on the use of so-called "conservation tillage."[24] In the decade following the adoption of Roundup Ready crops , tillage intensity and soil erosion declined very little. In fact, use of soil-eroding tillage has increased in response to glyphosate-resistant weeds.[25] In any case, organic farming methods build and conserve soil better than conservation tillage.[26]

GM Corn and Monoculture

Even those GM crops that appear to provide some benefits have contributed in broader terms to the unsustainability of American agriculture. Corn rootworm is a root-feeding insect that primarily afflicts corn that is grown year after year in the same fields ("corn-on-corn"). Farmers have traditionally "rotated" corn with soybeans (planted the crops in alternating years) to keep rootworm at bay, increase soil fertility, and reduce fertilizer use, among other benefits. GM corn engineered for resistance to rootworm, together with perverse subsidies promoting corn for ethanol, has facilitated much more "corn-on-corn" acreage over the past decade,[27] an unsustainable practice promoted by Monsanto and other seed companies.[28] One result has been a dramatic rise in rootworm that are resistant to the toxin in Monsanto's GM corn,[29] which in turn has led to a resurgence of chemical insecticide use.[30] Another consequence is increased use of nitrogen fertilizer on corn to all-time highs,[31] since fields of corn-on-corn do not benefit from soil-replenishing soybeans. Because corn grown every year also increases disease risks, fungicides are now sprayed on more corn than ever before.[32] Nitrogen fertilizer runoff is the major cause of "dead zones" in the Gulf of Mexico and other bays. Many fungicides are quite toxic to people and other life forms.[33]

Farming without Farmers

Herbicide-resistant GM crops are popular with larger growers because they simplify and reduce labor needs for weed control. They have thus facilitated the worldwide trend to concentration of farmland in fewer, ever bigger, farms. According to the Argentine Sub-Secretary of Agriculture, this labor-saving effect means that only one new job is created for every 1,235 acres of land converted to GM soybeans. This same amount of land, devoted to conventional food crops on moderate-size family farms, supports four to five families and employs at least half-a-dozen workers.[34] Small wonder that family farmers are disappearing and food security is declining. In Paraguay, the area planted to soybeans (95 percent now GM) has nearly tripled since 1997, while per capita food production has declined; 45 percent of the rural population is in poverty,[35] and 90,000 rural people move to urban slums each year.[36] The rapid expansion of "labor-saving" GM soybeans in South America has led to "*agricultura sin agricultores*" ("farming without farmers").[37]

"Glyphosate Babies"

GMOs do more than displace labor. They also erode farmers' skills and knowledge, making them ever more reliant on chemical companies. A striking illustration of this is provided by the Roundup Ready crop system, which has dumbed down the art of weed management to the single practice of spraying glyphosate. As University of Missouri weed scientist Jason Weirich put it, with reference to Roundup Ready crops, "My generation is known as the glyphosate babies. It's all we grew up with—glyphosate, glyphosate, glyphosate."[38] One result is the glyphosate-resistant weed epidemic discussed above. More insidiously, these "glyphosate babies" have lost the knowledge and skills to manage weeds sustainably, with practices such as cover cropping and judicious use of tillage. These farmers are easy prey for purveyors of toxic biotech "solutions"—such as crops resistant to 2,4-D and other herbicides—that will provide at best short-term relief at the cost of still more intractable weeds a few years down the line. In a similar manner, insect-resistant crops have fostered a decline in knowledge-intensive integrated pest management, whereby farmers use a variety of non-chemical methods to reduce use of and reliance on insecticides.

Conclusion

Under the cover of a massive and deceptive public relations campaign, GM crops have intensified some of the most pernicious aspects of industrial agriculture. By increasing pesticide use, they reduce biodiversity. By facilitating monocultures, they impoverish the soil and foster crop pests and disease. By reducing labor needs in farming, they enable wealthier farmers to expand at the cost of small farmers. Most insidiously, GM crops erode farmer self-reliance, and thus make them still more dependent on purveyors of expensive agrichemicals and patented seeds.

Agriculture must evolve beyond pesticide-intensive, monoculture-promoting GMOs, if there is to be any hope of eliminating hunger and preventing further environmental degradation. The path forward lies not in biotechnology, but rather in innovative agrocecology, defined as the application of ecological concepts and principles to design and manage sustainable food systems. Agroecological farming builds healthy soils, manages pests, and sustains small farmers through skillful farming practices such as complex rotations and cover crops, while minimizing hazardous and expensive inputs.[39] The increasingly evident failures of biotechnology make a transition to agroecology inevitable. The only question is whether the world will wake up from the biotech dreamland in time to avert the worst.

Science Interrupted: Understanding Transgenesis in Its Ecological Context

By Ignacio Chapela

Ignacio Chapela, PhD, *is an associate professor of microbial ecology at the University of California, Berkeley. In addition to his work on microbial ecology, he has engaged in research on the access, ownership, and stewardship of genetic resources. He advises national governments and multilateral institutions on policy-making on genetic engineering and sovereignty over genetic resources and assists indigenous organizations and NGOs in Latin America and elsewhere to meet challenges related to genetic engineering. He was a member of the National Academy of Sciences' special committee for the evaluation of the environmental impacts of the commercial release of transgenic crops, 2000–2001. He has also been an outspoken critic of the University of California's ties to the biotechnology industry.*

The year 2012 marked the fortieth anniversary of the first transgenic manipulation in history—human and evolutionary history. The geneticists Cohen & Boyer, Berg and their associates ushered this epoch with the publication of their key papers demonstrating the feasibility of DNA splicing across phylogenetic kingdoms, including bacteria, viruses, the amphibian *Xenopus*, and the insect *Drosophila*. This development was as momentous as it was irretrievable, given the fact that living things released into the world will tend to survive and reproduce. Quite suddenly, the gates were opened then for specific genetic materials (DNA and RNA) to move, at first aided by humans, into environmental contexts where they could never have been found before. How significant was this release? How far could a newly-released piece of DNA (a "DNA species") spread in space, in time, and over the phylogenetic landscape of the diverse species on the planet? What has happened with this process in the almost half-century since its first introduction into local, regional, and global ecologies?

Although the early years of transgenesis involved organisms under laboratory containment, it was clear from the beginning that the question of environmental release would become central as an increasing number of biological species became the target for manipulation. As time went by, it stood to reason and experience that we should expect to find DNA species in unexpected ecological contexts—out of place, at the unexpected historical moment, and being carried by unexpected organismic life-forms. This is the definition of what would be conceived and named as "genetic contamination."

Experiments have been carried out to determine how far transgenic DNA could move over a landscape via pollen, but in the highly charged political and economic atmosphere such experiments remain unreplicated and become often challenged in a manner that creates confusion around them. The plant geneticist Norman Ellstrand produced in 2003 what continues to be, in 2014, the most authoritative and comprehensive account of the movement of transgenic DNA through what is suspected as the most common route: cross pollination. Ellstrand's scholarly contribution, however, continued to be moored on Central Dogma (explanation of the flow of genetic information within a biological system) and traditional breeding, and had to be based on extrapolation from non-transgenic knowledge about pollination and Mendelian genetics, with only a few examples available with data specific to transgenic movement.

In tacit recognition of the dearth of data, or even the possibility of generating them, Ellstrand and a list of the most prominent commenters on the subject joined in 2008 to author a call for action, dressed as an academic paper on the pages of *Science* magazine, explaining that since we cannot derive the appropriate information from real field data, at least we could use the logs of seed sales in each US county to try to build a map of their distribution and abundance, the fundamental measurements in ecology (Marvier *et al.*, 2008). This is currently as close as anyone could hope to build a real map of the distribution of transgenic organisms, what Ellstrand et al. called the bare foundation of any possible understanding of the ecology and evolution of transgenesis—not very close at all to any reasonable standard of scientific norm, let alone social acceptability.

The Dogma Breaks Down

A more realistic way of approaching the question of contamination emerges when we acknowledge that the movement of transgenic DNA—and its consequences—is not really ever as simple as the Central Dogma ideology would require. DNA is a chemically homogenous molecule from the simple biochemical viewpoint, but not from its informatic configuration or its function in its reproductive environment. There are species of DNA (i.e. sequences of nucleotides in specific configurations) with characteristic and specific behaviors, reproductive dynamics, affinities and proclivities which make each quite unique. For example, not just any sequence of DNA can operate the functions required from a vector to transfer successfully from one genome to another, just as not just any sequence can act as a promoter. Many functions of individual DNA sequences, often surprising to the discoverers, are continuously being described, and the specific sequences chosen to act as promoters and vectors, for example, are always found by serendipitous discovery. Because specific fragments of DNA can be realistically envisaged as DNA species, the environment for such species begins with the immediate molecular milieu in the adjacent genome and the diversity of molecules with which it interacts (such as proteins, methyl-, acetyl- groups and innumerable others).

Furthermore, the function of a DNA species in its environment is now known to be nothing like the simple expectation of the Central Dogma. Far from the original model in which one "gene" mapped univocally to one "trait" of form or function, we now know that a multitude of possible outcomes can be expected from the behavior of individual DNA species in various environmental contexts. Barry Commoner's unravelling of the Dogma's myths continues to be as valid as it is overlooked (Commoner, 2002).

This means that an evaluation of the effects of movement of a DNA species in the environment must be re-cast with a re-definition of "environment"—that environment beginning at the ecological locus of the DNA species in the genome immediate to itself and beyond, out through the cellular, tissue, organ, organism, population, community, ecosystem, biome, and biospheric levels. At each one of these levels the introduced DNA species will have dynamic effects that, in most cases, will have little or nothing to do with the "trait" initially envisaged by the technician introducing the transgenic construct in the first place.

Finally, transposon-like DNA [a transposon is a segment of DNA that is capable of inserting copies of itself into other DNA sites within the same cell] used in vectors, promoters, and other DNA species of transgenic application shows a proclivity towards promiscuous re-association with genomes of various kinds. Such promiscuity operates at the small scale of molecular dynamics by producing genomes with multiple insertions of fragments and re-assorted transformants in addition to the expected "legitimate" insert desired by the technician.

Such promiscuous behavior in species of transgenic DNA accounts for the proclivity of transgenic constructs to reassert also outside of the immediate genomic context in which they are introduced by the first transgenic manipulation. Such movement, known as Horizontal Gene Transfer, makes transgenic DNA species quite unlike other fragments of DNA. Transgenic DNA fragments are endowed, by necessary design requirements, with viral and bacterial properties, which should be expected to increase their capacity to move through horizontal gene transfer.

We know from a few experiments in the laboratory and in controlled microcosm conditions, as well as from even fewer field-based observations, that horizontal gene transfer of transgenic DNA species occurs. This means that a DNA species introduced into the environment—as part of an organism carrying it—will move not only within that species over time and through space (such as through cross-pollination), but also across the breadth of the phylogenetic landscape formed by the variety of life forms with which the original organism comes into contact. Gene transfer occurs during sexual reproduction within a species, and is therefore considered "natural." As the transfer involves organisms which are less likely to exchange DNA without the increased promiscuity functions of transgenic DNA species, the term "horizontal" is used. Thus pollination between commercial crops and their wild relatives, the most widely considered form of "gene pollution" is a form of horizontal gene transfer. Important as it is, such transfer is only one example of the wider—and much less recognized or studied—transfer across broader distances in the phylogenetic landscape, such as for example transfers across genera of plants, from plants to insects or bacteria, and criss-cross in the wide domain of the Archaea, Eu-bacteria and eukaryotic microbes.

If the release of commercial crops carrying transgenic DNA species was conceptually difficult to grasp, the current drive to release other transgenic organisms such as fish and insects could not be more fraught with lack of understanding. The ecological and evolutionary dynamics of such organisms are much less well-understood than the equivalent for crop plants, let alone mammals like ourselves. Ignorance is even deeper around the dynamics of microbial life-forms, and yet there are innumerable proposals to release transgenic microbes for the most bewildering array of applications, in every possible ecological and evolutionary context: for bioremediation, for decomposition and biofuel generation, for carbon capture and many more purposes. All these applications derive their political and economic acceptability not from any scientific understanding of what they may bring to the open environment, but from the "exceptional era for science and for the public discussion of science policy" (to quote Berg once again) politically engineered in the 1980s.

How widespread horizontal gene transfers may be in time, in space, and over the phylogenetic landscape, continues to be an open question due to the lack of support for research on this subject. However, recently we have learned that such movements are already much more widespread, and of much more immediate significance than we imagined (Chen, 2012). Published in late 2012, a groundbreaking study found ampicillin-resistant bacteria already established in all of the six key Chinese rivers studied; the resistance to this antibiotic by such wild bacterial populations demonstrably originated in horizontal gene transfer from transgenically manipulated organisms. At a time when antibiotic resistance in bacteria is perhaps the most important question for public health practitioners, the finding that transgenic manipulation has likely boosted the generation of such threatening bacteria could not be more relevant. Yet, there was little or no media coverage of this finding, and perhaps not surprisingly, this key piece of ecological research has received, at the time of this writing in 2014 only one citation in the scientific literature record. Such is the level of inattention by the public as well as the scientific community to the releases into the environment of transgenic DNA species, their ecological spread and their consequences.

Disappearing Evidence: Muddying the Waters

If the dynamics of the spread of transgenic DNA species has received little attention, and the study of its consequences is but an orphan field of science, even the possibility of detecting the presence of transgenic DNA has been fraught with complication and confusion. Such was the experience derived from the first publication of the long-distance, inadvertent spread of transgenic DNA through the populations of industrial and non-industrial corn varieties by Quist & Chapela in 2001. This publication incited widespread interest and concern, since such populations were, until then, considered to be too remote in space from the nearest legal open-air release of commercial varieties of transgenic corn, and too near in time to the first commercial release of such varieties in 1996. What was more surprising was the degree of opposition to its message that the same publication unleashed, with both direct and elaborately veiled attacks on the results, the methods and the scientific messengers producing the evidence (Delborne, 2008).

Even more astounding was the appearance in 2005 of another scientific publication in the prestigious *Proceedings of the National Academy of Sciences,* where a team of scientists declared the *absence* of transgenic DNA sequences in the same geographical regions, and in the same corn populations as those sampled by Quist & Chapela Ortiz-García *et al.* in 2008. Far from enlightening, these contradictory studies should induce confusion in any reasonable observer. As it happens, the difference in statements between these two studies can be traced to a divergence in the *interpretation* of simple visual evidence. The methods used by both studies were very similar, allowing for their direct comparison yet their interpretation of the graphic display of their results could not be more opposed. Where Quist & Chapela saw a tell-tale band in a gel, indicating the presence of transgenic DNA in their samples, Ortiz-García *et al.* did not (see Figure 1 for direct evaluation of the raw visual data and the interpretation of such data through automated means). The question here is not whether the evidence is there to say one way or the other, but, to paraphrase Christophe Bonneuil and Jean Foyer from the Centre Koyré d'Histoire des Sciences et des Techniques at the Centre National de la Recherche Scientifique, referring to this confusing situation in Oaxaca: "To see or not to see: that is the question."

By 2008 a new study had emerged, published with much less media attention, helping to clarify the situation (Piñeyro *et al.*, 2008): not only did this paper declare again the presence of transgenic DNA in similar corn samples as the other two, in the same geographical location and at the same time, it made it clear that the willingness of the team led by Ortiz-García "not to see" could have also come from specific conflict of interest and conflict of commitment of some of its senior authors. More interestingly, however, Pinyero *et al.* (2007) pointed to the fact that the "reading" of Ortiz-García *et al*'s evidence was performed by a commercial company dedicated to the detection of transgenic DNA in cases of high commercial and legal importance.

In 2005 the company involved, Genetic ID, Inc., was already positioned in what has become a dominant and quasi-monopolistic role to provide technical evidence of the presence/absence of transgenic DNA in cases of commercial or legal conflict, particularly for international commerce across borders with different acceptance levels for its presence in their food and other products.

Because of such a role, it is understandable that a company exposed to steep liability claims would choose to set thresholds of detection relatively high, i.e. to choose to err on the side of calling a sample "negative" even with reasonable evidence of the opposite.

By using such a *commercial* criterion for a *scientific* publication, Ortiz-García *et al.* discounted the existence of transgenic DNA as "non-significant." A further layer of confusion is thus cast over the possibility of understanding, or even detecting, monitoring, or mapping the very existence of transgenic DNA species at relevant geographical scales. By maintaining a firm monopolistic grip on the acceptability of standard commercial methods which therefore cannot easily evolve, the industry of detection has grown to become yet another major challenge to the public understanding of transgenic contamination. Without access to inexpensive and easily accessible means of seeing the presence of transgenic DNA species in their environments, common people (including scientists unwilling to cooperate with commercial interests) are left in the dark unless they can afford hundreds of dollars charged by the twenty-first century detection industry for each determination, hardly the way to gain useful ecological information.

Conclusion

The release into the environment of transgenic organisms, carriers of complex re-arrangements of DNA species, raises many important questions of fundamental and practical interest. These questions, however, have never been addressed due to an alignment of coincidental and self-reinforcing actors, circumstances, and interests. Scientists, technicians, educators, media professionals, financiers, politicians, entrepreneurs, and well-established industries all saw their various interests aligned in the early 1980s and through the last two decades of the twentieth century in a spiral of self-reinforcing commitments to the growth and development of what was to become the biotechnology industry. The resonance of interests among these actors precluded the parallel growth of a science of precaution or even any serious critical engagement with possible drawbacks associated with the intentional release and the unintentional autonomous reproduction and recombination of transgenic DNA species in the environment. Most crucially, in the absence of critical scientific engagement, the basis of the biotechnology industry proposed in the 1980s remains to this day, for lack of critical checks on what was then understandable enthusiasm, tied at its very foundation to an outdated and remnant dogma, the Central Dogma of biology. On loud parade and dressed in such doctrinal finery, this forty-year-old complex can hear precious few voices calling out the scientific bankruptcy of its foundations.

As the age of Enlightenment was taking flight in the eighteenth century, Francisco de Goya was etching his deeply critical *Serie de los Caprichos*. Among the etchings in this series, one of the most famous is unusual in being inscribed: "*El sueño de la razón produce monstruos*." This inscription can be variously treated as meaning that "Reason, if asleep, allows monsters to come forth," but also that the dream of unchecked Reason is, ultimately, itself a source of monstrosity unleashed. The twenty-first century has indeed become a playground—not, however, of Enlightenment, but of its consequences. Nowhere is this reality more evident than in the environmental expression of the transgenic intervention begun almost half a century ago.

Both studies were performed using very similar methods, viz. an end-point PCR reaction designed to produce a band on a gel if a transgenic DNA sequence was found (band present) in a sample. For each panel, arrows on

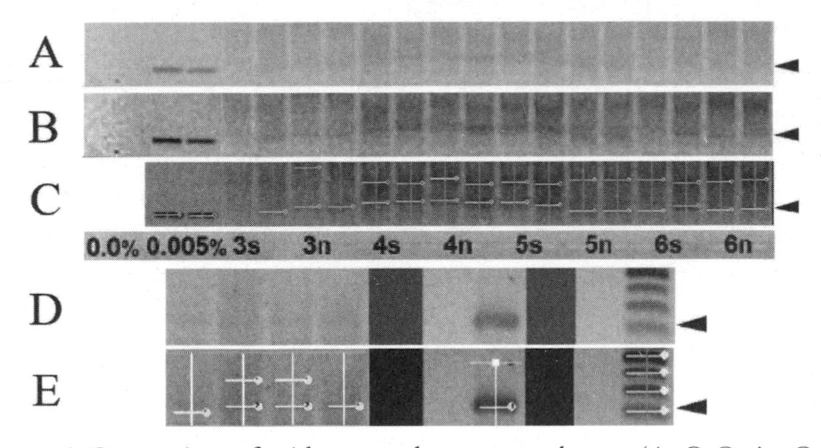

Figure 1. Comparison of evidence used to suggest absence (A–C; Ortiz–García *et al.*, 2005) or presence (D, E; Quist & Chapela, 2001) of transgenic DNA during the years 1999–2003 in local landraces of corn obtained from nearby locations in Oaxaca, Mexico. This comparison shows how very similar results have been interpreted to represent diametrally opposed conclusions concerning the fundamental question of presence/absence of transgenic contamination in a given place of interest. Needless to say, such contradictory interpretations lead to confusion by reasonable, intelligent observers.

Both studies were performed using very similar methods, viz. an end-point PCR reaction designed to produce a band on a gel if a transgenic DNA sequence was found (band present) in a sample. For each panel, arrows on the right indicate the location along the vertical axis where bands would be expected if transgenic DNA was present in the sample in question. Ortiz-García *et al.* (2005) published the image presented (in cropped form) in panel A, declaring that this photograph did not have any bands, and thus deducing that, to the best of their efforts, they were unable to detect transgenic DNA in their corn samples. Their study followed Quist & Chapela's (2001), who looked for the same transgenic DNA sequence, in the same geographical region, only a couple of years prior. Ortiz-García *et al.*'s image reproduced here (panel A), in cropped form, directly from its published version in *Proceedings of the National Academy of Sciences*; while Quist & Chapela's image is reversed (black-to-white; Panel D), for comparison, from the original published in *Nature*. In addition, images were sent for blind analysis at Kodak Molecular Imaging Systems, New Haven, Connecticut, where they were both independently and blind-analyzed in 2006 using Kodak's Molecular Image Software v.4.0. Panel A: cropped assembly from original in Ortiz-García *et al.* (2005), with two positive controls on the left end of the panel; Panel B: same as A, with contrast digitally increased to make bands more visually evident; Panel C: Kodak blind analysis of image on panel B, showing lines where the software automatically detected bands; Panel D: cropped from original in Quist & Chapela, 2001; Panel E: Kodak blind analysis of image E, showing lines where the software automatically detected bands.

Agricultural Technologies for a Warming World

By Lim Li Ching

Lim Li Ching, MPhil, *works in the biosafety and sustainable agriculture programs at Third World Network. This article originally appeared in* GeneWatch, *volume 26, number 1, January–March 2013.*

Climate change endangers the livelihoods and food security of the planet's poor and vulnerable, largely because it threatens to disturb agricultural production in many parts of the world. The Intergovernmental Panel on Climate Change (IPCC) projects that crop productivity would actually increase slightly at mid- to high latitudes for local mean temperature increases of up to 1 to 3° Celsius, depending on the crop. However, at lower latitudes, especially in seasonally dry and tropical regions, crop productivity is projected to decrease for even small local temperature increases (1 to 2°C). In some African countries, yields from rain-fed agriculture, which is important for the poorest farmers, could be reduced by up to 50 percent by 2020. Further warming above 3°C would have increasingly negative impacts in all regions.

Recent studies suggest the IPCC may have significantly understated the potential impacts of climate change on agriculture. New research suggests that production losses across Africa in 2050 (consistent with global warming of around 1.5°C) are likely to be in the range of 18 to 22 percent for maize, sorghum, millet and groundnut, with worst-case losses of 27 to 32 percent.[1] Other research suggests that rice production in South Asia, one of the most affected regions in terms of crop production, could decline by 14.3 to 14.5 percent by 2050, maize production by 8.8 to 18.5 percent and wheat production by 43.7 to 48.8 percent, relative to 2000 levels.[2] As such, unchecked climate change will have major negative effects on agricultural productivity, with yield declines and price increases for the world's staples.

The number of people at risk of hunger will therefore increase. Moreover, the impacts of climate change will fall disproportionately on developing countries, although they contributed least to the causes. The majority of the world's rural poor who live in areas that are resource-poor, highly heterogeneous and risk-prone will be hardest hit. Smallholder and subsistence farmers, pastoralists and artisanal fishermen will suffer complex, localized impacts of climate change. For these vulnerable groups, even minor changes in climate can have disastrous impacts on their livelihoods.

No wonder then that the world is desperately seeking solutions. Genetically modified organisms (GMOs) are one of the proposed options, for example through the development of drought-tolerant GM crops. There has been rather a lot of hype about these new GM crops, but closer examination reveals constraints. From the limited data supplied by Monsanto to the US Department of Agriculture, its drought-tolerant corn (recently deregulated in the US) only provides approximately 6 percent reduction in yield loss in times of moderate drought.[3]

Drought is a complex challenge, varying in severity and timing, and other factors such as soil quality affect the ability of crops to withstand drought. These complications make it unlikely that any single approach or gene used to make a GM crop will be useful in all or even most types of drought. Furthermore, genetic engineering's applicability for drought tolerance is limited insofar as it can only manipulate a few genes at a time, while many genes control drought tolerance in plants, raising questions as to whether the technology is fit for this purpose.

In contrast, conventional breeding has increased drought tolerance in US corn by an estimated 1 percent per year over the past several decades. According to the Union of Concerned Scientists,

> . . . that means traditional methods of improving drought tolerance may have been two to three times as effective as genetic engineering, considering the ten to fifteen years typically required to produce a genetically engineered crop. If traditional approaches have improved corn's drought tolerance by just 0.3 percent to 0.4 percent per year, they have provided as much extra drought protection as Monsanto's GE corn over the period required to develop it.[4]

While water availability during times of drought is also an important issue, there is little evidence that genetic engineering can help crops use water more efficiently, i.e. to use less water to achieve normal yields. Drought-tolerant crops typically do not require less water to produce a normal amount of food or fiber, and Monsanto has not supplied any data measuring water use by its drought-tolerant corn to suggest that it has also improved water use efficiency.

There are, moreover, biosafety issues with GMOs, and they have potential environmental, health and socio-economic risks. That is why there is an international law regulating GMOs: the Cartagena Protocol on Biosafety, ratified by 164 countries. Parties to the Cartagena Protocol have obligations to ensure that the risks are robustly assessed. There are also obligations in terms of risk management, monitoring, addressing illegal and unintentional transboundary movements, and public awareness and participation. Therefore, any decision to approve or release a GMO has to be weighed seriously and decision makers should consider the full range of options available.

With regard to climate change and its implications for poor farmers, a key question to ask then is whether the proposed option can meet the needs of small farmers with the least cost, most benefit, and lowest risks. What option can best contribute to resilience to deal with unpredictable climatic options? And given that the climate change challenge is so urgent, what can deliver results quickly?

The emerging consensus is that the world needs to move away from conventional, energy- and input-intensive agriculture, which has been the dominant model to date. Thus, the call has been for a serious transition towards sustainable and ecological agriculture. The International Assessment on Agricultural Knowledge, Science and Technology for Development (IAASTD),[5] stressed this in an extraordinarily comprehensive assessment of the global state of agriculture, involving more than 400 scientists.

The ecological model of agricultural production, which is based on principles that create healthy soils and cultivate biological diversity, and which prioritizes farmers and traditional knowledge, is climate-resilient as well as productive. Ecological agriculture practices and technologies are the bases for the adaptation efforts so urgently needed by developing-country farmers, who will suffer disproportionately from the effects of climate change. Many answers already exist in farmers' knowledge of their region and their own land—for example, how to create healthy soils that store more water under drought conditions or how to grow a diversity of crops to create the resilience needed to face increased unpredictability in weather patterns.

Ecological agriculture practices improve and sustain soil quality and fertility, enhance agricultural biodiversity and emphasize water management and harvesting techniques. Practices such as using compost, green manures, cover crops, mulching, and crop rotation increase soil fertility and organic matter, which reduce negative effects of drought, enhance soil water-holding capacity, and increase water infiltration capacity, providing resilience under unpredictable conditions. Moreover, cultivating a high degree of diversity allows farmers to respond better to climate change, pests, and diseases, and encourages the use of traditional and locally-adapted drought and heat-tolerant varieties and species.

There is also increasing evidence that ecological agriculture can increase yields where they matter most—in small farmers' fields—with low-cost, readily adoptable and accessible technologies that build on farmers' knowledge. A review of 286 ecological agriculture projects in 57 countries showed a 116 percent increase in yields for African projects and a 128 percent increase for East Africa.[6] During times of drought, scientific side-by-side comparisons at the Rodale Institute, USA, have demonstrated that organic yields are higher than both conventional and GM agriculture.[7]

While there is great potential in ecological agriculture, there has been little attention to it in terms of research, investment, training and policy focus. The challenge is to re-orient agriculture policies and significantly increase funding to support climate-resilient ecological agricultural technologies. Research and development efforts should be refocused towards ecological agriculture in the context of climate change, while at the same time strengthening existing farmer knowledge and innovation.

In conclusion, a comparison of genetic engineering with other technologies, such as conventional breeding and ecological agriculture, shows that the latter are more effective than the former at meeting the climate challenge, and at lower cost. An excessive focus on genetic engineering at the expense of other approaches is a risky strategy.

Down on the Farm: Genetic Engineering Meets an Ecologist

By David Pimentel

David Pimentel, PhD, *is a Professor Emeritus in the Department of Entomology, College of Agriculture and Life Sciences at Cornell University. He is an ecologist interested in resource management and environmental quality. This article was excerpted from Dr. Pimentel's talk on "Agricultural Biocontrol Methods" at the CRG National Conference "Creating a Public Agenda for Biotechnology, Health, Food and the Environment." This article originally appeared in* GeneWatch, *volume 4, number 3, May–June 1987.*

Ever since Gregor Johann Mendel crossed varieties of peas and discovered the basis of genetics in the 1800s, farmers have benefited from what might loosely be called genetic engineering. By patiently selecting and crossing plants, farmers and breeders have developed crop lines that do everything from produce larger vegetables, grains, and fruits; to survive harsh environments; to resist pest attack.

From this perspective, modern genetic engineering might seem to offer farmers little which is substantively different. The modern techniques, however, offer the opportunity to transfer genes from one kind of organism to another kind, such as bacterium to tobacco, and new strains can be created far more rapidly. Whereas it used to take up to twelve generations to produce an insect-resistant strain of tomato, gene splicing can cut the time to four generations, reducing the time in this case from a minimum of six years to two years.

The opportunities for gene transfer and other genetic engineering techniques to benefit agriculture are enormous. Scientists can develop products that will increase food production, reduce fertilizer use, and decrease the need for costly and environmentally dangerous pesticides. But genetically engineered agricultural products could also lead to certain sobering social, economic, and ecological problems. By itself, genetic engineering cannot solve all of the world's food production problems. Efficient irrigation schemes and proper soil care are also critical. As we enter an era when the products of genetic engineering are leaving the lab for the fields, we must head off their potential liabilities by thorough testing and careful decision-making.

Within a year, and continuing for decades into the future, genetically engineered farming products may be used for a broad range of purposes. Potential benefits include:

- pest control with engineered microorganisms;
- increase in crop yield with corresponding reduction in need for agricultural land;
- reduced susceptibility to frost ("ice-minus");
- ability of basic food grains to "fix" their nitrogen requirements;
- improve livestock production with vaccines and drugs;
- increase yield of plant byproducts such as ethanol;
- use of microbes to produce valuable new products.

Just as genetic engineering brings great advances to agriculture, it also includes economic and social liabilities. For example, engineered biocides for pest control, vaccines, and other technologies will increase food supplies and therefore reduce prices in the marketplace. It will also speed the demise of small farms.

Consider the mixed blessings associated with bovine growth hormone (BGH), which can now be produced using genetically engineered organisms. This research on BGH, started in the late 1970s by Prof. D. E. Bauman of Cornell University, is now being used experimentally and might be approved by the US Food and Drug Administration within the year. When injected into cattle, this hormone will increase milk production as much as 30 percent, allowing farmers to reduce production costs and the land needed for forage crops by approximately 10 percent.

But the advance comes at a time when milk production is at an all-time high. The milk surplus has prompted the US Department of Agriculture to pay some farmers to eliminate their herds of dairy cattle. The market has little or no capacity to increase its consumption of dairy goods, especially whole milk products. And because of the surplus of dairy products in the world, especially Europe, there is little opportunity for exporting milk products. The need for fewer cows translates into a need for fewer farms. This reduction could add to the number of farmers who are finding it difficult or impossible to stay in business. Bankruptcy represents both financial disaster and great social and emotional losses to those whose farms fail. Already farming communities, banks, and even some states such as those in the corn belt, are being severely hurt by the financial disaster in farming.

The full effects of biotechnology on society and the environment are difficult to predict. If some farmers and farm laborers who are forced out of agriculture because of growth hormones do not find work but turn to welfare, any benefits associated with lower food costs could be eliminated by increased taxes to support welfare. On the other hand, economic problems could be minimized if workable plans could be developed to assist in relocating displaced farmers to other gainful employment. Judging by the current lack of social policies to aid steel and textile workers displaced by automated equipment, the odds favor the development of another poor and disillusioned sector of American society.

Another set of economic and social problems concern the changes in communication that have taken place between public agricultural scientists and private agricultural scientists since the onset of genetic engineering. Scientific sharing between the public and private sector has become severely restricted. Some public geneticists now are keeping their genetic discoveries to themselves, to avoid losing potential financial rewards. In some cases these public scientists depend on research funds from private industry or consult for private firms. Today there are numerous lawsuits concerning the ownership of genes and genetic engineering techniques. There is growing distrust between genetic engineers employed by both the private and public sectors. If this continues, the anticipated benefits of genetic engineering will be diminished.

Even more disturbing is that some large companies are appropriating from poor farmers. In developing nations, seeds of valuable crop lines that have certain desirable traits that could be used in crop breeding. The collected plant types with the desirable traits from these seeds will be used to develop new commercial varieties, which will be sold back to the farmers and others, with no profit returning to the original seed owners. Ironically, it was those farmers and their ancestors who sensed the benefits of preserving and maintaining the special genetic traits in these crop varieties.

Environmental Risks

Many environmental questions should also be answered before new genetically engineered organisms are released for general use. These organisms may directly affect others in the environment. There is a chance that the engineered genes could be transferred to other organisms, transforming them into pernicious pests. In addition, genetically engineered agricultural products could lead to a decrease in genetic diversity, which may make the crops more susceptible to new pest types or to temperature and moisture extremes.

One of the first genetically engineered organisms, the ice-minus bacterium, illustrates some of the environmental factors that must be studied prior to release of an engineered organism. Last spring, a sharp controversy arose after Advanced Genetic Sciences, the developers of the frost-resistant *Pseudomonas syringae* strain, tested its effects outdoors without the approval of the US Environmental Protection Agency. A similar furor arose last year when the US Department of Agriculture approved the release of a genetically engineered live-virus swine vaccine without going through the established procedure of consulting its Recombinant DNA Committee. Fortunately, all present tests indicate that the vaccine, which controls pseudorabies, a serious disease of swine, is safe and appears not to be a threat to other organisms.

Let's take a look at an example from last year. Monsanto submitted a genetically engineered organism, which its scientists developed over a period of nearly five years, through extensive laboratory and greenhouse tests. Scientists at the company removed the toxic element In *Bacillus thuringiensis*, commonly called Bt, a bacterium used as a biocide or biocontrol agent for many crops. They inserted the toxic element of Bt into a modified Pseudomonas bacterium that has the ability to survive in the soil environment. There are potentially great benefits from using the engineered bacterium to control such major soil insects as the black cutworm, a serious corn pest that causes $50 to $100 million worth of damage each year. But we also need to be certain that Bt's toxic element is not a type that will kill beneficial insects and earthworms. It is possible that some beneficial soil insects may be damaged when the engineered Pseudomonas is released into the environment, since 99 percent of all insects are beneficial and a few are always more sensitive than others. Some soil insects, for example, carry out such important ecological activities as preying on pests, and burying livestock manures and other wastes. The soil organisms act in a similar manner to the organisms in a sewage plant by degrading wastes and recycling the nutrients for use by crops. The beneficial organisms also make holes that let water filter through. Bt sprayed on plant and soil surfaces seldom survives in the environment longer than a week or two, but the genetically engineered

Pseudomonas bacterium may persist and multiply in the favorable soil environment. This experiment has been delayed as a result of the EPA's request for further testing.

Another concern about how genetically engineered agricultural organisms might affect other species in the environment relates to the theoretical transfer of engineered genes. Scientists know, largely from laboratory studies, that engineered microorganisms can transfer plasmids—cellular replicating elements that can be transferred from parent to progeny—containing novel genes to other microbes. Instances are few, but occur often enough that there should be concern about what might happen in nature.

Since the potential exists for engineered genes to spread to indigenous organisms, ecologists and geneticists are concerned that some of the genetic characters added to crop plants could be transferred to weeds. If a gene that had been added to a cereal grain so that it could resist a plant pathogen, transferred by natural processes to a weed species of the same plant family, the weed might resist the pathogen and be able to spread more vigorously. The odds of this happening are extremely small, but such an occurrence could alter the ecosystems of either natural lands or farms.

Clearly, natural community niches are never full. Communities of plants and animals have tremendous flexibility to accommodate new genetic variations and species. After all, about 1,500 insect species have been introduced and established in the United States since 1640. Several of these, like the gypsy moth and Dutch elm disease, a fungal pathogen transmitted by a beetle, have become serious pests. These and other forest pests are reported to destroy about 25 percent of the annual production of US forests.

Some genetic engineers claim that their engineered organisms will be weaker than natural organisms. For this reason the engineered species are not supposed to be able to survive long in nature, and cannot cause ecological problems. But other genetic engineers claim that because their organisms were obtained from nature and only slight genetic modifications were made, the species will survive well in nature. The conflicting assessment highlights ecologists' concerns. Some truth exists in each viewpoint, depending on the particular organism and the specifics of the genetic engineering. The soundest policy is to thoroughly study the ecology of each engineered organism in the laboratory and greenhouse before releasing it into the complex, natural ecosystem.

"Pesticide Treadmill"

Genetically engineered agricultural products could have complicated effects on both the environment and economic structure, as demonstrated by the current development of herbicide-resistant crops. Certain crops are not affected by a particular broad-spectrum herbicide that can normally kill many types of green plants. Therefore, this chemical can be used to eliminate most weeds without damaging the crop itself. However, by using these herbicide-resistant crops, herbicides may become essential to the production of specific crops. The environmental implications are alarming.

Increased herbicide use would mean more chemical pollution of fields and water systems. Based on past experience, scientists can predict that the extensive use of herbicides would intensify the problem of weeds that can resist the compounds. Pesticide resistance contributes to the "pesticide treadmill" where pesticide use leads to resistance and destruction of natural enemies that, in turn, require more pesticides to deal with the newly created problems. In addition, the problem of herbicides and other pesticides drifting onto adjacent crop fields—which already destroys more than $70 million worth of crops annually—would intensify as herbicide use intensifies.

Increased chemical use would also prevent some crop rotations or the interplanting of herbicide-sensitive crops in soil contaminated by the compounds. This could intensify soil erosion and pest problems. Even now, certain crops cannot be planted in rotation or between crop rows if particular herbicides are used, despite many farmers' preferences to plant that way. For example, soybeans cannot be planted after a corn crop if there are residues of the herbicide atrazine. Clovers cannot be interseeded between corn, sorghum, or related crops when the triazine herbicides are used.

Furthermore, herbicide use has been demonstrated to make some crops more susceptible to certain insect pests and plant pathogens. For instance, when corn was treated with the recommended dosages of the popular herbicide 2 4-0, it was infested with three times as many corn-leaf aphids as normal and became significantly more susceptible to European corn borers, corn smut disease, and Southern corn-leaf blight. If ecological effects similar to this occurred after herbicide-resistant crop lines and more herbicides were used, insecticide and fungicide spraying would increase, thereby intensifying environmental problems.

An unanticipated, economic side effect of genetically engineered herbicide-resistant plants has been the move by some chemical herbicide producers such as Monsanto and Ciba-Geigy to buy seed companies. The chemical companies' goal is to sell the farmer a package of the genetically engineered seed and herbicide. The crops would be engineered to require the use of a specific herbicide produced only by that chemical company. This marketing strategy would benefit the companies, since the farmer using the seed has no option but to use their herbicide.

A final environmental problem, in which genetic engineering will be only one factor, is the inevitable increase in the size and industrialization of farms throughout the world. This is coming about in an attempt to produce the food and fiber for the growing population worldwide. Since 1954 the number of US farms declined from 4.8 to 2.3 million in 1985—less than half as many. At the same time the size of farms has about doubled from 242 acres to about 445 acres per farm. Without long-range planning that focuses on environmental protection, agricultural soils, water, and regional flora and fauna may be degraded by operations that put their efforts into short-term productivity.

Insuring Success: Recommendations

Ecologists can make some predictions that could be used to select appropriate organisms for genetic engineering. For instance, they can forecast that if an organism comes from the tropics, its chance of surviving the northern winters is zero. Therefore, I suggest that scientists should use tropical organisms for genetic engineering work, not species native to temperate zones. After a season, the engineered organisms would be eliminated by the winter cold. In addition, sound ecological protocols need to be followed before any genetically engineered organism is made and released. In the United States we already have several thousand pest species to contend with. We do not need genetic engineering to add any more to this list.

In June 1986, the federal government approved rules and guidelines that set the policy for regulating the biotechnology industry. Responsibility for assessing risks and benefits was divided among five federal agencies. These include the Food and Drug Administration, which is generally responsible for genetically engineered organisms that may be present in foods and drugs; and the Department of Agriculture, which is responsible for engineered organisms to be used in crop plants and animals. Also included are the Environmental

Protection Agency (EPA), which is responsible for engineered organisms released into the environment for pest control, pollution control, and related activities; and the Occupational Safety and Health Agency, which is responsible for engineered organisms that may affect workers. Finally, the National Institutes of Health are responsible for engineered organisms with the potential to affect public health.

Although setting up these regulations was a much-needed first step in controlling biotechnology, many scientists believe that the divisions of authority are cumbersome, even inadequate. Another concern is that agencies such as the Department of Agriculture are left both to promote and regulate the new technologies. This never worked when the Department of Agriculture handled both pesticide promotion and regulation. That is why pesticide regulation was transferred to EPA in 1970. In my opinion, it would be best for biotechnology to be regulated by only one agency, the EPA, with the other agencies having input into the regulatory process. For example, the other agencies could have a representative on the EPA review committee.

Although there is only a small chance that an environmental problem would be caused by a genetically engineered organism, a single mistake could be a major disaster. To reduce environmental risks, the government should require that before engineered plants are released into the wild, the companies and other institutions that develop the lines thoroughly test the organisms in both the laboratory and greenhouse to determine their potential for surviving and reproducing in nature.

As part of this process, companies and institutions should have to identify potential hosts of an engineered organism, and test its interactions with them. For example, the ice-minus bacterium should be tested on all major crops for which it is listed as an important pathogen. It should also be tested against several species of beneficial natural plants and insects. The beneficial organisms selected might be major species that are important in maintaining the quality of the environment.

Companies should also be required to investigate indoors any genetically engineered organisms' possible ecological effects on native plants and animals. And genetic engineers should document the potential that an engineered organism has to transfer genetic material to others.

Then companies should conduct field tests on islands and similarly isolated areas. This has already been done, for example, to test the effectiveness of the sterilized screwworm fly, which was not genetically engineered, in control of the screwworm pest.

For their tests, the companies and other institutions should have to use teams that include microbiologists, ecologists, plant breeders, agronomists, wildlife specialists, public health specialists, and botanists.

At the same time that careful environmental assessments are made of genetically engineered organisms, it is essential that economic and social impacts of the new biotechnology be assessed. Such an analysis might help federal and state governments determine the benefits and risks and what might be done to reduce potential economic and social costs. Granted, this kind of testing is time-consuming and costly. But if it insures the success of biotechnology, it is worth the extra caution.

Engineering Crops for Herbicide Resistance

By Sheldon Krimsky and Roger Wrubel

Sheldon Krimsky, PhD, *is a professor of urban and environmental policy at Tufts University and is board chair of the Council for Responsible Genetics.* **Roger Wrubel, PhD,** *is director of the Environmental Studies Program at the University of Massachusetts—Boston. From* Agricultural Biotechnology and the Environment: Science, Policy, and Social Issues *by Sheldon Krimsky and Roger P. Wrubel. Copyright 1996 by the Board of Trustees of the University of Illinois. Used with the permission of the University of Illinois Press. This article originally appeared in* GeneWatch, *volume 11, number 1–2, April 1998.*

The creation of crops that are resistant to herbicides is among the most controversial applications of biotechnology to agriculture. A prominent theme of the agricultural biotechnology industry is that genetically engineered crops will reduce the use of pesticides and are thus environmentally beneficial. Crops engineered with protein products to kill insects or that resist disease have been cited by industry as examples.

But how can it be argued that herbicide-resistant crops (HRCs) will reduce herbicide use and benefit the environment? Environmental and alternative agricultural groups that have been generally critical of industry and government efforts to develop biotechnology products for agriculture have seized on this apparent contradiction to publicize their case that the goal of companies developing biotech products is short-term profit and not the long-term health of agriculture and the environment. However, the companies developing these products have promoted HRCs as consistent with the responsible and wise use of biotechnology to solve pest problems in an environmentally compatible manner.

A company will gain substantially if it can increase the market share for one of its herbicides. By creating crops resistant to its herbicides, a company can expand markets for its patented chemicals. The US agricultural market for herbicides is now more than $3.9 billion annually and dwarfs sales of other pesticides, such as insecticides and fungicides.

Two main strategies have been used to create HRCs. The first strategy is to alter the active site of the herbicide in the plant, reducing the sensitivity of the plant to the herbicide while maintaining the plant's normal biochemical functioning. A second strategy involves introducing genes into crops for metabolic detoxification of the herbicide. Plants and bacteria that are naturally tolerant to herbicides often have enzymes that convert the herbicide to a nontoxic metabolite.

Opposing opinions regarding the merit of herbicide-resistant crops center on different perceptions regarding the effects of herbicides on the environment and on our health and the necessity of using herbicides in crop production. There is a clear dichotomy of opinion regarding the safety of herbicides. As a general matter, environmentalists and alternative agriculturalists believe that all pesticides carry undue risk. Some risks are known, but many others remain to be identified.

In contrast, most weed scientists distinguish herbicides from other pesticides and evaluate the safety of each herbicide individually. Weed scientists acknowledge that some herbicides have detrimental environmental or health effects and should be phased out. But most also believe that the majority of herbicides are safe, especially when used as directed.

Two examples in the ongoing research and development of HRCs are the herbicides glyphosate and bromoxynil. Glyphosate is the active component in Roundup, a widely used herbicide that can be used to control weeds while the crop is growing in the field. Monsanto, which holds the patent on and markets Roundup in the United States, is actively pursuing commercial development of genetically engineered resistance to glyphosate in a number of crops, including cotton, corn, soybeans and potatoes. Field tests of glyphosate-resistant tomatoes, sugar beets and canola have also been conducted by Monsanto and other companies. Glyphosate-resistant soybeans were the first HRC to be commercialized fully.

Bromoxynil is the active ingredient in Buctril, an herbicide manufactured by Rhone-Poulenc. Calgene, in association with Rhone-Poulenc, has been developing cotton resistant to bromoxynil. Meanwhile, the US Environmental Protection Agency (EPA) continues to evaluate the potential human health hazards of bromoxynil. Applied at low rates, this herbicide degrades rapidly in the soil, does not accumulate in groundwater, and does not leave residues in food. However, environmental and alternative agriculture groups have staunchly opposed any expanded use of bromoxynil, citing evidence of birth defects in laboratory animals, possible human developmental effects, potential cancer-causing linkages, and toxicity to fish.

The real unknown about herbicides and for that matter almost all xenobiotics (foreign chemicals introduced into the environment) is the health effects of low-level chronic exposures. Most research on the long-term effects of pesticides has focused on cancer risk, and far less attention and low funding priority have been given to neurological, immunological, developmental, and reproductive effects. The bottom line is that we do not presently have the tools to evaluate accurately the risk of long-term exposure to many chemicals, including herbicides.

If resistance is conferred on a crop by making the active site insensitive to an herbicide, residues of the herbicide as well as degradation products may build up in the plant. If herbicide resistance is conferred by transferring a gene to a plant that detoxifies the herbicide, the degradation products will build up in the plant. Before an herbicide is approved for registration, residue levels of the herbicide in the new crop and any metabolic products of the herbicide in the plant must be determined and the data presented to the EPA, which is responsible for establishing acceptable levels of pesticide residues in plants and plant products.

Critics of efforts to develop herbicide-resistant plants have argued that the acceptable residue levels established by the EPA underestimate the true danger of herbicides to human health. Critics point out that the inert ingredients in herbicides, including surfactants and solvents added in formulations to improve application and spreading on plant surfaces, can be more acutely toxic than the active ingredients. Industry and weed scientists reason that any crop that is resistant to an herbicide, whether it be natural tolerance or engineered resistance, has to store, process, or excrete the herbicide or its metabolic products. This is not a new problem, and they argue that the regulations regarding registration of herbicides for the new HRCs are stringent and should not be any different than they are for non-genetically-engineered crops.

The factor that most threatens the success and agronomic usefulness of HRCs is the potential for weeds to develop resistance to the associated herbicides. The extensive and continuous use of herbicides since the 1950s has resulted in the evolution of more than a hundred weed species resistant to one or more herbicides. While the use of HRCs may not greatly alter the total number of acres treated with herbicide, it will shift the number of acres treated with particular herbicides. If HRCs are widely accepted, there will likely be increased reliance on a few of the newer herbicides, with the phasing out of older, less desirable ones. The widespread use of HRCs developed for resistance to single herbicides will accelerate the selection pressure on weeds to evolve resistant biotypes.

Intuitively, the development of HRCs would seem to make it easier to use herbicides on more crops and over longer periods of the growing season and thus increase the overall amount of herbicide applied. However, because the current use of herbicides is so prevalent—more than 90 percent of all corn, soybean and cotton acreage receives at least one herbicide treatment each year—it is unlikely HRCs will increase the overall use of herbicides. Rather, the likely impact of HRCs will be to shift the types of herbicides that are used, and farmers will increasingly rely on a few broad-spectrum herbicides.

For some crops, several different herbicides are now used through the growing season to control the range of infesting weeds. With HRCs, multiple herbicide treatments might be replaced by use of a single broad-spectrum herbicide, resulting in a net reduction of the number of applications and the quantity of herbicide applied. If HRCs are successful in reducing the use of herbicides, farmers should not expect to reap the full savings, because companies will likely charge a premium for the seed. However, farmers will accept HRCs only if they believe there is a significant economic benefit in their use compared to ordinary seed. Because there are many weed control options for all major crops, farmers need not switch to HRC varieties without convincing evidence of their effectiveness. Seed and chemical companies must prove that their products have a significant economic advantage derived from lower overall weed-control costs and/or more effective weed control leading to higher yields and improved crop quality.

Although HRCs may not necessarily increase herbicide-treated acreage in the United States, HRCs do encourage continued dependence on these chemicals, which critics view as harmful to the environment and human health. Alternative agriculturalists argue that nonchemical methods of weed-control can be implemented immediately to decrease dependence on herbicides and that research into better alternatives needs to be intensified. Among the alternative techniques are the reintroduction of crop rotations where they have been abandoned, cultivation designed to minimize erosion, timing of planting, high-density plantings, cover cropping, intercropping, and biological control.

In the development of herbicide-tolerant crops, new techniques in molecular genetics fostered breakthroughs in developing HRCs when classical selection methods had largely failed. Agrichemical companies viewed this as a way to increase the value of certain herbicides by expanding their range of uses. Few in industry expected the harsh criticism of HRCs from the public interest community and international nongovernmental organizations.

Clearly, HRCs are part of high-input, chemically intensive agriculture in the United States. As such, the technology is rejected outright by those demanding a shift to low-input agriculture. For the critics, herbicide-resistant crops increase the dependence of farmers on chemicals, delay the development of alternative weed-control methodologies, and are heavily promoted because of industry self-interest to the detriment of environmentally sound agriculture. Unfortunately, most farmers rely on their agrichemical dealers for pest control advice. Industry pressures to increase herbicide sales conflict with the wise use of herbicides and the long-term health of agriculture.

Both the agrichemical industry and its critics need to examine HRCs thoughtfully. Herbicide-resistant crops could help phase out environmentally damaging herbicides, reduce overall herbicide use, and make integrated weed management more attractive to farmers. But HRCs could just as easily result in reliance on a few chemicals, eventually leading to herbicide-resistant weeds.

Genetic Engineering for Biological Control: Environmental Risks

BY DAVID PIMENTEL

David Pimentel, PhD, *is a Professor Emeritus in the Department of Entomology at Cornell University. He is an ecologist who is interested in resource management and environmental quality. This article originally appeared in* GeneWatch, *volume 2, numbers 4 and 6, November–December 1985.*

For several decades naturally occurring organisms have been used effectively in biological control programs to limit major pests of agriculture and forestry. Now with the development of genetic engineering technology, scientists will be able to modify genetic systems of biocontrol agents to increase their effectiveness against pests. Although great potential awaits the adoption of this technology, it might fail because we neglect to follow careful test protocols and develop sound regulations.

Ecological Basis of Biological Control

Nearly 110 pest species have been effectively controlled by organisms brought from a distant region of the world and released into the environment where the pest is located. However, predicting whether a released biocontrol parasite will be pathogenic to a species that is not its usual target has always been extremely difficult. This is so even when the parasite's host and host distribution is well known. For instance, of the forty natural enemies of the gypsy moth introduced into the United States from Europe for biological control, only ten became established. No one knows why the others failed to do so. Also unknown is why none of the ten established parasites has provided effective control of the gypsy moth under what appear to be favorable environmental conditions for the parasite.

On the other hand, ecologists can make some predictions with certainty. For example, most tropical parasitic organisms will not become established if released into a northern temperate region. A wasp parasite of the coconut moth, for instance, would not survive in New York State. Also, parasites from a humid environment introduced into a highly arid habitat usually do not become established. These predictions are based on intensive knowledge of the ecology of an organism and the environment into which it is to be introduced.

In addition, when several parasitic species are released into one habitat, with a potential host and favorable climatic conditions, some of the parasites will become established in the environment. The most reliable data confirming that new parasite species can adapt to new hosts relatively easily are based on ecological information about pests already associated with world crops. For example, of the nearly 1,700 species of sugarcane pests in tropical regions of the earth, about 1,000 are native insects that are unique to a specific ecological region where sugarcane is grown. Each country and region has a unique set of pest species that has adapted to its own sugarcane crop. Here in the United States, about 60 percent of all insect pests associated with US crops are native insects that have moved from local plants onto the introduced crop plants. Thus, parasitic organisms can and often do adapt easily from one plant host type to a new host.

Clearly, insects and other organisms can move from one food host to another. A food host can usually tolerate another species feeding on it.

Niches are never full. Community systems, in nature, have tremendous flexibility to accommodate new genotypes and species. Consider the fact that about 1,500 new insect species have become established in the United States since about 1640, most since 1900. Although a few of these species were introduced intentionally, most gained entrance and spread along with humans as they expanded into the new territory.

Genetically Engineered Biocontrol Agents

Although past biological control programs have proven effective in limiting some major pests of agriculture, great potential exists for the application of genetic engineering technology to improve biocontrol agents. For example, the genetic makeup of parasites can be altered to make them virulent and pathogenic to pests, thus improving their effectiveness as pest control agents. Specifically, a parasite could be made more virulent by increasing its rate of reproduction, transmission, infective ability, seasonal survival (diapause characteristics), or its pathogenicity with toxins.

Still another way to increase parasite virulence would be to add or delete genes in the parasite, thus allowing it to either overcome or avoid resistant characters in the host. In a sense, this is the equivalent of producing a *new* parasite. The genetically engineered parasite could become highly virulent to its host and in this way it would be similar to any natural, newly associated parasite.

Using genetic engineering technology, it should be possible to make a naturally avirulent parasite, virulent. Also it should be possible to transfer a toxin-producing unit from one organism to a related, but nonpathogenic type organism. The toxin-producing unit would thus convert an organism into a potential biological control agent.

Assessing Ecological Risks

Just as with biocontrol programs using naturally occurring organisms, the use of genetically engineered organisms must be based on knowledge of the entire ecosystem and how the altered organism will interact in it. Benefits, as well as risks, must be weighed to ensure successful control programs.

Parasite-host systems in nature are well integrated genetically, that is, the parasite and host evolved a balanced "demand/supply economy." This enables the parasite and host to coexist in relative stability or "natural balance." Although altering this balance by genetic engineering could improve biological control, it could also alter the dynamics of the entire natural system. Serious new environmental problems and even an increase in pest problems could result.

For example, the "ice minus" bacterium, *Pseudomonas syringae*, which was developed to reduce the susceptibility of potatoes, tomatoes, strawberries and other crops to frost in California, is reported to displace the natural *P. syringae* type that serves as a nucleus for ice formation in plants. In nature, the unaltered organism is itself an important pathogen of several major crops, including beans, alfalfa, tomatoes, plums, pears, peaches, apples, almonds, cherries, oranges, and grapefruit. The unaltered form of *P. syringae* changes the natural resistance of some insects, making them more susceptible or resistant to freezing, depending on the genotype.

The genetically altered form of *P. syringae* may cause many unexpected problems after their release. At this point, there is not a priori way of knowing what value these changes will have to the environment. What if the new *P. syringae* adversely affects the honeybee, which is the major crop pollinator, responsible for $20 billion worth of crops, as well as diverse native plants? Although the deletion of certain genes in this organism reduces its ability to cause frost injury to certain crops, its activity in the natural ecosystem should be carefully assessed by sound ecological protocols prior to its release into the natural habitat. This procedure is essential before the release of genetically altered biocontrol organisms. Without proper tests we will not know whether the genetically altered *P. syringae* has retained or lost its pathogenic characteristics.

Another genetically engineered biocontrol agent has been developed by inserting the toxic element of *Bacillus thuringiensis (Bt)* into *Pseudomonas* sp. for insect control in soil. Using this engineered organism for insect control can be beneficial, but because it is similar to *Bt*, it can be expected to be pathogenic to various beneficial insects and earthworms. Since approximately 99 percent of all insects are beneficial, there is a good chance that many beneficial soil insects will be destroyed when the genetically engineered *Pseudomonas* sp. is used. Therefore, the risks of using this organism should be carefully studied in the laboratory before it is released into the natural ecosystem and allowed to spread.

Another potential risk of inserting a pathogenic unit into a nonpathogenic organism is the likelihood that pathogenic genes will be transferred to other microbes. Scientists know it is possible for engineered microorganisms to transfer plasmids containing novel genes to other microorganisms in the environment. While nonconjugative plasmids are generally considered "safe" for genetic engineering of organisms, it has been demonstrated that some transfer of plasmids to indigenous bacteria does occur. Data on gene transfer in microbes are based primarily on laboratory studies, and little is known about this response in nature and whether undesirable engineered genes could spread to indigenous microbiota.

With such great potential awaiting the use of genetically engineered biocontrol agents, it would be a disappointment if they failed because widespread release into the natural ecosystem. Although *no set of protocols will ever be 100 percent effective* in preventing ecological catastrophes, the risks to the environment as well as hazards to public health can be minimized if suitable ecological protocols and regulations are adopted.

The following ecological procedures would reduce environmental risks:

- Thoroughly test the organism in the laboratory and greenhouse to determine the potential the engineered organism has for surviving and reproducing in nature;

- Identify all potential hosts of the organism and test the hazards of the parasite to these hosts in the laboratory and greenhouse;
- Investigate possible direct and indirect ecological effects (e.g., parasitizing native plants and animals) of genetically engineered organisms in controlled environments before they are released into the environment;
- Document the potential of engineered organisms to transfer genetic material to other organisms;
- Conduct field tests on islands and similar "contained" areas before widespread release;
- Investigate possible environmental risks of using multidisciplinary teams that include microbiologists, ecologists, and other specialists.

The benefits that genetically engineered organisms can make in pest control can only become a reality if they are carefully tested, regulated, and monitored prior to widespread release. Lessons learned from the introduction of pest organisms like the gypsy moth and the indiscriminate use of pesticides should be heeded so the ecological risks of the release of genetically engineered organisms into the natural environment can be minimized. Only in this way will genetically engineered organisms develop into reliable and safe biocontrol agents for agricultural and forest pests.

GM Mosquitoes: Flying through the Regulatory Gaps?

By Lim Li Ching

Lim Li Ching, MPhil, *works in the biosafety program at Third World Network and is deputy editor of* Science in Society. *This article originally appeared in* GeneWatch, *volume 25, number 3, April–May 2012.*

In December 2010, 6,000 genetically modified mosquitoes were released in my country, Malaysia. This followed releases of large numbers of mosquitoes engineered with the same modification—a dominant lethal gene—in the Cayman Islands, where over 3.3 million GM mosquitoes were released in 2009 and 2010. Since February 2011, more than 3 million of these mosquitoes were released in the city of Juaziero in northeastern Brazil. The release of these same mosquitoes is currently being considered in the Florida Keys in the United States. Many other countries are reportedly evaluating the GM mosquitoes for laboratory research and possible future field releases.

The genetic modification in question targets *Aedes aegypti*, commonly known as the yellow fever mosquito, which is a vector of dengue fever and other diseases. The so-called RIDL technology involves a genetic regulation that, in the absence of the antibiotic tetracycline, causes death at the larval stage of the offspring. The release of mainly male GM mosquitoes carrying this lethal gene is intended to result in mosquito population suppression, with the consequent aim of reducing the incidence of dengue fever.

The GM mosquitoes were developed and the associated technology patented by the UK-based company Oxitec, which appears to be approaching many countries and offering the mosquitoes as a potential solution to the dengue problem. Dengue fever is a serious problem in many countries, and authorities are increasingly looking for alternatives, as tools such as pesticides are rendered ineffective due to resistance development.

However, the release of these GM mosquitoes into the environment raises many scientific, social, ethical and regulatory concerns. Even while these issues are still being debated, it seems that there is a headlong rush to release the GM mosquitoes.

The situation is compounded by the fact that the international regulatory and risk assessment frameworks governing GM insects in general, and GM mosquitoes in particular, are still immature. So much so that in the US, discussion is on-going as to which agency should regulate the proposed release of GM mosquitoes in Florida, since this is a completely new area which the regulatory world is unfamiliar with.

Moreover, under the Cartagena Protocol on Biosafety—the only international law dealing exclusively with genetic engineering and genetically modified organisms—a technical expert group revised its guidance last year for GM mosquito risk assessment. This guidance, part of a larger package of guidance on risk assessment, will be forwarded to the Parties of the Cartagena Protocol for consideration in October 2012. To my knowledge, a corresponding group which convened under the World Health Organization to develop guidance principles for GM mosquito evaluation has yet to finish this task.

At the national level, the first release of GM mosquitoes in the world, which occurred in the Cayman Islands, was conducted in the absence of a biosafety law. While the release was approved by the authorities concerned, the Cayman Islands only had a draft biosafety bill at the time. Moreover, the provisions of the Cartagena Protocol did not apply to the Cayman Islands, even though the UK, under which the Caymans are a British Overseas Territory, is a Party to the Protocol. This meant that specific biosafety questions may not have been fully considered nor evaluated, because of the absence of a detailed and comprehensive biosafety regulatory framework.

Indeed, the risk assessment that was used to support the approval of the releases in the Cayman Islands has been roundly criticized. Scientists at the Max Planck Institute for Evolutionary Biology in Germany conducted a thorough examination of the regulatory procedures and documents. They concluded that the risk assessment was incomplete, with no provision of experimental data on the releases; that there was poor referencing (unlikely to meet peer review standards); and worst of all, that there was a marked absence of discussion of the potential health or environmental hazards specific to the GM mosquito in question.

This trend of substandard regulatory oversight is regrettably not a one-off. The Max Planck scientists assessed the regulatory process in the first three countries (US, Cayman Islands, Malaysia) permitting releases of GM insects (including GM mosquitoes in the latter two countries) in terms of pre-release transparency and scientific quality, and found the process wanting. They suggest deficits in the scientific quality of the regulatory documents and a general absence of accurate experimental descriptions available to the public prior to the releases.

Worryingly, they judged the world's first environmental impact statement on GM insects, produced by US authorities in 2008, to be scientifically deficient. This assertion is made on the basis that 1) by and large, the consideration of environmental risk was too generic to be scientifically meaningful; 2) it relied on unpublished data to establish central scientific points; and 3) despite the approximately 170 scientific publications cited, the endorsement of the majority of novel transgenic approaches was based on just two laboratory studies of only one of the four species covered by the document. However, the environmental impact statement appears to be used as the basis for regulatory approvals around the world, including that of the GM mosquitoes.

One of the most obvious questions to ask is whether humans can be bitten by the GM mosquitoes. In public information available on the Cayman Islands and Malaysian trials, however, this question is either conspicuously ignored or it is implied that there is no biting risk, "as only male mosquitoes are released and they cannot bite."

However, as detailed by the Max Planck scientists, it is probable that transgenic daughters of the released males will bite humans. This is because the males are only partially sterile as the technology is not 100 percent effective. Furthermore, if the mosquitoes encounter tetracycline contamination in the wild, the numbers of survivors could increase. The likely presence of transgenic females in the environment requires the consideration of a more complex series of potential hazards, but this does not appear to have been done.

Public information, consultation, and participation have been also lacking. In the case of the Cayman Islands, while Oxitec and the local Mosquito Research and Control Unit claim that adequate information was provided to the public prior to the release of the GM mosquitoes, the video information provided by MRCU for outreach does not once mention that the mosquitoes in question are genetically modified. Moreover, given the significance of the first release of GM mosquitoes in the world, it is puzzling as to why Oxitec only announced the fact of the release more than a year after they occurred, catching even scientists in the field of transgenic insects off guard.

It is clear that the regulatory processes that have governed the release of GM mosquitoes into the environment so far have been lacking. While international guidance may have recently been completed, the implementation at national level still suffers from a lack of adequate experience in dealing with this novel application of genetic engineering, a lack of rigorous risk assessment and robust investigation of unanswered questions and a lack of effective and meaningful public consultation and participation. In light of this, the push to release the GM mosquitoes in various countries is grossly premature.

43

Why Context Matters

By Craig Holdrege

Craig Holdrege, PhD, *is the founder and director of the Nature Institute, a research and education nonprofit organization dedicated to developing new qualitative and holistic approaches to seeing and understanding nature and technology. He is the author of numerous books and articles including* Genetics and the Manipulation of Life: The Forgotten Factor of Context *(1996) and* Beyond Biotechnology: The Barren Promise of Genetic Engineering *(2010).*

For most of the twentieth century, a main branch of genetics focused on distinct "traits" that appear to remain constant over generations—think of Mendel's peas, the eye color of the fruit fly, or the coat pattern of guinea pigs that you learned about in school. A kind of atomization of the organism into myriad separate characteristics occurred. The hypothesis that such traits are caused by discrete and stable agencies called genes led to the search for genetic material. DNA was found to fit the requirements for genetic material, and by the 1960s everything seemed pretty clear: there are segments of DNA called genes that are stably inherited, and they cause, via enzymes and proteins, an organism's manifold traits.

This elegant and simple picture suggested that if scientists could take DNA from one organism and "insert" it into another, the host organism would take on characteristics of the donor. This is the idea of genetic engineering and was pursued in practice with great effort, intelligence, and financial backing from the 1970s on. By the early 1990s, genetically modified (GM) food crops had been produced in biotech company labs and universities around the world. The technique worked. For example, biologists found bacteria that did not die when exposed to the herbicide glyphosate. With the aim of conveying this herbicide resistance to plants, they contrived to insert relevant bits of the bacterial DNA into embryonic plants. A small percentage of the developing plants both grew normally *and* survived treatment with glyphosate. GM crops were born.

What could be a neater solution to agricultural problems than GM crops (GMOs) that allow farmers to spray a relatively fast degrading herbicide to control weeds? Or crops that produce their own pesticide (so-called "Bt" crops) to ward off hungry and harmful grubs? Today hundreds of millions of acres of cropland are planted with GM crops possessing such traits—soybeans, corn, canola, cotton, sugar beets, alfalfa and more.

So what's the problem? It's often the case that technologies are "successful" because they emerge out of a promising and yet narrow perspective that has been pursued with single-minded intensity in order to produce specific results. Along the way much gets ignored that turns out to be important when the technology is finally released from the clearly circumscribed confines of conceptual models and labs into the rich, unpredictable, and complex realities of the world.

The Trouble with Genes

When a canola plant grows in a soil to which compost has been added, it sends a multitude of branching roots into the soil (Figure 1, left). The above-ground plant grows large and robust, with differentiated leaves, and over time it develops many flowers and fruits. In contrast, seeds of the same variety that have grown in a sandy loam soil develop into a very different plant (Figure 1, right).

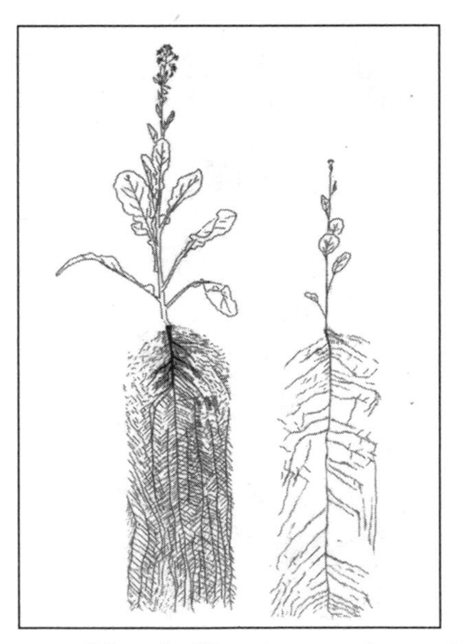

Figure 1. Two specimens of Canola (*Brassica napus*) growing in different soils.

Many fewer roots form and the root system is less branched. This contained root growth is mirrored by the small above-ground plant that has a thin main stem, small leaves with little differentiation, and few small flowers and fruits. In each case the canola plant is interacting intensively with its environment, and this interaction becomes embodied in its growth, its forms and structures, and its substances.

What we can learn from such a seemingly simple and yet fundamentally important example is that organisms are not fixed. They have a high degree of plasticity, a potential to interact with their concrete circumstances and to develop in relation to those circumstances. There are, in reality, no fixed traits that do not vary in relation to circumstances. To find the most stable and fixed aspects of inheritance, geneticists had to focus on those characteristics that vary least and, in addition, they had to control the growing conditions as much as possible in order to achieve uniform results. They bracketed out plasticity as far as possible; the more you isolate something, the more fixed it will appear. But when you subject an organism to different conditions, the trait will show plasticity. Of course, some traits are more plastic than others, but the fact remains that organisms are fundamentally adaptive.

So, if the term "trait" is to reflect reality, it must be thought of as a spectrum of possibilities. Within the spectrum's fluid boundaries there are an infinite number of possible expressions that could be realized, depending on the conditions of the organism and the environment. This is true even of Mendelian traits.

And it is also true of genes. While this fact could have been obvious to anyone who thought carefully about an organism's plasticity, it has taken the sheer force of myriad findings within the field of molecular biology over the past few decades to motivate scientists to begin to move beyond a misconceived gene-centered view of heredity. A gene as a discrete entity that causes a discrete trait is a figment of the imagination. To get a sense of how scientists have left behind the straightforward idea of the gene—the idea that a specific sequence of DNA is responsible for the structure and therefore function of a specific protein molecule or part thereof—consider a couple of recent definitions of the gene:

Genes might be redefined as fuzzy transcription clusters with multiple products.[1]

A gene is a statistical model to help interpret and provide concise summarization to potentially noisy experimental data.[2]

As leading geneticist Walter Gelbart stated already in 1998,

> We may have well come to the point where the use of the term "gene" is of limited value and might in fact be a hindrance to our understanding of the genome . . . . Genes are not physical objects but merely concepts that have acquired a great deal of historic baggage over the past decades.[3]

DNA has been found to be interwoven and to interact with its cellular environment in a remarkably dynamic and context-dependent way.[4]

Current science is re-contextualizing what was conceived of in an atomistic way in order to do justice to the realities of the living organism. For this reason, the idea of the discrete gene becomes an obstacle to evolving understanding.

Paradoxically, during the decades in which the gene concept has been dissolving and genetic processes have been viewed in increasingly dynamic terms, genetic engineering has operated with the older, much more static view of the gene as the cause of distinct traits ("gene for" herbicide resistance, etc.).

And just as the idea of distinct traits and genes "works" as long as you focus on particular characteristics studied under controlled conditions and ignore the larger context, so also does genetic engineering "work" as long as you focus solely on whether an organism exhibits the desired trait. The moment you look at the larger genetic, organismic, environmental, economic, and social contexts, the picture shifts.

The Organismic and Ecological Context of GMOs

From an organismic perspective that takes the dynamics of heredity into account, you would expect that GMO should not behave as the mechanisms they are considered to be. And this is the case. Often scientists find so-called "unintended effects"[5]

A desired trait ("herbicide resistance") is the intended effect, and when other effects arise they are viewed as side effects. Of course, this is an anthropocentric perspective. From the perspective of the organism, the side effects may be more essential than the intended ones.

For example, researchers wanted to improve the digestibility of alfalfa by reducing its lignin content or altering its lignin composition. They succeeded, but the plants showed other characteristics as well: Many were only 25 to 50 percent as tall as the parents; flowering was delayed, sometimes by as much as twenty days; in some of the GM lines, the flowers were white instead of their normal purple-blue color; the researchers even reported a different floral scent for one of the GM lines.[6]

In commercialized GM crops unintended effects often show themselves only after they are growing in farm fields. GM crops that are resistant to the herbicide glyphosate have been grown in the US since 1996. These crops (mainly commodities like soybeans, cotton and corn) can be sprayed with the herbicide and survive, while weeds in the fields around them will die. This makes weed management much simpler on huge farms, and over the last twenty years most of the commodity crops grown in the USA are GM.

Gradually farmers and weed scientists noticed that some weeds were becoming resistant to glyphosate—they didn't die when sprayed. This problem has increased, and to date in the US fifteen different species of weeds have become glyphosate-resistant.[7]

Farmers now have to spray other herbicides on their fields and the costs of using herbicides have increased dramatically. In fact, many weeds are now showing resistance to multiple herbicides. As one weed specialist recently warned, "US farmers are heading for a crisis".[8]

Any plant ecologist—as someone who considers phenotypes by their environmental and genetic relationships—knows that weeds will become resistant to herbicides if farmers plant only herbicide-resistant crops, discontinue crop rotation practices, and spray mainly glyphosate. In other words, the monolithic approach of current GM agriculture is doomed to fail. Amory Lovins wisely remarked that "if we don't understand how things are connected, quite often the cause of problems is solutions."[9]

In the case of weed resistance, we do have some idea of how things are connected, and what agribusiness and many farmers have done is to blatantly ignore this knowledge. Monsanto and other companies have earned billions of dollars from GM crops during the past eighteen years, a "successful" strategy that makes profits from short-term efficacy. Perhaps they hope that they can develop new herbicides and herbicide-resistant GM crops to provide the next "solution" to the problems they've already created, while sustaining their earnings.

The Paradigm of Context

Such are the consequences of a world view and the practices that emanate from it. When you envision the organism as a complex mechanism and aim to produce distinct and encapsulated effects by altering the parts, you may imagine you can achieve your goal. But as genetically engineered crops show, this is an illusion. You have to ignore both the internal and external ecology of the organism with all the context-dependent interactions, and you have to treat "side effects" as just more problems to solve with the same mindset that produced them.

There is another paradigm, the paradigm of context that you learn from studying the dynamics of life, and not life in isolation[10]

In this paradigm you are keenly aware of the dangers of isolating phenomena in order to understand them, since such procedures lead to overly simple and static concepts, and to the illusion that there are magic-bullet solutions to complex problems. You are committed to looking at living processes in their complexity and dynamism. Along the way you realize you will not understand everything, and this awareness of ignorance—ignorance that is based on substantial knowledge—leads you to be more cautious and circumspect in your thinking and doing. You are in awe of the intricacies of life and realize that fruitful approaches to interacting with the world need to take the form of a dialogue. In this way human action can begin to reintegrate itself into the ecology of life.

PART 7

The Ethics of GMOs

Many consumers are concerned with the corporate ownership of seeds and genetically engineered food plants. Monopolistic business practices and the extreme consolidation of the seed industry, coupled with the patenting of seeds has led many consumers to question whether agribusiness corporations have public welfare as one of their priorities.

Too often, corporate profit motive is cloaked with benevolent intentions. Ending world hunger is a prime example. Hunger is not caused by a global shortage of food. Nearly a billion people suffer from malnutrition because they do not have access to the world's abundance of food, they lack the money to buy it, and they are excluded from the land to grow it. Private biotech corporations prevent small farmers from reusing their seeds, a traditional practice that provides food security for 1.4 billion people. Furthermore, since improvements on staples that feed the world's poor, such as cassava and potato, do not have much potential for profit, the majority of genetically engineered crops are aimed at helping large-scale farmers boost their yields and profits. Genetically engineered crops also reduce food security in the long run by decreasing biodiversity and increasing the use of ecologically damaging chemicals.

In the same vein, the creation of fortified foods through the use of genetic engineering technology is another example of a quick techno-fix to a larger, more complex problem. Opponents of genetic engineering have questioned the effectiveness of the technology associated with the production of Golden Rice, proving that products fortified with vitamin A through GE will not end vitamin A deficiencies. Deficiency of a single micro-nutrient like vitamin A seldom occurs in isolation but is one aspect of a larger context of deprivation and multiple nutrient deficiencies. In addition, the same issues of distribution and access remain with genetically engineered rice as with non-GE rice—the obstacles of access and distribution must still be overcome to get the rice to those who need it. The use of this technology in this manner provides good publicity for the industry, even while it falls short of reducing hunger and blindness. The moral rhetoric of industrial biotechnology is in stark contrast with its actual ethical impact.

The essays in this section explore how GMOs fit into a larger ethical framework.

—Jeremy Gruber

The Biopiracy of Wild Rice

BY BRIAN CARLSON

Brian Carlson, MS, *is the Health and Environmental Affairs Manager at Municipality of Provincetown and a former environmental community organizer for the White Earth Land Recovery Project.* *This article originally appeared in* GeneWatch, *volume 15, number 4, July–August 2002.*

Manoomin (wild rice), which covers a vast portion of North America, is a sacred food to the Anishinaabeg and other indigenous peoples. "Anishinaabeg" means literally "original people," whose prophecy and migration story directed them to settle in the region "where the good berry grows on water." This "good berry" is Manoomin, or wild rice. The heart of the Anishinaabeg culture and traditions lies in the geographical center of biodiversity for wild rice. Located in the northern Midwestern states of Minnesota, Wisconsin, and Michigan, and southern Canada, the Anishinaabeg have honored and cared for wild rice for thousands of years. The Anishinaabeg culture believes the intrinsic identity of wild rice is such that it holds within it a spirit given by the Creator to the Anishinaabeg for their sustenance, both spiritual and traditional. This characteristic precedes economics and laws. It stands as a fundamental human right of the Anishinaabeg and as a source of dignity in their culture. Manoomin binds the Anishinaabeg together as a community and instills a relationship to their land that nurtures their spirituality and their livelihoods. Unfortunately, the Manoomin is now being threatened by the completed mapping and potential genetic manipulation of the wild rice genome.

The fight to save Manoomin is directly related to the national and international movement against the globalization of the world's food supply, the corporate take-over of our common foods, and the introduction of genetic pollution into our ecosystem, food, and bodies. To respond to these threats the Anishinaabeg are organizing to educate their community and to stop the biopiracy and biocolonialization of their culture.

The people gather every harvesting season for ceremonies and to harvest that which the Creator has given them. While giving thanks for the gift of Manoomin, the Anishinaabeg are also seeking support for their battle to save their culture and traditions. With the genetic research by the University of Minnesota that has produced the genomic map of wild rice, the possibility of further patents on wild rice varieties seems inevitable. There are also fears that the natural stands of wild rice on protected and ceded lands may be infiltrated by hybridized or genetically modified wild rice varieties, as happened with corn in Mexico.

Here in northwestern Minnesota, on Gahwahbahbahnikag, the "White Earth" Reservation, wild rice grows abundantly. Each individual lake contains its own stand of wild rice, making each a center of biodiversity. Wild rice from the White Earth Reservation in the western portion of Minnesota tastes and looks different from wild rice from the Fond DuLac Reservation in eastern Minnesota. Winona LaDuke, founder of the White Earth Land Recovery Project, stated, "The abundant biodiversity of wild rice in Indian country assures that there will be a crop somewhere for the Anishinaabeg. This is the way the Creator designed it." The biodiversity of wild rice must be protected from the hybridization programs, genetic research, and patenting to ensure that Anishinaabeg and other people can continue harvesting the natural wild rice for generations to come.

Biocolonialism, widespread in Western societies, is based on the premise of domination over nature, and is realized in the actions of large agribusiness. Debra Harry, Executive Director of the Indigenous Peoples Council on Biocolonialism, states that ". . . colonization is an age- old process of theft and control facilitated by doctrines of conquest . . . that claim the land as empty and nonproductive (in its natural state) . . . and as the self-proclaimed 'discoverers' of crops, medicinal plants, genetic resources, and traditional knowledge, these bioprospectors become the new 'owners.' The quest for knowledge by 'bioprospectors' through exploration and research has led to the mapping of the wild rice genome, creating threats to the sacredness of Manoomin."

The commodification and commercialization of common foods and agriculture can be seen in the continued research on wild rice. Since the 1960s, the University of Minnesota's hybridization programs (they have four known experiment stations) have produced nine different hybridized varieties of paddy wild rice. Scientists like to state that they are improving the plant by making it more efficient or conducive to mechanical harvesting technologies and standardized for ease of collection and processing. However, they do not take into account the adverse and damaging effects of their so-called improvements on the Anishinaabeg people and others who rely on natural wild crops. The private and corporate takeover of wild rice production destroys our environment by increasing the frequency of chemigation (chemical irrigation) on paddy wild rice. It also turns desert land in northern California into aquaculture habitats to grow a plant that is native to wetland and monsoon type ecosystems.

The biopiracy of a sacred food and destruction of a spiritual livelihood has infringed upon the sovereignty and spiritual dignity of the Anishinaabeg. In a genome research lab, the bioengineer Ken Foster has engineered a method, in which, by so-called Cytoplasmic Male Sterility, he produces a hybrid strain of wild rice. Cytoplasmic Male Sterile wild rice exists in nature; however the cross breeding of two varieties allowed Foster to claim his "inventiveness" and secure two patents on wild rice. The previous existence and knowledge of this trait created a strong argument by the Anishinaabeg that these patents are infringing on indigenous cultural, intellectual, and sovereign treaty rights.

Universities and private research labs are moving to profit from their work on wild rice. It is urgent that tribes, along with the environmental and food safety community, address these issues and take action. The corporate control and claims of ownership over wild rice using these patents is another form of colonialism and infringes on tribal sovereignty and treaty rights granted by the United States. A treaty is made with sovereign nations; these treaties need to be honored and upheld. The University of Minnesota, one of the collaborators and large supporters of wild rice genetic research, is located in the region of origin for the species *Zizania paiustris*. The center of wild rice biodiversity and germplasm in the northern Midwest and southern Canada is of great interest to the University of Minnesota and the paddy farmers they support. The establishment of a $20 million paddy wild rice market in the late 1990s is one of the economic and moral reasons that the University of Minnesota uses to continue to genetically modify wild rice. Ironically, the majority of paddy wild rice is now grown in Northern California and shipped back to the Midwest.

Out of concern for the well-being of their members, the president of the Minnesota Chippewa Tribe, Norman Deschampe, wrote a letter to President Mark Yudof of the University of Minnesota asking him to halt wild rice genetic research and objected to the research of wild rice as "imprudent and even provocative." He wrote, "This rice from these waters holds a sacred and significant place in our culture . . . we urge the greatest possible level of caution before the University proceeds too far in this process . . ." Though the plea to halt genetic research was sent in September 1998, almost a year later, in August 1999, the genome map of wild rice was published, showing the Minnesota Chippewa Tribe and the Anishinaabeg that their request was obviously ignored. Additionally, upon the publication of the wild rice genome study it became known that two of the researchers that worked on this genomic study now work for Monsanto and Pioneer Hi-Bred, the world's two largest seed companies.

President Deschampe ended his plea to the University of Minnesota president with the demand that "We (the Anishinaabeg) are prepared to undertake every legal and lawful measure to protect our interests in this matter. I hope you do not feel we do so merely to stop the progress of our general society. (We are all too aware of the historical outcome for Indians when the general society feels we are in the way of their progress.) I assure you, our interest is only in protecting the few rights and advantages that we have granted at such great cost."

The biopiracy of a sacred food and destruction of a spiritual livelihood has forever altered the intrinsic identity of Manoomin and has infringed upon the sovereignty and spiritual dignity of the Anishinaabeg. Actions are underway to secure the rights and sovereignty of Native peoples and their beloved Manoomin. The right and responsibility to protect wild rice for future generations is an inherent right of the Anishinaabeg, and is further protected by their self-governance, sovereignty, and treaty rights.

Conflicts of Interest Undermine Agricultural Biotechnology Research

By Susan Benson, Mark Arax, and Rachel Burstein

Susan Benson *is a San Francisco freelance writer and former editor of* Farmer to Farmer. **Mark Arax** *is a reporter for the* Los Angeles Times. **Rachel Burstein** *is an investigative reporter for* Mother Jones. *Adapted from "Growing Concern," originally published in* Mother Jones *magazine.* Mother Jones's *reporter Jeanne Brokaw contributed additional research for this story. This article originally appeared in* GeneWatch, *volume 11, number 1–2, April 1998.*

Farmers trying to decide whether to take the leap into biotech crops lean heavily on university researchers for scientific analysis of a product's pros and cons. But if that recommendation and governmental regulation have been influenced by industry dollars—and university scientists working for companies lose sight of farmers' interests—the process becomes tainted in a way that can harm agriculture and ultimately affect consumers.

In fields across the country, genetically pumped-up crops—from a virus-resistant yellow crookneck squash to transgenic wheat—are being groomed for market. And hundreds more are in the pipeline, some implanted with the genes of animals, such as new varieties of corn, soybeans, oats, rice, apples, broccoli, lettuce, melons, and strawberries. There's even transgenic seafood in the works, including genetically altered salmon, prawns, and catfish.

There may be serious side effects to messing with Mother Nature. There are concerns about the long-term impacts of releasing new genes into the food supply. For example, little is known about the potential of transgenic foods to provoke allergic reactions in human beings. And there are environmental concerns, such as the possibility that crops genetically altered to contain toxins from the bacterium *Bacillus thuringiensis* (Bt) will encourage strains of "super-bugs." Pests that become resistant to Bt could render it useless as a natural topical pesticide.

Preventing pest resistance is the responsibility of the US Environmental Protection Agency (EPA), which, along with the Food and Drug Administration and the Department of Agriculture, is charged with regulating new crops. But critics say the EPA has regularly caved in to pressure from Monsanto and other seed companies and approved genetically engineered products without taking adequate measures to guard against pest resistance and other dangers.

Furthermore, the university scientists who initially test the seeds in the field are often faced with conflicts of interest. (Cuts in public funding for research and regulatory oversight have left the door open for industry—specifically a handful of companies including Monsanto, Northrup King, American Cyanamid, Ciba-Geigy, Rhone- Poulenc, and Dow—to lead the headlong rush into genetic engineering and control the entire lab-to-field-to-table process.) By giving money to a specific scientist at a specific university, a company can fund the field research it needs to determine the potential of a specific product. In many cases, the company then uses the scientist to attest to the virtues of the biotech product.

Congress has helped pave the way for corporate biotech programs, passing a series of laws in the 1980s that pushed federally funded research at universities into the eager hands of agrochemical companies. Congressional specialty grants, which are designed to let Congress respond to pressing agricultural concerns, are generally awarded to researchers who already have industry sponsors in place. "[Universities] don't necessarily say who their other funders are, but they will tell us if the project is leveraged four-to-one by private dollars," says Tim Sanders, a staff member of the House Appropriations Agriculture Subcommittee. Industry support is important, he says, because committee members "want to see everyone participate."

Under a banner of global competitiveness, this new relationship between academia, business, and government encourages universities to waste no time converting their science into patent rights. Previously, such research had been considered public property. Any patents that emerged typically were held by the government. Indeed, so ingrained was this public ethos that when Jonas Salk was asked who owned the patent to his polio vaccine, he responded incredulously, "The people, I would say. Could you patent the sun?"

Today, however, universities are quick to license patent rights to companies for profit-making. These same companies, meanwhile, award grants to university entomologists and geneticists to conduct research on future products.

Often, critics say, it doesn't take a great deal of money to entice a university department or scientist over to the corporate side, particularly in this time of state and federal funding cuts. "Universities are more than ever hunting for corporate money, and while that money may be a small percentage of the overall budget, it's often enough to influence the direction of public science," explains Kathleen Merrigan of the Henry A. Wallace Institute for Alternative Agriculture, a nonprofit research and education organization based in Washington, D.C. "Corporate money can be the tail that wags the dog."

For example, in 1985, Cornell University agreed to do research on bovine growth hormone (BGH) for Monsanto. Tess Hooks, a sociologist at the University of Western Ontario whose graduate work at Cornell dealt with scientific ethics, reviewed the agreement between Cornell and Monsanto. According to Hooks, the university would test BGH on dairy cows and report the findings to Monsanto, which would present its case to the FDA. The government agency would then decide if the hormone—which increases a cow's milk production—created any health risks to cows or milk consumers. But before Cornell received the $557,000 grant from Monsanto, Hooks says, it essentially had to agree to hand over control of its research to the biotech company.

Computers in the university's dairy barn sent the raw data directly to Monsanto in St. Louis. According to Hooks, the company, rather than the university's principal research scientist, controlled and interpreted the data. "I couldn't believe that a university would agree to such restrictions," says Hooks.

Monsanto's efforts to get BGH approved in the United States were dogged by controversy. Current and former FDA employees accused the agency of overlooking important safety concerns in its review of the product and of committing ethics violations because several recently hired FDA officials had worked on BGH for Monsanto. In the end, the FDA was cleared of misdoing. But questions about the hormone persisted. In 1994, several British scientists charged that Monsanto had suppressed their independent analysis of the company's data because it showed a higher rate of infection for cows treated with BGH than Monsanto had acknowledged.

At North Carolina State University, a mini-scandal erupted three years ago when several professors were found to be moonlighting as paid consultants to Rhone-Poulenc, Monsanto, and American Cyanamid—at the same time the professors were evaluating the companies' biotech products for the university. One distinguished weed science professor, Harold Coble, appeared in a Rhone-Poulenc marketing brochure singing the virtues of the company's genetically engineered cotton plant and its companion herbicide, bromoxynil (BXN). "There isn't a downside to the BXN," he says in the brochure.

As a result of the controversy, the university instituted a policy requiring faculty to report on a yearly basis any potential conflicts of interest, such as consulting for a chemical company. Other scientists who have done research for biotech companies dismiss these examples as anomalies. "Practically all of my money for research comes from industry, but I've never done anything to help a company promote its product," says Daniel Colvin, a University of Florida agronomist. "If you manipulate the truth, it takes only one season on the farm to find out that the product doesn't work like you said it would. After one bad season, your credibility with the farmer is shot."

But in some cases it is difficult to tell where public research ends and the company's marketing begins.

Take, for example, the August 25, 1996, letter from Ron H. Smith, an entomologist at Auburn University, that Monsanto faxed to *Mother Jones* magazine in support of its Bt cotton. "Weeks from now," Smith wrote, "when the last bale of the 1996 cotton crop is harvested . . . producers finally will have time to pause and reflect on the revolution that has gripped their profession. The results, so far, have been astonishing. . . . The proof, as they say, is in the pudding—or, in this case, the [farmer's] pocketbook."

Although the letter bore Smith's signature, an Auburn public relations official actually wrote it for him. When asked if he received any funding from Monsanto for his research, Smith replied, "No, not directly." However, *Mother Jones* found university records indicating that Monsanto gave $500,000 to Auburn University between 1991 and 1996; $26,000 was earmarked for projects listing Smith's name. Asked again, Smith confirmed the information, saying he had misunderstood the original question.

Genetically Engineered Foods Changing the Nature of Nature

By Martin Teitel and Kimberly Wilson

Martin Teitel, PhD, *is former CEO of the Cedar Tree Foundation and a former executive director of the Council for Responsible Genetics.* **Kimberly Wilson** *is an author and activist. Excerpted from the book of the same title published by Park Street Press, 1999 (innertraditions.com). This article originally appeared in* GeneWatch, *volume 12, number 5, October 1999.*

Imagine yourself one morning on a modern jetliner, settling into your seat as the plane taxis toward the active runway. To pass the time you unfold your morning newspaper, and just as the plane's rapidly building acceleration begins to lift the wheels from the ground, your eye catches a front page article mentioning that engineers are beginning a series of tests to determine whether or not the new model airplane that you arc in is safe.

That situation would never happen, you say to yourself. People have more foresight than that. Yet something we entrust our lives to far more often than airplanes, our food supply, is being redesigned faster than any of us realize, and scientists have hardly begun to test the long-term safety of these new foods. The genetic engineering of our food is the most radical transformation in our diet since the invention of agriculture 10,000 years ago. During these thousands of years, people have used the naturally-occurring processes of genetics to gradually shape wild plants into tastier, more nutritious, and more attractive food for all of humanity.

At the end of the twentieth century enough genetically engineered crops are being grown to cover all of Great Britain plus all of Taiwan, with enough left over to carpet Central Park in New York. With this abrupt agricultural transformation, humanity's food supply is being placed in the hands of a few corporations who practice an unpredictable and dangerous science.

As we eat genetically altered food and read about new safety tests, we may start to realize that we are the unwitting and unwilling guinea pigs in the largest experiment in human history, involving our entire planet's ecosystem, food supply, and the health and very genetic makeup of its inhabitants. Worse yet, results coming in from the first objective tests are not encouraging. Scientists issue cautionary statements almost weekly, ranging from problems with monarch butterflies dying from genetically modified corn pollen to the danger of violent allergic reactions to genes introduced into soy products, as well as experiments showing a variety of actual and suspected health problems for cows fed genetically engineered hormones and the humans who drink their milk.

Three features distinguish this new kind of food. First and most important, the food is altered at the genetic level in ways that could never occur naturally. As genes from plants, animals, viruses, and bacteria are merged in novel ways, the normal checks and balances that nature provides to keep biology from running amok are nullified. The second novel feature of the revolution in our food is that the food is owned. Not individual sacks of wheat or bushels of potatoes, but entire varieties of plants are now corporate products. The term "monopoly" takes on new power when one imagines a company owning major portions of our food supply, the one thing that every single person now and into the future will always need to buy. Finally, this new technology is "globalized." This means that local agriculture, carefully adapted to local ecology and tastes over hundreds and thousands of years, must yield to a planetary monoculture enforced by intricate trade agreements and laws.

Biotech's commandeering of our food is widespread, but hardly inevitable. Tens of thousands of natural seeds still exist to form the basis of a diverse, healthy, and locally controlled food system in our world. With proper attention from ordinary people, our food supply will be put back into the hands of farmers and food suppliers and all the rest of us, for the sake of our health and our environment, and for the future that we leave to our children's children.

Lessons from the Green Revolution: Do We Need New Technology to End Hunger?

By Peter Rosset, Frances Moore Lappé, and Joseph Collins

Peter Rosset, PhD, *is an agricultural economist with the Center for the Study of Rural Change in Mexico and the Land Research Action Network. He is the former executive director of Food First/The Institute for Food and Development Policy, which was founded by Joseph Collins and Frances Moore Lappé in 1975. Frances Moore Lappé is an activist and author of over eighteen books including* Diet for a Small Planet. *She is the cofounder of three national organizations that explore the roots of poverty, hunger, and environmental distress. Joseph Collins, PhD, is an author and researcher who writes about world hunger and other issues in related to inequality. This article is substantially similar to an article by the same name published in* Tikkun, *volume 15, number 2, pages 52–56.* *This article originally appeared in* GeneWatch, *volume 13, number 2, April 2000.*

Faced with an estimated 786 million hungry people in the world, cheerleaders for our social order have an easy solution: we will grow more food through the magic of chemicals and genetic engineering. For those who remember the original "Green Revolution" promise to end hunger through miracle seeds, this call for "Green Revolution II" should ring hollow. Yet Monsanto, Novartis, AgrEvo, DuPont, and other chemical companies who are reinventing themselves as biotechnology companies, together with the World Bank and other international agencies, would have the world's anti-hunger energies aimed down the path of more agrochemicals and genetically modified crops. This second Green Revolution, they tell us, will save the world from hunger and starvation if we just allow these various companies, spurred by the free market, to do their magic.

The Green Revolution myth goes like this: the miracle seeds of the Green Revolution increase grain yields and therefore are a key to ending world hunger. Higher yields mean more income for poor farmers, helping them to climb out of poverty and more food means less hunger. Dealing with the root causes of poverty that contribute to hunger takes a very long time and people are starving now. So we must do what we can—increase production. The Green Revolution buys the time Third World countries desperately need to deal with the underlying social causes of poverty and to cut birth rates. In any case, outsiders—like the scientists and policy advisers behind the Green Revolution—can't tell a poor country to reform its economic and political system, but they can contribute invaluable expertise in food production. While the first Green Revolution may have missed poorer areas with more marginal lands, we can learn valuable lessons from that experience to help launch a second Green Revolution to defeat hunger once and for all.

Improving seeds through experimentation is what people have been up to since the beginning of agriculture, but the term "Green Revolution" was coined in the 1960s to highlight a particularly striking breakthrough. In test plots in northwest Mexico, improved varieties of wheat dramatically increased yields. Much of the reason why these "modern varieties" produced more than traditional varieties was that they were more responsive to controlled irrigation and to petrochemical fertilizers, allowing for much more efficient conversion of industrial inputs into food. With a big boost from the International Agricultural Research Centers created by the Rockefeller and Ford Foundations, the "miracle seeds quickly spread to Asia, and soon new strains of rice and corn were developed as well.

By the 1970s, the term "revolution" was well deserved, for the new seeds—accompanied by chemical fertilizers, pesticides, and, for the most part, irrigation—had replaced the traditional farming practices of millions of Third World farmers. By the 1990s, almost 75 percent of Asian rice areas were sown with these new varieties. The same was true for almost half of the wheat planted in Africa and more than half of that in Latin America and Asia, and about 70 percent of the world's corn as well. Overall, it was estimated that 40 percent of all farmers in the Third World were using Green Revolution seeds, with the greatest use found in Asia, followed by Latin America.

Clearly, the production advances of the Green Revolution are no myth. Thanks to the new seeds, tens of millions of extra tons of grain a year are being harvested. But has the Green Revolution actually proven itself a successful strategy for ending hunger? Not really.

Narrowly focusing on increasing production—as the Green Revolution does—cannot alleviate hunger because it fails to alter the tightly concentrated distribution of economic power, especially access to land and purchasing power. Even the World Bank concluded in a major 1986 study of world hunger that a rapid increase in food production does not necessarily result in food security—that is, less hunger. Current hunger can only be alleviated by "redistributing purchasing power and resources toward those who are undernourished," the study said. In a nutshell, if the poor don't have the money to buy food, increased production is not going to help them.

Introducing any new agricultural technology into a social system stacked in favor of the rich and against the poor, without addressing the social questions of access to the technology's benefits, will lead to an even greater concentration of the rewards from agriculture, as is happening in the United States.

Because the Green Revolution approach does nothing to address the insecurity that lies at the root of high birth rates—and can even heighten that insecurity—it cannot buy time until population growth slows. Finally, a narrow focus on production ultimately defeats itself as it destroys the very resource base on which agriculture depends. We've come to see that without a strategy for change that addresses the powerlessness of the poor, the tragic result will be more food and yet more hunger.

More Food and Yet More Hunger?

Despite three decades of rapidly expanding global food supplies, there are still an estimated 786 million hungry people in the world in the 1990s. Where are these 786 million hungry people? Since the early 1980s, media representations of famines in Africa have awakened Westerners to hunger there, but Africa represents less than one-quarter of the hunger in the world today. We are made blind to the day-in-day-out hunger suffered by hundreds of millions more. For example, by the mid-1980s, newspaper headlines were applauding the Asian success stories. India and Indonesia, we were told, had become "self-sufficient in food" or even "food exporters." But it is in Asia, precisely where Green Revolution seeds have contributed to the greatest production success, that roughly two-thirds of the undernourished in the entire world live.

According to *Business Week* magazine, "even though Indian granaries are overflowing now," thanks to the success of the Green Revolution in raising wheat and rice yields, "5,000 children die each day of malnutrition. One-third of India's 900 million people are poverty-stricken." Since the poor can't afford to buy what is produced, "the government is left trying to store millions of tons of foods. Some is rotting, and there is concern that rotten grain will find its way to public markets." The article concludes that the Green Revolution may have reduced India's grain imports substantially, but did not have a similar impact on hunger.

Such analysis raises serious questions about the number of hungry people in the world in 1970 versus 1990, spanning the two decades of major Green Revolution advances. At first glance, it looks as though great progress was made, with food production up and hunger down. The total food available per person in the world rose by 11 percent over those two decades, while the estimated number of hungry people fell from 942 million to 786 million, a 16 percent drop. This was apparent progress, for which those behind the Green Revolution were understandably happy to take the credit.

But these figures merit a closer look. If you eliminate China from the analysis, the number of hungry people in the rest of the world actually increased by more than 11 percent, from 536 to 597 million. In South America, for example, while per capita food supplies rose almost 8 percent, the number of hungry people also went up, by 19 percent. In south Asia, there was 9 percent more food per person by 1990, but there were also 9 percent more hungry people. Nor was it increased population that made for more hungry people. The total food available per person actually increased. What made possible greater hunger was the failure to address unequal access to food and food-producing resources.

The remarkable difference In China, where the number of hungry dropped from 406 million to 189 million, almost begs the question: which has been more effective at reducing hunger—the Green Revolution or the Chinese Revolution where broad-based changes in access to land paved the way for rising living standards?

Whether the Green Revolution or any other strategy to boost food production will alleviate hunger depends on the economic, political, and cultural rules that people make. These rules determine who benefits as a supplier of the increased production—whose land and crops prosper and for whose profit—and who benefits as a consumer of the increased production who gets the food and at what price.

The poor pay more and get less. Poor farmers can't afford to buy fertilizer and other inputs in volume; big growers can get discounts for large purchases. Poor farmers can't hold out for the best price for their crops, as can larger farmers whose circumstances are far less desperate. In much of the world, water is the limiting factor in farming success, and irrigation is often out of the reach of the poor. Canal irrigation favors those near the top of the flow. Tubewells (a type of water well), often promoted by development agencies, favor the bigger operators, who can better afford the initial investment and have lower costs per unit. Credit is also critical. It is common for small farmers to depend on local moneylenders and pay interest rates several times as high as wealthier farmers. Government-subsidized credit overwhelmingly benefits the big farmers. Most of all, the poor lack clout. They can't command the subsidies and other government favors accruing to the rich.

With the Green Revolution, farming becomes petro-dependent. Some of the more recently developed seeds may produce higher yields even without manufactured inputs, but the best results require the right amounts of chemical fertilizer, pesticides, and water. So as the new seeds spread, petrochemicals become part of farming. In India, adoption of the new seeds has been accompanied by a six-fold rise in fertilizer use per acre. Yet the quantity of agricultural production per ton of fertilizer used in India dropped by two-thirds during the Green Revolution years. In fact, over the past thirty years the annual growth of fertilizer use on Asian rice has been from three to forty times faster than the growth of rice yields.

Because farming methods that depend heavily on chemical fertilizers do not maintain the soil's natural fertility and because pesticides generate resistant pests, farmers need ever more fertilizers and pesticides just to achieve the same results. At the same time, those who profit from the increased use of fertilizers and pesticides fear labor organizing and use their new wealth to buy tractors and other machines, even though they are not required by the new seeds. This incremental shift leads to the industrialization of farming.

Once on the path of industrial agriculture, farming costs more. It can be more profitable, of course, but only if the prices farmers get for their crops stay ahead of the costs of petrochemicals and machinery. Green Revolution proponents claim increases in net incomes from farms of all sizes once farmers adopt the more responsive seeds. But recent studies also show another trend: outlays for fertilizers and pesticides may be going up faster than yields, suggesting that Green Revolution farmers are now facing what US farmers have experienced for decades—a cost-price squeeze.

In Central Luzon, Philippines, rice yield increased 13 percent during the 1980s, but came at the cost of a 21 percent increase in fertilizer use. In the Central Plains, yields went up only 6.5 percent, while fertilizer use rose 24 percent and pesticides jumped by 53 percent. In West Java, a 23 percent yield increase was virtually canceled by 65 and 69 percent increases in fertilizers and pesticides respectively.

To anyone following farm news here at home, these reports have a painfully familiar ring. And why wouldn't they? After all, the United States—not Mexico—is the true birthplace of the Green Revolution. Improved seeds combined with chemical fertilizers and pesticides have pushed corn yields up nearly three-fold since 1950, with smaller but still significant gains for wheat, rice, and soybeans. Since World War II, as larger harvests have pushed down the prices farmers get for their crops while the costs of farming have shot up, farmers' profit margins have been drastically narrowed. By the early 1990s, production costs had risen from about half to over 80 percent of gross farm income. So who survives today? Two very different groups: those few farmers who choose not to buy into industrialized agriculture and those able to keep expanding their acreage to make up for their lower per acre profit. Among this second select group are the top 1.2 percent of farms by income, those with $500,000 or more in yearly sales, dubbed "superfarms" by the US Department of Agriculture. In 1969, the superfarms earned 16 percent of net farm income; by the late 1980s, they garnered nearly 40 percent.

Superfarms triumph not because they are more efficient food producers or because the Green Revolution technology itself favored them, but because of advantages that accrue to wealth and size. They have the capital to invest and the volume necessary to stay afloat even if profits per unit shrink. They have the political clout to shape tax policies in their favor. Over time, why should we expect the result of the cost-price squeeze to be any different in the Third World? In the United States, we've seen the number of farms drop by two-thirds and average farm size more than double since World War II. The gutting of rural communities, the creation of inner-city slums, and the exacerbation of unemployment all followed in the wake of this vast migration from the land. Think what the equivalent rural exodus means in the Third World, where the number of jobless people is already double or triple our own.

Not Ecologically Sustainable

There is also growing evidence that Green Revolution-style farming is not ecologically sustainable, even for large farmers. In the 1990s, Green Revolution researchers themselves sounded the alarm about a disturbing trend that had only just come to light. After achieving dramatic increases in the early stages of the technological transformation, yields began falling in a number of Green Revolution areas. In Central Luzon, Philippines, rice yields grew steadily during the 1970s, peaked in the early 1980s, and have been dropping gradually ever since. Long-term experiments conducted by the International Rice Research Institute (IRRI) in both Central Luzon and Laguna Province confirm these results. Similar patterns have now been observed for rice-wheat systems in India and Nepal. The causes of this phenomenon have to do with forms of long-term soil degradation that are still poorly understood by scientists. An Indian farmer told *Business Week* his story:

> Dyal Singh knows that the soil on his 3.3-hectare [8-acre] farm in Punjab is becoming less fertile. So far, it hasn't hurt his harvest of wheat and corn. "There will be a great problem after five or ten years," says the sixty-three-year-old Sikh farmer. Years of using high-yield seeds that require heavy irrigation and chemical fertilizers have taken their toll on much of India's farmland. So far, 6 percent of agricultural land has been rendered useless.

Where yields are not actually declining, the rate of growth is slowing rapidly or leveling off, as has now been documented in China, North Korea, Indonesia, Myanmar, the Philippines, Thailand, Pakistan, and Sri Lanka.

The Green Revolution: Some Lessons

Having seen food production advance while hunger widens, we are now prepared to ask: under what conditions are greater harvests doomed to failure in eliminating hunger?

First, where farmland is bought and sold like any other commodity and society allows the unlimited accumulation of farmland by a few, super-farms replace family farms and all of society suffers.

Second, where the main producers of food—small farmers and farm workers—lack bargaining power relative to suppliers of farm inputs and food marketers, producers get a shrinking share of the rewards from farming.

Third, where dominant technology destroys the very basis for future production, by degrading the soil and generating pest and weed problems, it becomes increasingly difficult and costly to sustain yields.

Under these three conditions, mountains of additional food could not eliminate hunger, as hunger in America should never let us forget. The alternative is to create a viable and productive small farm agriculture using the principles of agroecology. That is the only model with the potential to end rural poverty, feed everyone, and protect the environment and the productivity of the land for future generations.

Successful Examples

That sounds good, but has it ever worked? From the United States to India, alternative agriculture is proving itself viable. In the United States, a landmark study by the prestigious National Research Council found that "alternative farmers often produce high per acre yields with significant reductions in costs per unit of crop harvested," despite the fact that "many federal policies discourage adoption of alternative practices." The Council concluded, "Federal commodity programs must be restructured to help farmers realize the full benefits of the productivity gains possible through alternative practices."

In South India, a 1993 study was carried out to compare "ecological farms" with matched "conventional" or chemical-intensive farms. The study's author found that the ecological farms were just as productive and profitable as the chemical ones. He concluded that if extrapolated nationally, ecological farming would have "no negative impact on food security," and would reduce soil erosion and the depletion of soil fertility while greatly lessening dependence on external inputs.

But Cuba is where alternative agriculture has been put to its greatest test. Changes underway in that island nation since the collapse of trade with the former socialist bloc provide evidence that the alternative approach can work on a large scale. Before 1989, Cuba was a model Green Revolution–style farm economy, based on enormous production units, using vast quantities of imported chemicals and machinery to produce export crops, while over half of the island's food was imported. Although the government's commitment to equity, as well as favorable terms of trade offered by Eastern Europe, meant that Cubans were not undernourished, the underlying vulnerability of this style of farming was exposed when the collapse of the socialist bloc joined the already existing and soon to be tightened US trade embargo.

Cuba was plunged into the worst food crisis in its history, with consumption of calories and protein dropping by perhaps as much as 30 percent. Nevertheless, by 1997, Cubans were eating almost as well as they did before 1989, yet comparatively little food and agrochemicals were being imported. What happened?

Faced with the impossibility of importing either food or agrochemical inputs, Cuba turned inward to create a more self-reliant agriculture based on higher crop prices to farmers, agroecological technology, smaller production units, and urban agriculture. The combination of a trade embargo, food shortages, and the opening of farmers' markets meant that farmers began to receive much better prices for their products. Given this incentive to produce, they did so, even in the absence of Green Revolution–style inputs. They were given a huge boost by the reorientation of government education, research, and extension toward alternative methods, as well as the rediscovery of traditional farming techniques.

As small farmers and cooperatives responded by increasing production while large-scale state farms stagnated and faced plunging yields, the government initiated the newest phase of revolutionary land reform, parceling out the state farms to their former employees as smaller-scale production units. Finally, the government mobilized support for a growing urban agriculture movement—small-scale organic farming on vacant lots which, together with the other changes, transformed Cuban cities and urban diets in just a few years.

The Cuban experience tells us that we can feed a nation's people with a small-farm model based on agroecological technology, and in so doing we can become more self-reliant in food production. A key lesson is that when farmers receive fairer prices, they produce, with or without Green Revolution seed and chemical inputs. If these expensive and noxious inputs are unnecessary, then we can dispense with them.

The Bottom Line

In the final analysis, if the history of the Green Revolution has taught us one thing, it is that increased food production can—and often does—go hand in hand with greater hunger. If the very basis of staying competitive in farming is buying expensive inputs, then wealthier farmers will inexorably win out over the poor, who are unlikely to find adequate employment to compensate for the loss of farming livelihoods. Hunger is not caused by a shortage of food, and cannot be eliminated by producing more.

This is why we must be skeptical when Monsanto, DuPont, Novartis, and other chemical-cum-biotechnology companies tell us that genetic engineering will boost crop yields and feed the hungry. The technologies they push have dubious benefits and well-documented risks, and the second Green Revolution they promise is no more likely to end hunger than the first.

Far too many people do not have access to the food that is already available because of deep and growing inequality. If agriculture can play any role in alleviating hunger, it will only be to the extent that the bias toward wealthier and larger farmers is reversed through pro-poor alternatives like land reform and sustainable agriculture, which reduce inequality and make small farmers the center of an economically vibrant rural economy.

Let Them Eat Promises: The Fight to Feed the World is Being Betrayed from Within

By Devinder Sharma

Devinder Sharma *is an Indian journalist, writer, and thinker. Trained as an agricultural scientist, Devinder quit active journalism to research policy issues concerning sustainable agriculture, intellectual property rights, environment and development, biotechnology and hunger, and the implications of the free trade paradigm for developing countries. He was the founding member of the Chakriya Vikas Foundation (Foundation for Cyclic Development), is on the Asia Rice Foundation's board of directors, and chairs the Forum for Biotechnology and Food Security in New Delhi. This article originally appeared in* GeneWatch, *volume 16, number 2, March–April 2003.*

It was in 2001 that hundreds of people in the United States, most of them agricultural scientists, signed an AgBioWorld foundation petition asking multinational seed giant Aventis CropScience to donate some 3,000 tons of genetically engineered experimental rice to the needy rather than destroy it. Beyond feeding the hungry, the appeal was a public relations exercise to demonstrate the concern of biotechnology's proponents for feeding the world's poor.

Aventis had expressed worry about global hunger, stating that it is "working hard to ensure that US farmers can grow abundant, nutritious crops and we hope that by contributing to that abundance all mankind will prosper." At the same time, the AgBioWorld Foundation conveyed its "disapproval of those who, in the past, have used situations similar to this one to block approved food aid to victims of cyclones, floods and other disasters in order to further their own political (namely, anti-biotechnology) agendas."

Eradicating global hunger is certainly a worthy intention. But when told that India had 65 million tons of non-genetically modified surplus food in 2001, and has a staggering population of some 320 million people whose food requirements are unmet, those who signed the appeal were suddenly not interested. All the concern and "humanitarian intentions" vanished into thin air. Food surplus in India has now been reduced by eleven million tons, with the balance exported in the past year.

If you are wondering whether the international community is in any way genuinely concerned at the plight of the hungry, don't hold your breath. At the time of the first World Food Summit (WFS) in Rome in 1996, heads of atate from all the countries of the world "reaffirmed the right to have access to safe and nutritious food, consistent with the right to adequate food and fundamental right of everyone to be free from hunger." They considered it unacceptable that more than 840 million people throughout the world did not have enough food to meet their basic nutritional needs.

These leaders vowed to halve that number by the year 2015, meaning that they would need another twenty years to provide food to the other 400 million hungry people. In other words, they postponed the job. By 2015, the world's hungry will have multiplied to 1.2 billion—so, essentially, the heads of state actually expressed their helplessness in tackling hunger and malnutrition.

The heads of state met in Rome again for the "WFS plus Five" in June 2002, to take stock of the efforts made to reduce hunger since they met five years earlier. That, too, at a time when some 24,000 people die every day from hunger, starvation and related diseases. And by the year 2015, by which the United Nations' Food and Agriculture Organization aims at reducing the number of hungry by half, more than 122 million people would succumb to mankind's greatest shame—hunger during times of plenty.

Not acknowledging that the lack of political will exacerbates hunger and destitution, or that political power is being used to promote technologies and strategies that deepen the imbalances at the heart of this human debacle, politicians have joined hands with industrialists and agricultural scientists to chant the mantra of the potential of genetic engineering to boost food production and abolish hunger and malnutrition.

At the same time, free trade and globalization are being used to push highly subsidized agricultural products from the Organization for Economic Cooperation and Development (OECD) into the Global South. The richest trading block in the world, the OECD provides agricultural subsidies of $1 billion a day to its farmers, resulting in underpriced food imports being dumped on developing countries. These imports have already resulted in further marginalization of small farmers and the loss of supplementary livelihoods and other poverty-coping mechanisms for millions of agricultural workers in the developing world. Since importing food is like importing unemployment, the world is fast heading towards a situation when the developing countries will be left with little option but to remain dependent upon the West for their basic food requirement.

Free trade and market domination of food and agriculture will quite clearly be inadequate to make food reach those who need it most. Market interventions in the developing countries are geared towards supporting more commercial farmers and export crops. However, projections indicate that market demand for food can be met mostly from within developing countries. Reliance on food production by those who need it is far healthier than imports from the developed world. The food-insecure populations need income through employment generated in the production of food, not just its physical availability.

The twin engines of economic growth—the technological revolution and globalization—will only widen the existing gap. Biotechnology will, in reality, push more people into the hunger trap. With public attention and resources being diverted from the ground realities, hunger will only grow in the years to come.

What is also not being accepted, for obvious reasons, is the startling fact that South Asian warehouses are overflowing when the average productivity of cereals hovers around 2 tons per hectare—amongst the lowest in the world. India has more than 54 million tons of food surplus at present. Till recently, Pakistan and Bangladesh, too, were overflowing with foodgrains. These are the countries inhabited by nearly 40 percent of the world's poor and hungry. These countries have the potential to raise production at least by three times with the available technology.

Launching a frontal attack on hunger to ensure that food reaches those who need it most could drastically reduce the number of the world's hungry. The world does not have to wait until 2015 for everyone to have enough to eat.

PART 8

Modifying Animals for Food

From the beginning, bioengineering animals has been a controversial issue. Unlike plants, animals have feelings; they sense pain and distress. They cannot give or withhold their consent. We should be concerned about their welfare. Genetically engineering animals for food is an outgrowth of domestic animal breeding methods, which has been carried out for thousands of years. Concerns over transgenic animals have been part of longstanding criticisms of factory farming and the suffering of animals caught up in agribusiness. Transformations that would address animal welfare in industrial production are often opposed on the basis that it would cost the consumer more for food or it would reduce American competitiveness.

It's easy to get distracted by strikingly unusual bioengineered animals, but the majority of the most important ones don't appear bioengineered at all. Genetically modified salmon, for example, look just like regular salmon only larger, but they pose very real risks to natural fisheries and ocean ecosystems. Pigs modified with a gene from spinach to lower levels of saturated fat still look like pigs but raise the question, "Why not just eat spinach?" Cows treated with bovine growth hormone (BGH) have raised serious concerns over the potential human health effects of consuming their milk and meat. Indeed a number of individual animals have gained fame—or infamy—for being bioengineered including "Dolly," the first cloned sheep, and "Enviropig," which was developed to have reduced phosphorous levels in its manure.

As Paul Root Wolpe asks in his essay, "When does the use of biotechnology on animal bodies step over an ethical line, or are their bodies open platforms for our biomechanical tinkering?"

This section explores several authors' responses to that question.

—Jeremy Gruber

Ethical Limits to Bioengineering Animals

By Paul Root Wolpe

Paul Root Wolpe, PhD, *is professor of Bioethics and director of the Center for Ethics at Emory University. This article originally appeared in* GeneWatch, *volume 24, number 2, date April–May 2011.*

The history of humanity's use of animals for its own purposes is not a particularly benevolent one. We have used animal flesh for our meals and fur for warmth, bones for weapons and skin for parchment, blubber for oil and horns for medicine. We have harnessed their bodies for labor and domesticated them as companions. We have killed them for sport, worked them to death, and used them for experiments without anesthesia.

Today, we believe we are better. We have laws against animal cruelty and strong public sanctions for the mistreatment of animals. At the same time, though, we have transformed the individualized exploitation of animals into industrial animal processing. Each year, billions of chickens and millions of turkeys have their beaks partially cut off so that they can be crowded into warehouse-sized barns without cannibalizing each other. Cattle are fed unnatural diets and sometimes castrated without anesthesia, while geese are force-fed to fatten them up for pâté.

Of course, one could go on. No need here to review the litany of our current cruelties or to lament our lack of concern for the suffering of the animals we consume; there is a vast literature on those things and their ethical implications (from Peter Singer's *Animal Liberation* to Jonathan Safran Foer's *Eating Animals*). Yet, despite our general inattention to the quality of life of our livestock and poultry, we still maintain that, as a society, we care about the treatment of animals. We are outraged when a famous quarterback is caught with fighting dogs; we give money to save endangered species; we individually wash oil-soaked birds in the gulf.

We are deeply conflicted about our treatment of animals as a society; so when we see the use of animal bodies as platforms for genetic experiments, it is little wonder that we are confused about how to react.

Our conflicted attitude toward animals expresses itself in many ways. I often present the idea of using transgenic pigs to provide heart valves and whole heart transplants to my undergraduate students as provocatively as possible, by saying, "Imagine! You can create a drove of transgenic pigs whose hearts are not as immunoreactive to humans, or, perhaps, even engineer a pig to grow a heart using some genes from the intended recipient. Then, when the heart is needed, you can choose the best pig, slaughter it—maybe right in the hospital—and transplant the heart directly into the human being!"

The reaction is immediate and passionate, and predictable. Students object that keeping pigs at the ready to be slaughtered for hearts is wrong. "Why?" I ask. The best response I usually get from students (or at least, those who have taken Intro to Philosophy) is that it is using the animal entirely as a means, and that is wrong. My next question, of course, is whether they eat meat.

Why is it that students who had bacon for breakfast react so strongly to the idea of slaughtering a pig for its heart? After all, no one is going to die without bacon, and we are sacrificing the transgenic animal to save a person's life.

The answer lies in our tendency to personalize morality. We are far more likely to feel a responsibility to one person who is suffering than to groups who are suffering. Charities raising money for poverty in developing nations know it is far more effective to tell one person's story than to cite statistics. It is also manifested in what Albert Jonsen referred to as the "rule of rescue."[1] If you ask people whether insurance companies have a right not to reimburse experimental, unproven treatments, a majority say yes. However, if you then give them a particular case of a particular person—a woman with treatment-resistant breast cancer, for example—the same people think it unconscionable that the insurance company will not pay for any and all experimental treatments, even those with only a remote chance of helping her. Personalize the case and our general principles often go out the window.

Students feel sorry for that particular chicken, or pig, but not so much for chickens and pigs in general. The thought of slaughtering "pigs" for bacon seems not to offend some visceral sense of fairness, while slaughtering that particular pig to put its heart in that particular person seems to create an ethical calculus that we do not put into play with masses of animals. One death a tragedy, a million deaths a statistic.

The same reaction seems to be operative in our response to laboratory animals. The same mice we can glue-trap or snap-trap in our basements must be euthanized consistent with the recommendations of the American Veterinary Medical Association Panel on Euthanasia when killed in a lab. Vermin in the home turn to protected subjects in the lab.

The reactions to biotechnologically engineered animals, then, become complicated. Is there anything really wrong with using green fluorescent protein to create a glowing rabbit, fish, or monkey? Should we protest the construction of an ear-shaped polymer scaffolding on the back of a mouse? Do we cross a line when we create flocks of transgenic animals as bioreactors, or to be "pharmed" for drug molecules? Is it ethically questionable to wire electrodes into animal brains to control behavior, or to keep disembodied animal brains alive in nutrient media? With all the suffering that we bring to animals caught up in agribusiness, is it not better to create an animal that is well taken care of, even if it has been engineered with genes that make it glow or express a protein in its milk?

When does the use of biotechnology on animal bodies step over an ethical line, or are their bodies open platforms for our biomechanical tinkering?

The conflict becomes clear in cases such as the experiments of Sanjiv Talwar and John Chapin and their colleagues at SUNY, who wired rats with electrodes in their sensorimotor cortex and again in their pleasure centers (medial forebrain bundle), and then controlled the rats' movements by stimulating them to turn right or left as the experimenter desired.[2] Animal rights groups and ethicists often complained that these animals were denied their autonomy, turned into "ratbots" or "roborats." Even Sanjiv Talwar admitted that the "idea is sort of creepy."[3] Yet we use animals for work and recreation purposes every day in ways that significantly restrains their autonomy, whether they are drug-sniffing dogs or plow oxen or aquarium dolphins or canaries in a cage. Are roborats really any worse off?

Clearly, from an ethical perspective, the suffering of animals in general does not free us from the obligation to treat our animals ethically in the laboratory or biotech industries. So what is wrong with the roborat?

The difference between roborats and drug-sniffing dogs, for example, is that in the case of the dogs, they are trained to exhibit certain behaviors, then rewarded with affection or food for those actions. They are taught to understand, and are rewarded for understanding. Dogs unable or unwilling to conform are not used as work dogs. The roborats, on the other hands, do not really understand what they are asked to do, and the rewards, delivered directly to the brain, do not conform to the nature of rewards we generally expect to give animals—food, for example—which connect an understandable action to a primary biological function, in the hands of a recognizable human master. The roborat does not understand the behavior being asked and does not participate in the reward being offered, and it has no potential relationship to the master. All is done remotely, both request and reward, through impulses sent directly to the brain. In that sense, the roborat is instrumentalized beyond the pet or work animal, and is denied a level of autonomy given even to farm animals and caged pets. The violation in the case of the roborat is taking that last step: removing all relationship with the animal and truly treating it as a mechanism to be controlled rather than a living creature.

The lines are not clear, and the standards shift with time and place. As we begin to create animals whose very bodies are pharmaceutical manufacturing plants, whose organs provide life-saving transplants, and whose bodies have altered genetics and physiology to provide experimental platforms for our science, we have to scrutinize our motives and our actions.

Animals are both commodities and autonomous beings, and in order to maintain our rights to consider them as the former, we are obligated at the same time to honor the latter. Clearly, there is a certain amount of self-contradiction involved. We do not treat all animals the same, and we have different standards even for the same animal in different contexts. Our pretenses, however, do not absolve us of our ethical responsibilities. As we create new ways to integrate biotechnologies into animal bodies, we must constantly revisit and redefine the line between the use of animals and their exploitation, the control of behavior and their right to a certain level of autonomy, and the instrumental use of animals for the general welfare and the manipulation of animals for our curiosity or entertainment. There may be no better measure of our humanity than how we treat our animals.

Back on "The Farm"

By Rob DeSalle

Rob DeSalle, PhD, *is a curator in the American Museum of Natural History's Division of Invertebrate Zoology and co-director of its molecular laboratories and a member of CRG's board of directors. This article originally appeared in* GeneWatch, *volume 24, number 2, April–May 2011.*

About a decade ago, I had the great pleasure to spend some time in the studio of a well-known New York City artist who was interested in the then burgeoning and often over-publicized science of genetic modification of animals and plants.

This inquisitive artist was Alexis Rockman, a painter with a reputation among his colleagues for paintings "depicting nature and its intersections with humanity," and "painstakingly executed paintings and watercolors of the phenomena of natural history."

The interaction was as timely as it was interesting. Alexis knew little about the techniques of genetic modification or genomics, but was—and continues to be—a superb natural history artist. Meanwhile, I had just begun to curate an exhibition on the human genome and genetic technology at the American Museum of Natural History entitled "Genomic Revolution." Thus began our relationship: an artist and a scientist talking every Thursday afternoon about genetic technology over coffee in his studio.

Some of the topics that Alexis wanted to discuss seemed pretty bizarre. However peculiar the topic, I would first try to explain the technology to Alexis and then would add an extra layer of science once we felt comfortable with the primer and its jargon. Throughout this process, it became immediately obvious to me that as he was soaking up the material, Alexis was worried that some of the genetically modified versions of animals would impact our natural world.

After about two months of my genomics "tutorial," Alexis dismissed me and began work on a piece of art he was later to call "The Farm." I left the last meeting with some apprehension about the art that might come from our conversations. While I felt that Alexis possessed a firm general understanding of genetic technology, I was—and continue to be—wary of how an artist or an author might take creative license with science.

While walking in the SoHo neighborhood of New York City one fall day in 2000, I looked up at a huge billboard at the intersection of Lafayette and Houston Streets. The billboard stunned me. Erected by an organization called DNAid, it featured Alexis's "The Farm" in all of its glory. Since I had only seen sketches of some of Alexis's ideas, I was blown away by the immensity of the piece, by its vividness and candor. Through his strong understanding of natural history, Alexis strove to create an awareness about the existence of plant and animal ancestral forms among his audience. More specifically, Alexis wanted his audience to understand that all living organisms—not just plants and animals—have ancestral forms. To facilitate this understanding, Alexis painted certain domestic animals while including the "ancestral" versions of them. Hence, we see chickens, swine, cattle, wild mice, and domestic crops in the background of the painting. It is purposefully ironic that each of these domestic forms has its wild form that existed probably at most 20,000 years ago, when domestication began.

In the foreground, a slew of genetically modified organisms lurk near a barbwire fence. All of these peculiar creatures came from Alexis's thinking about the extent and limits of genetic modification. The painting includes an interesting menagerie with plants, such as tomatoes, grown in cube-like shapes; a mouse with an ear growing off of its back; a rather porcine pig with human organs growing inside it; and a large cow I can only describe as "Schwarzeneggerish." As bizarre as the painting's modified organisms look, they were, as Alexis suggests in the description that accompanied the piece, informed by reality.

I thought it might be interesting to look at these four modified organisms a decade later to see how well-informed the artist was in drawing them and what their status is now. Let's start with the geometrically bizarre domestic plants. I recall from our conversations that Alexis was already aware of "Flavr Savr," one company's attempt to genetically modify tomatoes to maintain freshness, but he was particularly taken aback by the possibility of genetically modifying things to change their shape. The square tomatoes in "The Farm" would be much more easily and efficiently packaged. Alas, Flavr Savr went bust around the time Alexis produced the piece, and to my knowledge no genetically modified cube tomato has been produced.

Perhaps the most peculiar animal in the piece is the mouse with a human ear attached to it. I recall that Alexis and I had discussed the potential of using non-human animals as culturing media for human organs. In 2000, this idea was very prevalent in the news, so I didn't label this fascination as "bizarre"; rather, I thought that his questions about the topic were timely and warranted. In fact, the pig with the human organs growing inside it was also a popular news story at that time.

The "earmouse" (also known as the Vacanti mouse) actually did not have a human ear growing out of it. The "ear" consisted of a gob of cow cartilage grown in the shape of an ear. It was produced not by genetic engineering, but by inserting a polyester fabric that had been soaked with cow cartilage cells under the skin of an immunocompromised mouse. Pigs as donors of human organs, meanwhile, are something we may see in our lifetime. Some pig organs, including their hearts, are about the same size and have the same general plumbing as human organs, and scientists have suggested that pigs might be a good source of organs for human transplants. Like any transplantation, though, tissue rejection is an important consideration, and some genetic modification of the pigs to overcome rejection would be needed. While this might seem farfetched—especially when Alexis produced "The Farm," over ten years ago—the possibility has been resurrected as a result of work in 2009 in China producing pig stem cell cultures. Such cultures can be used as an easier method to genetically modify pigs to circumvent the rejection problem. A year later, Australian scientists genetically modified a line of pigs by removing a stretch of a single chromosome in order to alleviate the rejection problem and allow experiments with lung transplants to proceed.

The last animal in Alexis's menagerie is the Incredible Hulk Cow. "Double-muscled" cows do exist, such as the Belgian Blue and the Piedmontese. These breeds have a defective myostatin gene which would otherwise, when expressed normally, slow muscle cell growth in a developing cow. However, these breeds came about through conventional breeding, not genetic modification. To my knowledge, super-muscular cows have not been successfully engineered, although researchers are working on several other bovine genetic engineering tricks, including cows with altered genes to improve the conversion of milk to cheese or engineered for resistance to mad cow disease.

While only one of the bioengineered animals in Alexis's menagerie still exists today—pigs being developed as organ donors—ten years later, Alexis's "The Farm" still makes a compelling statement about technology and nature. We are in the midst of our second generation of genetic modification (the first being the initial domestication of plants and animals), and we have not yet figured out how to proceed. Alexis draws attention to the fundamental novelty of genetic engineering in depicting the progression of domesticated animals as beginning with already-domesticated forms, subtly omitting their wild ancestors.

All of this reminds me of a conversation that I had with an uncle who is a farmer in upstate New York. Over a beer in the shade of a barn, we were talking about how farming has changed since he was a young man. During the conversation, I excitedly tried to explain to him the possibility of using plants with a genetically engineered gene involved in stress tolerance. This gene would be linked to a luminescent beacon, so that when the plants were under drought stress, the beacon would glow, telling the farmer to water it. My uncle looked at me incredulously and said, "Now how good of a farmer would I be if I couldn't tell my crops needed water?"

Biotechnology and Milk: Benefit or Threat?

By Michael Hansen

Michael K. Hansen, PhD, *is a senior staff scientist with Consumer's Union where he works primarily on food safety issues. This article is excerpted from the "Executive Summary of Biotechnology & Milk: Benefit or Threat?" copyright © 1991 by Consumers Union of the United States, Inc., Yonkers, NY 10703-1057, a nonprofit organization, written by Michael Hansen, PhD. Used with permission from Consumers Union for educational purposes only. No commercial use or reproduction permitted (www.ConsumerReports.org). This article originally appeared in* GeneWatch, *volume 17, number 1–2, March 1991.*

Within the next year, Americans may be drinking milk "with a difference." A synthetic growth hormone produced by genetically engineered bacteria that stimulates milk production is already being used experimentally. Its approval for commercial use is expected soon. It will be the first major and—according to many—most important agricultural biotechnology product to hit the market. It is expected to have annual sales of $300 to $500 million in the US and $1 billion worldwide.

The product, however, has generated heated controversy from the start. Opponents and proponents have different names for it, the former calling it bovine growth hormone (BGH) and the latter calling it bovine somatotropin (bST). We will get around this problem by using both names, "BGH/bST."

A cow's pituitary gland normally makes BGH/bST and secretes it into the cow's bloodstream. The hormone has a variety of effects, one of which is to stimulate milk production.

Scientists have now developed techniques whereby, through genetic engineering, bacteria can be "programmed" to produce BGH/bST. The current consensus is that injecting cows with synthetic BGH/bST can raise milk yields by an average of 10 percent to 25 percent. How much it would increase the total US milk supply depends on how fast farmers begin to use it. Any increases in milk production would add to existing milk surpluses, which are bought up by the federal government under the dairy price support system.

At present, four US companies are seeking approval from the US Food and Drug Administration (FDA) to commercially market synthetic BGH/bST: Elanco (a division of Eli Lilly), Upjohn, American Cyanamid, and Monsanto. These companies have contractees and/or consultants in some twenty-two US university dairy science departments, and twenty thousand cows have been treated experimentally.

Synthetic BGH/bST has been banned in three Scandinavian countries and parts of Canada. The US FDA decided in 1985 to allow milk from experimental BGH/bST-treated herds to enter the general milk supply. However, in April 1990, two major dairy-producing states—Wisconsin and Minnesota—enacted temporary bans (of at least one year) on the commercial use of synthetic BGH/bST. As of early 1990, several dozen major supermarket chains, dairy cooperatives, and other food industry firms said they would not accept milk from BGH/bST-treated herds, because of consumer concerns.

Consumer Impact

Proponents of BGH/bST argue that it will lead to lower milk prices for consumers. They theorize that BGH/bST will increase production enough to cause a drop in the federal dairy price support which, in turn, will lead to cheaper retail prices. However, although the support price for milk has fallen because of dairy surpluses, since 1985, the average price of a half-gallon of milk has increased, from $1.11 in 1986 to $1.43 in 1990, a 29 percent price rise. Because of a myriad of intervening factors, including marketing orders, state and local regulation, and actions and decisions by wholesalers and retailers, lower support prices do not seem, at the present time, to lead to lower milk prices at the supermarket. Furthermore, since the 1990 farm bill set a floor on dairy price supports, it is even less likely that increased milk supplies will lead to reductions in retail price. Thus, use of BGH/bST cannot be said to lead to lower prices for the consumer.

Meanwhile, at least eight studies have found significant consumer concern about use of BGH/bST in the dairy industry. Most of the consumers surveyed in a 1986 study financed by the National Dairy Board said that, if given a chance, they would buy milk from untreated cows over milk from BGH/bST-treated cows, even if they had to switch brands or pay more. The vast majority of consumers surveyed favored labeling of dairy products from BGH/bST treated cows in all of the studies that posed that possibility.

In late 1985, the FDA decided to permit milk and meat from experimental herds to be sold for human consumption based on a finding that there would be no adverse human health effects. This finding was delivered prematurely, and the FDA did not adequately address several major human health questions regarding BGH/bST use. The most critical of these relate to possible elevated levels of a substance called IGF-I and antibiotics in the milk and meat of treated cows. (See "Effects on Human Health.")

Effects on Human Health

The potential human health effects of consuming milk and meat from BGH/bST-treated cows are hotly debated. Bovine growth hormone is secreted into the cow's bloodstream and acts on a number of different cell types throughout the cow's body. It increases overall metabolism, stimulates bone growth in young animals, muscle growth in growing animals, milk production in adult females, and feed intake. Its mode of action on many of the cells involved in these processes is indirect: BGH/bST acts by changing the levels of another class of hormones, the somatomedins. Among these is insulin-like growth factor, or IGF-I. This hormone is known to be increased in BGH/bST-treated cows and in the milk produced by them. Furthermore, IGF-I from cows and humans are identical biochemicals and can act interchangeably.

Many important questions raised by these facts remain untested and unanswered. Once ingested by humans, are these hormones resistant to digestion by enzymes and physiological conditions of the human digestive tract? It has been shown that IGF-I can affect the metabolism of epithelial cells grown in the laboratory; does IGF-I in the human intestinal tract cross the epithelial cell layer and affect the metabolism of those cells which line the intestinal tract? Such questions as these can be addressed experimentally.

Another set of concerns rises from indirect effects of the treatment of cows with BGH/bST. It has been observed already that high-producing milk cows, bred over the last fifty years, show increased susceptibility to infectious diseases. BGH/bST could lead to increased antibiotic use, which in turn could pose direct or indirect risks to human health. Trace levels of antibiotics have been detected in milk and meat from such animals, and these levels pose hazards to human health in two ways: many people are allergic to certain antibiotics at extremely low levels, and cows can accumulate antibiotic-resistant microorganisms which can be transferred to humans through milk and meat and cause serious infections which are difficult to treat due to multiple antibiotic resistance. Indeed, several incidents have been documented which illustrate such a scenario. For example, *Salmonella* resistant to ampicillin, carbenicillin, and tetracycline, was traced to meat from animals slaughtered in South Dakota, after causing illness in a number of people, one of whom died.

To our knowledge, FDA has not collected the data necessary to answer the questions raised by IGF-I. The FDA should therefore reopen its investigation of the human food safety of BGH/bST use.

The other major human health concern is whether BGH/bST injections will lead to increased health problems in cattle, with consequent increased use of antibiotics. Widespread antibiotics use in food-producing animals is of concern because it leads both to more antibiotic residues in milk and to the development of antibiotic resistance in bacteria that can infect cows and people. Since the evidence has not all been published, we cannot say for sure whether synthetic BGH/bST will increase disease rates in cattle. Some published studies have not found an increase; others have. Unfortunately, virtually none of these studies were designed specifically to look at disease rates; rather, the studies were designed to assess efficacy.

FDA should not approve BGH/ bST on animal safety grounds if in fact data shows that BGH/bST test herds have higher disease rates than untreated herds (leading to more antibiotic use) or have other significant health problems.

Economic and Social Impact

The US dairy industry has been so successful that during 1980-85, the US government milk price support program was spending an average of $2.1 billion a year to purchase milk and dairy products for which dairy farmers could find no other market. The glut of milk led Congress to enact a voluntary dairy herd termination program in 1985. The government paid farmers to slaughter their dairy cows and agree to refrain from dairy farming for at least five years. During 1986 and 1987, 14,000 farmers participated in the program and 1.55 million cows were slaughtered. The whole-herd buyout program cost

$1.8 billion. During 1987-89, the government paid between $600 million and $1.3 billion a year to purchase surplus milk.

A number of studies have examined what will happen to dairy farmers if BGH/bST is approved. Although the studies vary significantly in the magnitude of the effects they predict, virtually all have found that BGH/bST use will increase the milk supply, which will drive down support prices paid to farmers, which will drive some farmers out of business. Most have found that small farmers, who are located principally in the Northeast and upper Midwest, will suffer disproportionately.

The conclusions of the various studies all depend on the assumptions made in the models. Basically, those studies that assumed larger effects on yield, and widespread and rapid adoption, were the ones that predicted the largest impact on dairy farmers. Changes in the federal dairy price support program in 1990 might lessen the impact on small dairy farmers of BGH/bST use if the resulting dairy surpluses are small, although not if surpluses are significant. On the other hand, these changes could increase the costs to the federal government. Under the 1990 provisions, a 10 percent surplus in the milk supply induced by BGH/bST use could cost taxpayers $1.75 billion in the first two years after FDA approval, beyond what taxpayers would otherwise pay for the dairy price support program.

In view of the possible economic and social costs of approval of BGH/ bST and the absence of any governmental body charged with considering such costs, Congress could commission studies of: 1) the possible costs to federal programs, including the dairy price support and social welfare programs, of approval of BGH/bST, as well as any possible benefits (such as to the tax base); 2) the impact on dairy farmers and rural economies in dairy states of approval of BGH/bST, taking into account changes in 1990 in the dairy price support program; 3) the impact of introduction of BGH/bST on the consumer price of milk.

Environmental Impact

BGH/bST use also raises environmental impact questions. It appears fairly likely that BGH/bST use will result in increased pollution of surface and ground water from urine and manure generated on large-scale confinement feedlots.

When the FDA decided to allow large-scale testing of BGH/bST in 1985, it prepared no environmental impact statement, based on a "Finding of No Significant Impact" to its action. It kept secret this "finding" until forced to make it public by a court suit brought under the Freedom of Information Act. This "finding" completely ignored some issues, and dealt wholly inadequately with others. Before giving its final approval to BGH/bST, the FDA should prepare and seek public comment on a full-fledged environmental impact of approving BGH/bST, including impacts related to possible increases in the number of confinement-type farms in the dairy industry, and possible environmental release of BGH/bST-producing bacteria from manufacturing facilities via waste water or worker exposure.

Impact on World Hunger

Corporations developing BGH/ bST claim that it will help feed the developing world's growing population. The marketing and testing of BGH/bST in the developing world is inappropriate for many reasons, however. Increased production of milk in the developing world may subvert cheaper traditional protein sources, leading to conversion of land to pasture and the feeding of grain to cattle, rather than people. It should also be noted that a large portion of the world's adult population is presently unable to drink milk due to lactose intolerance.

Animal Welfare

Animal welfare advocates have raised ethical questions about treating cows as "milk machines." One area of concern is the overall treatment of the animal and the management techniques encouraged by BGH/bST. Economic studies predict that BGH/bST will accelerate a shift in production from smaller farms in the Northeast and Midwest to larger farms in the Southwest. Such farms are far more prone to use the capital-intensive confinement systems characteristic of "factory farming" than are smaller-scale farms.

Freedom of Information

The FDA has so far kept confidential all the studies on which it is basing its evaluation of BGH/bST's safety. An article in *Science* magazine defending the FDA's decision-making presented only summary data, which precludes independent review. The law requires the FDA to keep confidential only bona fide trade-secret information. To retain the public's trust, the FDA should make all the human and animal safety data on which it is basing its decisions on BGH/bST available to the public.

Broader Implications

Much of the public anger over BGH/bST comes from a sense that the public has little control or even say over what new technologies are developed and introduced. The consumer surveys to date show that people do not want milk from BGH/bST-treated cows. Many dairy farmers—even large ones who stand to benefit from BGH/ bST—are concerned that BGH /bST use will seriously erode consumer confidence in milk as a natural, safe food and lead to reduced sales. It seems as though the only ones to benefit from synthetic BGH/bST will be the chemical companies that sell it and some of the farmers who are first to use it. Yet no one is looking at who loses and who gains from this product's introduction.

There is also the basic question of whether society needs more milk in the first place, and if so, whether other techniques would be more desirable.

Critics have raised similar questions about the social and economic impacts of other biotechnology products, asking whether bioengineering is the best way to meet particular societal needs, or whether a more traditional approach might be more appropriate. Such critics have been labeled anti-technology Luddites who are undermining US efforts to remain competitive in the world market. But this does not necessarily follow. In some cases a new technology may improve our lives; in other cases it may not.

In the Bullpen: Livestock Cloning

By Jaydee Hanson

Jaydee Hanson *is senior policy analyst at the Center for Food Safety. This article originally appeared in* GeneWatch, *volume 26, number 1, January–March 2013.*

On February 22, 1997, a team of Scottish scientists announced that the world's first cloned mammal, a sheep cloned from an adult cell, had been born the previous July.[1] The scientists cloned a ewe by inserting DNA from a single sheep cell into an egg and implanted it in a surrogate mother. Actually, cloned embryos were placed in 277 surrogate ewes and the sheep, called Dolly, was the only success. Dolly was cloned as a way to help copy sheep that had been genetically engineered to produce human proteins in their milk or blood. In effect, cloning was intended to help make pharmaceutical drugs.

By 2006, it was clear that a number of cattle, pig, and goat breeders were planning to introduce cloned animals and their offspring into the food market. The US Food and Drug Administration drafted a risk assessment on the use of clones for food in 2006 and eventually approved their use for food in 2008. However, the FDA risk assessment relied on minimal research for their approval of the meat and milk of clones for food.[2] The FDA assessment found no peer-reviewed studies on meat from cloned cows or on milk or meat from their offspring. No peer-reviewed studies were found on meat and milk from cloned goats or their offspring either, nor for pigs. The three peer-reviewed studies on milk from cloned cows showed marked differences in milk from clones that should have prompted further research. Nonetheless, based on submissions of data from two cloning companies, the FDA approved food and milk from cloned cattle, pigs, and goats as safe for human consumption. Ironically, while the first mammal cloned was a sheep, the FDA recommended against approving meat or milk from sheep due to lack of data.

After the approval of meat and milk from cloned cattle, goats, and pigs, the US Department of Agriculture asked cloners to voluntarily keep such products off the market. The US Congress requested a study of the economic effects of cloning on livestock producers, but USDA has yet to release the report. Our research at the Center for Food Safety has found that at least six US sellers of bull semen for the artificial insemination of cattle are advertising that they have cloned bull semen for sale. While there is no tracking of the number of cloned bull semen straws that have been sold, we assume that a small portion of the cattle sold in the US are the offspring of clones. While cloning is sold as a way to improve cattle and pig genetics, most of the US pedigrees now require the cloning status of an animal to be listed in its pedigree. The three breed associations with the most cloned animals—Angus, Hereford, and Texas Longhorn—all require cloning status to be listed in the pedigree. In Europe and Canada, most breeding associations simply prohibit the listing of a cloned animal in the pedigree of the breed.[3]

Who Is Doing the Cloning?

Most of the cloned animals produced in or for the US market are cloned by two companies: Viagen, a Texas company owned by billionaire John Sperling,[4] and Cyagra, formerly a US company but now owned by Argentina's richest man, Gregorio Perez Companc.[5] Viagen claims that it has produced a total of 1,000 clones, mostly cows but also racing horses and pigs. Japanese company Kirin Pharmaceuticals now owns another US cloning and animal genetic engineering company, Hematech. Its scientific officer, James Robl, told me that while the company has an application before the FDA for approval of its cloned and genetically engineered "mad cow" resistant cow, it plans to use that animal only to produce pharmaceutical drugs, not for food. The next generation of clones could come from wealthy Chinese corporations. The Beijing Genome Institute, a powerhouse in DNA sequencing, has set up a pig cloning company that is working on improving the efficiency of the cloning process.[6]

Regulating Animal Cloning for Food

Only the United States has explicitly approved animal clones for food. The European Union has had a significant discussion about whether food from clones and their offspring should be approved. The agriculture ministers of most EU countries have called for approving meat and milk from clones and their offspring.[7] The European Parliament, on the other hand, has called for an outright ban of clones and their offspring in food and forced a stop to efforts by the European Commission to approve clones for food. The Health and Consumers agency of the European Commission conducted a public comment period last fall and is expected to issue a report soon.[8] That report will likely engender another international discussion of animal cloning for food.

Conclusion

Comparatively few cloned animals or their offspring have entered the US or European Union markets. It is likely that imports of meat from Argentina or China will be the source of cloned meat products in the future. The US should take steps to be able to track clones and their offspring. At the very least, the National Organic Program should implement the 2007 recommendation of the National Organic Standard Board that clones and their offspring should be excluded from USDA certified organic products, so that consumers who want to avoid clones and their offspring in meat and milk products would have a viable option.

Canada Banned BGH!

By Kimberly Wilson

Kimberly Wilson *is an author and activist. For more information about the Canadian Health Reports visit http://www.hc-sc.gc.ca/english/archives/rbst. This article originally appeared in* GeneWatch, *volume 12, number 1, February 1999.*

Canadian milk consumers can breathe a little easier now that Health Canada, a regulatory branch of the Canadian government, has decided not to allow dairy cows to be treated with Monsanto's Bovine Growth Hormone. Monsanto's Bovine Growth Hormone (BGH), or Bovine Somatotropin (BST) is a genetically engineered hormone which tricks a cow's body into producing more milk than it otherwise would. Milk from cows treated with BGH has been known to be contaminated with pus from udder infections, with antibiotics administered to stem those infections, with foreign growth hormones, and with high levels of insulin-like growth factor (IGF-I) which has been linked to human breast and gastrointestinal cancers.

The decision to ban the use of BGH was based on animal-welfare issues such as the increased occurrence of udder infections and decreased lifespan of cows treated with BGH. Monsanto's application was controversial from the start more than eight years ago, when consumers and farmers alike opposed the use of BGH. Since then, profuse claims of corruption have circulated, including a recent claim that Monsanto offered HealthCanada scientists research money in exchange for their approval of BGH. The issue was brought to a head in late 1998 when six scientists at HealthCanada produced a report aptly titled the "Gaps Report," which chronicles the history of Monsanto's application within HealthCanada, and cites the many missing documents, or "gaps" in the normal approval process, including faulty experiments, and incomplete data.

GeneWatch obtained a copy of the Canadian "Gaps Report," which states that "The manufacturer of the product [Monsanto] did not subject it to any of the normally required long-term toxicology experimentation and tests for human safety."[1] The report also suggests that Canadian regulatory officials, made an exception for Monsanto by not requiring that it submit the appropriate tests to make sure that BGH was safe for human consumption. The report has called into question not only Monsanto's unconscionable application, but the conduct of senior officials at both the US FDA and the Bureau of Veterinary Drugs (BVD-Canada) who seemed to show the application unusual favor. The gaps report ends with the following statement, "Duly arising from this particular issue certain senior officials of both these agencies have allegedly been asked to be investigated for employing unauthorized influence against subordinate staff and a personal conflict of interest." Upon its release, the Gaps Report caused such a furor that senior officials at HealthCanada refused to release the report in its entirety and legally silenced the scientists who created the report. In response, the Canadian Senate Agricultural Commission began an investigation of the issue, and shortly thereafter HealthCanada released its decision to ban BGH.

Food and Drug Amalgamation

By Eric Hoffman

Eric Hoffman *is the former Biotechnology Policy Campaigner at Friends of the Earth US. This article originally appeared in* GeneWatch, *volume 24, number 2, April–May 2011.*

The environmental dangers posed by the FDA's approval of genetically engineered salmon for human consumption were highlighted in the fall issue of *GeneWatch*.[1] Unfortunately, a whole herd of genetically engineered animals are in the works. If we do not fix our inadequate regulatory system now, we could face a host of irreversible risks in the future.

In the US, the process of regulating biotechnology comes from the "Coordinated Framework for Regulation of Biotechnology," which was created in 1986 to prevent biotechnology-specific regulations from being written. In their place, agencies were asked to use current rules and find ways to apply them to biotechnology products. The coordinated framework also designates which US agencies will oversee which products. For example, the USDA approves genetically engineered (GE) crops before they can be planted, while the FDA governs GE crops once they leave the farm. The FDA also regulates GE animals for the production of food and pharmaceuticals. This has led to an absurd status quo, with the FDA approving GE animals not as new foods, but as new animal drugs.

The FDA defines a "drug" as something that is intended to affect an animal's structure or function. For GE animals, recombinant DNA (the engineered genes) qualifies as a "drug" under the FDA's definition. It is important to note that in the case of GE salmon, this "drug" does not improve the health of the salmon or the people consuming it. In fact, the rDNA construct that produces growth hormone year round is responsible for a number of adverse effects, such as jaw erosion and other physical abnormalities in the salmon, the potential for increased allergenicity among human consumers, and lower ratios of Omega-3 and -6 fatty acids in the meat—even though the presence of these healthy fats is one of the primary reasons many people choose to eat salmon in the first place.

The only reason GE salmon is being proposed as a "drug" is to allow the company that produces it to avoid the more stringent regulations that it may be subjected to if the GE salmon were defined as "food." Yet the salmon's recombinant DNA makes for a poor drug, as it does not appear to provide any benefit to consumers or the environment. Like trying to fit a square peg into a round hole, the FDA is trying to force a genetically engineered food product into a regulation written for drugs.

Using the New Animal Drug (NAD) process to approve GE animals is also inappropriate, as it only compares the risk of GE salmon compared to eating non-GE salmon. The process fails to properly look at the major impact GE salmon will have on the environment or human health. It does not look at the impact GE salmon farming will have on fishing communities or the impact expanding use of GE salmon will have on the production and consumption of these fish. The NAD process does not require a proper cost-benefit analysis, nor does it look at alternatives to using GE fish entirely. At the heart of the issue, the NAD process fails to look at GE salmon approved for food as food and slyly tries to get these fish onto our plates without the proper precautions or environmental review.

While this "frankenfish" is unsettling to most Americans (91 percent of whom don't want this fish to reach their plates), the trouble with GE animals does not end there. The same company that is marketing this GE salmon, AquaBounty Technologies, also has GE tilapia and GE trout waiting in the pipeline for approval. Another company, Hematech, is working on GE cows that theoretically cannot get mad cow disease, which begs the question: Is it really easier to alter the genome of a cow to stop producing prions in its brain than to simply stop feeding cows dead animal brains (which is how mad cow disease is spread)?

Other companies are looking into developing GE chickens that are unable to transmit bird flu, and recent reports from China revealed that researchers have created a GE cow that makes "human breast milk" instead of cow milk. And of course there is the "Enviropig" from a Canadian university, a transgenic pig that produces less phosphorus in its waste—allowing industrial animal factories to shove even more pigs into their tightly confined feedlots while continuing to pollute as much as they do today.

If the FDA approves the GE salmon for human consumption, it will open up a floodgate of GE animals onto our plates—and it is likely GE food products will not even be labeled as such. These approvals will all happen under the illogical regulatory framework of approving GE animals for food as "animal drugs" and not the foods they truly are.

So where are we now? The US Food and Drug Administration's Veterinary Medicine Advisory Committee held a public hearing on the approval of GE salmon last September. Since then, the FDA has remained silent on the issue as it finalizes its Environmental Assessment, which will be posted for a thirty-day comment period sometime soon (and will likely be completely inadequate).[2] Approval will shortly follow this Environmental Assessment if the FDA does not find any significant environmental harms.

Since the FDA's hearings on the GE salmon, two federal bills were introduced in both the House and Senate by Senators Mark Begich (D-Alaska) and Lisa Murkowski (R-Alaska) as well as Representative Don Young (R-Alaska). These bills would ban the approval of GE salmon or require labeling if the fish is indeed approved. The current legislation proposed in Congress is important but the bills, if passed, are only a temporary fix for the larger problem of a completely inadequate system surrounding GE animal regulation.

What would the proper oversight of GE animals look like? Below is an outline of what regulations on GE animals that protect human health and the environment would entail.

Legislation on the approval, use, and commercialization of genetically engineered animals must:

1. Encompass all GE organisms for human or animal food, or food and feed containing GE organisms.
2. Require independent and comprehensive risk assessment of GE organisms, including:
   a. Analysis of the engineered genes and their long term stability over multiple generations;
   b. Safety of eating the GE animal or products from the GE animal;
   c. Comprehensive and independent environmental impact reviews, including whether approval of any animal with a wild relative that lives in or could ever enter US jurisdiction;
   d. Assess the economic harms likely to be caused by any approval; and

   e. Assess the health and welfare of all GE animals over the lifespan of the animals.

3. Require labeling of GE animals for:
   a. All products intended for human or animal consumption that have been genetically engineered;
   b. Any foods containing any amount of GE product; and
   c. The creation of a tracking system of GE animals and products through the food supply.
4. Mandate transparency in the approval process with adequate time and ability for public participation.[3]

Until such regulations are put in place, all decisions on the approval of GE animals for food must be stopped. Our current way of regulating GE animals as "new animal drugs" is nonsensical and does not require the proper analysis of risks to human health, the environment, the health of wild-type populations related to the GE animals, the economic impact these GE animals may have, or the level of transparency needed to guarantee a reasonable level of public participation in the decision-making process.

Food Unchained

By Samuel W. Anderson

Samuel W. Anderson *is editor of GeneWatch. This article originally appeared in* GeneWatch, *volume 24, number 2, April–May 2011.*

More than 10,000 years ago, humans began to domesticate animals. Livestock—sheep, goats, and cattle in Southwest Asia, pigs in current-day China—began to replace hunting as the primary source for meat and skins. Humans found new uses for animals: collecting milk from lactating ruminants and eggs from poultry; harnessing cattle, donkeys, and water buffalo to plows; shearing the wool from sheep and alpacas; climbing onto the backs of horses and camels for personal transportation; and, of course, producing biopharmaceuticals using proteins extracted from the milk of transgenic goats.

If that last item sounds a bit abrupt in the chronology of animal domestication, that's because it is. Even after thousands of years of selective breeding, farmers and animal science researchers in the twentieth century found plenty to improve upon. The establishment of new and enhanced livestock breeds moved at an astonishingly rapid pace, as developments that might have taken centuries were accomplished in decades. Yet, while modern farmers and researchers possessed tools unavailable to their forbears, they still relied on selective breeding to achieve genetic upgrades.

In the 1980s, the advent of recombinant DNA technology appeared to herald a new age in animal agriculture, allowing the engineering of specific genes and selective introduction of novel traits. Scientists began creating transgenic animals that served as better lab subjects such as "knockout mice," which were engineered not to express a certain gene. They enhanced animal-generated biotherapeutics, such as pharmaceutical proteins from sheep's milk (first at Roslin Institute in 1989). Scientists developed animals into future organ donors (usually attempts at pigs modified to grow specific organs compatible for transplant into humans). They even created novelty pets such as the fluorescent GloFish.

This period also marks when researchers began attempting to apply germ-line engineering to improve animals' food production. Unlike other types of transgenic technologies, most of which aim to adapt animals for novel uses, attempts to upgrade livestock-based food production through transgenics pits genetic engineering against the time-tested mechanisms of selective breeding that had gradually honed those same characteristics over millennia. Humans have been developing specialized varieties of sheep for at least six to eight thousand years, selecting for some of the same traits—fast growth rates, feed conversion efficiency—that the Roslin Institute has tried, mostly in vain, to improve through genetic engineering.

Meanwhile, when other researchers at Roslin began engineering transgenic sheep that produce biotherapeutic proteins in their milk, they were attempting something that couldn't be done through conventional methods. And unlike bioengineered food animals, "pharming" reached the market and is being adopted in new forms. GTC Biotherapeutics was the first to get regulatory approval for a pharmaceutical produced by a transgenic animal with ATryn, an antithrombrin derived from the milk of transgenic goats. Other companies have since developed goats, cows, and even rabbits that can produce various therapeutic proteins in their milk. The end product may not be novel, but the means certainly is. Most of the biotherapeutics being produced through animal pharming are already commercially available through other production means, but the strength of pharming is its potential to manufacture the same product at a significantly lower cost. A few hundred of GTC's goats can produce as much antithrombrin proteins as a lab that costs millions of dollars to set up and millions more to scale up.

The same cannot always be said of transgenic food animals. In many cases, germ-line engineering of a food-producing animal may be used to attempt to extend the aims of conventional breeding; in other words, to improve food production. These traits can be improved without genetic engineering, of course, and have been for thousands of years, but genetic interventions can produce much more drastic results—for better or worse. AquaBounty's genetically modified salmon grows far more rapidly than its conventional counterpart, and although this "improvement" raises a host of serious concerns, from a strictly production-oriented standpoint, it could be a boon to some fish farmers (and certainly to AquaBounty). On the other hand, when the United States Department of Agriculture funded the development in the 1980s of pigs carrying the human growth hormone in an attempt to create a faster growing, leaner meat animal, the results yielded only a sickly litter afflicted with an array of odd conditions, including pneumonia, peptic ulcers, and arthritis. Of the nineteen now infamous "Beltsville pigs," seventeen died before reaching one year of age.

These ventures may have resulted in an animal welfare fiasco and the very real threat of a catastrophic disruption of ocean ecosystems, but we can at least see why they were carried out. The achievement—or intent, in the Beltsville pigs' case—was to more efficiently convert grain to meat, doing so with one radical improvement that may have taken many years to accomplish through conventional breeding.

Many other endeavors to improve livestock food production through germ-line engineering appear redundant or superfluous, which may explain their tendency to either fail or fizzle out. Ironically, many of these projects also receive the most press, if only for their novelty:

- "Enviropig," developed by researchers at Ontario's University of Guelph, is able to digest a form of phosphorous in feed grains that it would normally excrete, reducing the phosphorous levels of its manure by 30 to 70 percent, with the aim of diminishing the environmental impact of large-scale hog production.
- Akira Iritani, a scientist at Japan's Kinki University, reported in 2002 that his team was the first to successfully add a functioning plant gene to an animal, in the form of pigs that carried a spinach gene. As a result, Iritani said, the pigs' carcass held 20 percent less saturated fat, converted by the novel gene into linoleic acid.
- In the 1990s, British researchers began attempts to develop transgenic sheep resistant to scrapie, a prion disease similar to bovine spongiform encephalitis (or mad cow disease), which is 100 percent fatal in sheep. Research has also been undertaken to develop cows resistant to mad cow disease, so far with no published success.

The prevailing flaw in these technologies—presuming they were successfully developed—is that while the method may be novel, the result is not. A surprising amount of research on transgenic food animals has been bent on achieving what can already be accomplished more gracefully through conventional breeding, altered production practices, or human behavioral adjustments.

In many cases, all of the time and money spent developing a new transgenic livestock breed serves only to replace an existing solution, even if it is more efficient, effective, and sustainable than a genetically engineered silver bullet.

Take the above examples:

- "Enviropig" is awaiting approval for human consumption in Canada and the US, and already has the green light from Canada's Department of the Environment and the blessing of swine industry groups. Yet, it represents an incomplete solution (phosphorous is not the only problem nutrient in pig manure) to a problem that already has solutions at hand. Unfortunately, those alternatives—changing the pigs' rations by adding an enzyme (phytase, which can reduce phosphorous in manure by over 50 percent) or using different grains, being more careful and strategic when spreading manure on fields as fertilizer, and most of all, ditching the vast 10,000+ hog confinement operations in favor of smaller, diversified farms—require behavioral changes in the hog industry, as opposed to maintaining the status quo with new pigs. Enviropig has been framed by its creators and the swine industry as an environmental breakthrough, but from the perspective of environmental protection, it addresses a problem that already had known solutions. In reality, despite the name, Enviropig was designed to solve hog industry problems. It reduces the amount that producers need to spend on mineral supplements, but more importantly, it allows the hog industry to appease regulators and scale up operations without changing the prevailing practices.
- Iritani's "Popeye pig," with apparently 20 percent of its saturated fat converted to healthier unsaturated fats by an inserted spinach gene, never resurfaced after it was announced in 2002. At the time, Iritani essentially admitted that he did not expect the pig to be commercialized due to lack of public acceptance, but he also expressed his hope that "safety tests will be conducted to make people feel like eating the pork for the sake of their health." The notion of encouraging people to eat pork for their health may raise an eyebrow; beyond that, one need not think too hard to come up with simpler ways to cut down on saturated fat (trim it off of your pork chop, or simpler yet, cut back on the meat).
- Attempts to engineer transgenic scrapie-resistant sheep appear to have fallen by the wayside, presumably because many sheep already carry a

dominant gene for scrapie resistance, allowing producers, after sending biopsies to a lab, to select against scrapie susceptibility. Mandatory scrapie ID tags in the US and other countries have also helped track and control the spread of scrapie. The best argument that conventional breeding and well-executed containment practices preempt any usefulness of transgenic scrapie-resistant sheep is their success: Australia and New Zealand have officially eradicated the disease, and the US has reduced it to 0.03 percent of the nation's entire flock.

- Like transgenic scrapie resistance, genetically altering cows' genomes to grant them mad cow resistance is essentially a superfluous advance. While some research suggests that genes can influence mad cow resistance or susceptibility, the disease is most often believed to be contracted when brain tissue from another mad cow disease carrier (or scrapie, some studies say) enters into a cow's feed. In the US and Canada, stringent steps to keep ruminant tissue out of ruminant feeds helped essentially eradicate mad cow in North America—no transgenic influence needed.

Despite the headlines about cows that produce something akin to human breast milk or cattle genetically engineered for immunity to sleeping sickness, the most marked shortcoming of transgenesis as a means of improving food animals is evidenced by the host of experiments that don't make headlines. Attempts to create healthy, fast-growing transgenic sheep carrying the human growth hormone began in the mid-1980s. In early experiments, the transgenic lambs grew at average rate until reaching fifteen to seventeen weeks, at which point "over expression" of the growth hormone resulted in two rather counterproductive side effects: "reduced growth rate and shortened life span." Fifteen years later, growth hormone experiments in sheep had only managed to yield larger than normal sheep. At twelve months, transgenic rams were only 8 percent larger than the control rams, and with no significantly increased feed efficiencies noted.

AquaBounty's success bringing genetically modified salmon to market is, so far, an anomaly; to date, no other animal has been commercialized carrying a transgene that increases the amount of food it produces or the efficiency with which it converts feed to meat, milk, or eggs. Meanwhile, the goats, cows, chickens, and even rabbits that have been developed to produce human biopharmaceuticals are, in some cases, already proving to be the most economical producers of some therapeutic proteins. Pharming is not without its drawbacks, and the need to carefully test and regulate all products of transgenic animals is evident. Nonetheless, if a genetically modified animal is to deliver significant benefits for humans, there certainly seems to be a surer path for those using genetic engineering to coax an entirely new use out of an animal which has been selected on the basis of its existing advantageous traits; as opposed to those projects taking on conventional breeding programs at their own game—attempting, with a single transgenic silver bullet, to outshine thousands of years of purposeful selection.

Just Say No to Milk Hormones

By John Stauber

John C. Stauber *is a political activist and author of books such as* Trust Us, We're Experts *and* Mad Cow USA. *He is the founder and former executive director of the Center for Media and Democracy. This article originally appeared in* GeneWatch, *volume 7, number 1–2, March 1991.*

In the fall of 1986, dairy farmer John Kinsman was joined by environmentalist and author Jeremy Rifkin in a protest action on the University of Wisconsin campus in Madison. Rifkin, a leading critic of genetic engineering, shared Kinsman's concerns regarding bovine growth hormone, or BGH, that was being tested at the university. When Kinsman discovered that the university's dairy plant was selling ice cream made with milk from cows injected with BGH, he picketed and organized a student boycott of the ice cream. Eventually, the university dairy refused to use any more milk from BGH-treated cows.

What began in 1986 as a small protest in "America's Dairyland" has evolved into an international movement that unites family farmers, consumers, and animal welfare, environmental, and public health activists. By working together at the grassroots, these groups have legislated bans on BGH in key dairy states such as Wisconsin and Minnesota and within the European community.

These critics say that recombinant bovine growth hormone has no redeeming social value and is bad for cows, family farmers, and consumers. They have ignited a debate which extends far beyond any single product to the question of who should control animal husbandry in the twenty-first century.

Four multinational drug and chemical companies (Monsanto, Eli Lilly-Elanco, American Cyanamid, and Upjohn) have invested a half-billion dollars in developing and promoting bovine growth hormone. They view the issue differently, intending BGH to be just the first of many bioengineered agricultural products in a multibillion-dollar-a-year global market.

Healthy Cows or Healthy Profits for Biotechnology?

BGH represents much more than just the first billion-dollar product for the agricultural biotechnology industry. To begin with, there is the question of animal health. What does it do to a milk cow when she is routinely injected with a drug that forces her to produce more than she possibly could without hormone shots? According to Dr. Richard Burroughs, a former FDA veterinarian whose job it was to evaluate confidential industry tests of BGH, the hormone is detrimental to the health of the cow.

In 1989, Burroughs blew the whistle and revealed the FDA's cover-up of cow health problems. Burroughs asserts that the dosage of BGH required to produce more milk is toxic to the cow, resulting in the udder infection mastitis, developmental and reproductive problems, enlarged internal organs, and a shortened life-span.

Even if BGH had no adverse effects on cows, it would still have severe and devastating economic impacts upon America's dairy farmers. This is because BGH would increase milk production by as much as 25 percent in individual cows at a time when the price farmers are paid for their milk is already too low to cover their costs of production.

The federal government now spends billions of tax dollars purchasing surplus dairy products, yet dairy farmers are going bankrupt and the family farm is being displaced by huge milk factories in Texas and California. Studies at Cornell and the University of Wisconsin predict that tens of thousands of additional family dairy farmers would be forced out of business as BGH pushes milk production higher and farm prices lower.

BGH use would undermine consumer confidence in dairy products, thereby decreasing milk sales and further reducing farm income. Surveys conducted during 1990 at the universities of Wisconsin and Missouri show that consumers would reject milk from hormone-treated cows. Ninety-five percent of consumers in the University of Missouri study want BGH milk to be labeled. In the Wisconsin study, 67 percent of regular milk buyers would pay an average of $.22 more per half-gallon to be sure the milk they buy is not from hormone-treated cows.

These surveys reveal that the public views milk as one of the few natural and untainted foods in the grocery store. Milk is trusted. Genetically engineered milk hormones are neither trusted nor wanted. In the Wisconsin study,

71 percent of the respondents felt that, down the road, experts would discover human health problems in consuming milk from BGH-treated cows. Apparently the American consumer of the 1990s, having seen many food additives recalled for safety reasons, has developed a street-wise attitude and would rather avoid any unnecessary risks. Surveys show that unless grocery shoppers can choose between natural milk and BGH milk, they will significantly cut back their milk consumption.

Boycott BGH

The struggle to keep milk pure, natural, and free of synthetic BGH has really just begun. Consumers Union, the highly respected publisher of *Consumer Reports* magazine, evaluated BGH in December 1990. It called upon the Food and Drug Administration to halt the sale of milk from cows being treated experimentally with BGH while the agency reevaluates its 1985 decision that such milk is safe for human consumption.

But despite all the criticism it has received, the FDA is still expected to approve BGH for commercial farm use sometime in the next year. With the federal government poised to give BGH the go-ahead, a coalition of consumer, farmer, and animal welfare groups has announced a proactive boycott campaign to insure that the dairy hormone will be rejected in the marketplace.

A project of the Foundation on Economic Trends, in coalition with the National Family Farm Coalition and other groups nationwide, the Boycott BGH campaign is organizing farmers and consumers in local communities to contact dairy and grocery companies directly. Activists are asking dairy plants, dairy product companies, and grocery stores to establish a "pure milk policy," stating that they will not accept or sell any dairy products from cows injected with BGH, even if the FDA approves the drug.

The BGH boycott is using tactics that successfully removed BGH test milk from grocery shelves in Vermont and Wisconsin. In 1989, two dozen major companies—including Kraft, Borden, Dannon, Yoplait, Safeway, Krogers, Ben & Jerry's, and Dean Foods—refused to sell any BGH test milk. Now these companies are being asked to take the next step by establishing a permanent "BGH-free" policy.

Later this year, the Boycott BGH campaign will release its first Pure Milk Advisory, listing which companies have adopted a BGH-free policy and which have not. Some of the latter will be targeted for picketing, boycotts, and advertising campaigns in localities around the country.

Suing the National Dairy Board

On November 29, 1990, the Foundation on Economic Trends and fourteen dairy farmers filed a lawsuit against the Department of Agriculture. They charged that the department's National Dairy Promotion and Research Board, established by Congress to promote dairy consumption, has joined BGH manufacturers and dairy trade associations in a multimillion-dollar public relations campaign to reassure consumers that the hormone is safe.

"It really upsets me that these people deceived us and used our money against our wishes," says dairy farmer John Kinsman, one of the plaintiffs in the suit. "We are very angry they would use our money to work against us."

Last March, letters from the National Dairy Board to concerned farmers claimed that they were not promoting BGH, nor would they begin a consumer campaign until after the drug was approved, to maintain compliance with federal law. But a Freedom of Information Act request filed by Rifkin's foundation turned up documents showing that the board awarded a $1 million contract to a public relations firm last February to promote BGH. And notes from meetings held over the past two years show that dairy trade association and BGH industry representatives met with agriculture department officials to discuss ways to alleviate public fears about the hormone and to make BGH palatable to farmers and consumers.

Who will decide if milk comes from cows routinely injected with biosynthetic growth hormones? BGH is a "first," and as such is establishing critical precedents. As the *New York Times* noted on April 29, 1989, "For the first time in this century, farmers and industrial leaders are battling to control how a new agricultural technology is deployed." BGH is an important test case because its success or failure in the marketplace will affect other products in the pipeline of agricultural biotechnology and, says the *Times*, ultimately the structure of American agriculture.

Gene Technology in the Animal Kingdom

PAUL B. THOMPSON

Paul B. Thompson, PhD, *is the W. K. Kellogg Chair in Agricultural Food and Community Ethics at Michigan State University in East Lansing, Michigan. His research has centered on ethical and philosophical questions associated with agriculture and food, and especially concerning the guidance and development of agricultural techno-science. He is the author of thirteen books and editions, such as* The Spirit of the Soil: Agriculture and Environmental Ethics; The Ethics of Aid and Trade; Food Biotechnology in Ethical Perspective, *and co-editor of* The Agrarian Roots of Pragmatism. *He has served on many national and international committees on agricultural biotechnology and contributed to the National Research Council report "The Environmental Effects of Transgenic Plants."*

Novel genetic and reproductive technology is used on non-human animals, but it is not easy to arrive at a firm conclusion as to what we should think about this. On one side, there are commentators who are quite simply horrified by it. Their reaction tracks the repulsive shock and condemnation of Nazi death camps. On the other side are legions of scientific researchers and would-be entrepreneurs who clearly regard the manipulation of animals and animal tissue through genetic engineering, cloning and stem-cell technology as so utterly routine as to be without the barest trace of moral significance. I should confess at the outset that I am mystified as much by the extremity of these reactions as anything.

What Is Animal Biotechnology?

Animal biotechnology is hard to get one's mind around in part because it encompasses a vast array of different applications and techniques. At one end of the spectrum, microorganisms have been genetically manipulated in order to produce animal drugs or to simulate biochemical processes normally performed in animal tissue. Recombinant chymosin is an example of the latter. It is an enzyme essential to making cheese that was traditionally derived from rennet, which was harvested from the entrails of slaughtered calves. Recombinant chymosin is produced by a genetically engineered bacterium, and no one (so far as I know) has objected to the fact that it has virtually displaced rennet in the vast majority of the cheese that has been made throughout the industrial world. The use of recombinant chymosin is not labeled, even in the European Union.[1] If you have eaten cheese, you have eaten recombinant chymosin.

The fact that animal species have been cloned is widely known due to the celebrity of Dolly, the sheep cloned at the Roslin Institute and announced to the world in 1997. Cloning does not in and of itself involve the transfer of genetic material from one species to another. Nevertheless, cloning is widely regarded as virtually essential to the commercialization of genetically engineered food animals because of the relatively low reproductive rates of cattle, pigs and sheep, and the length of time it takes for these species to reach sexual maturity. There is thus a close tie between cloning of food animal species and the use of genetic engineering (about which, more will be said soon).

Mice are a different story. As city dwellers are likely know, the fecundity of the mouse is legendary, and the biotechnology industry has had no difficulty in commercializing genetically engineered mice and rats. These animals are used primarily for medical research. Researchers can custom order laboratory rats and mice with specific genetic disorders or other characteristics that are tailored to particular types of biomedical experimentation. Sometimes these animals are intentionally transformed so that they will suffer from a disease condition of interest to the researcher in question, and as such these laboratory animals may endure significant suffering during the course of the experiment.[2]

There have also been a number of animal biotechnology projects that produced a single genetically modified individual, rather than a line of transformed animals intended for commercial applications. These one-off animal transformations often seem to have been done just to attract attention. The artist Kac has produced a series of works in the medium of "bioart"—plants or animals transformed and displayed as expressions of aesthetic creativity. In 2000, Kac displayed Alba, a green fluorescent rabbit produced through genetic engineering. In 2012, a Royal College of Art graduate student named Koby

Barhad exhibited a mouse that had been genetically engineered so that DNA from the long-dead pop singer Elvis Presley was in its genome. Such projects illustrate the breadth of potential for genetic technologies applied to animals, but they are often justified by their creators in terms of the public discussion or comment that they generate, as opposed to any intrinsic aesthetic merit.

Animal genomes have also been transformed for a number of food and industrial applications. A few of these experiments have been widely reported. Founded in 1998, Hematech, a South Dakota–based company, has been producing antibodies that are fully compatible with human immune systems in cattle. Goats have been genetically engineered to make spider silk in their milk, with the hope that this ultra-strong material could have industrial applications if it could be obtained in larger quantities. Similarly, animals have been transformed to manufacture drugs or other biologically active substances in their blood or milk. There are also a few transformations that have food or agricultural purposes. The "Enviro-Pig" has been genetically transformed to produce phytase in its saliva, which will allow it to metabolize the phytic acid that would otherwise contribute to the pollution impact of pig manure. Neither the goats nor the pigs have so far made it through regulatory approval or found a commercial application.

The most widely known product of food animal biotechnology is the Aqua Bounty salmon, which has been transformed so that it will grow extremely quickly, if food is provided at an optimum rate. The fish is not, at this writing, approved for food consumption by the regulatory agency of any industrial country. However, the United States Food and Drug Administration has released a number of preliminary documents, including environmental risk assessments, that apparently clear the way for regulatory approval. Indeed, there has been widespread speculation that approval of this fast-growing fish is imminent for nearly ten years. Another fish biotechnology—the GloFish, a zebra fish engineered to fluoresce under ultraviolet light—has been available in pet stores since the mid-2000s. But of course, zebra fish are not to eat.

Transformation can theoretically be undertaken to address issues of animal welfare that arise in industrial production settings. In 2005, results were announced for a genetic transformation that would potentially reduce the rate of mastitis in high producing industrial dairy cows, for instance. The most recent trend has been toward applications of biotechnology that would alleviate the problems of animal welfare by producing animal products without the bother of keeping animals, at all. In 2013, Dutch scientist Mark Post announced the world's first synthetic hamburger. It was composed of meat that had been produced using a stem-cell technology under laboratory conditions. Post's approach further challenges our understanding of animal GMOs by introducing a product that is biologically classifiable as animal—it is composed of animal tissue—but that is not derived from *an* animal. There was no sentient cow or steer that was the source of Post's artificial hamburger.

The Ethics of Animal Biotechnology

These applications of gene technology raise all the same ethical issues that are raised by any other genetic technology. Some observers think that any kind of genetic manipulation crosses a moral threshold, and certainly the genetic engineering of animals would not be an exception. Other ethical concerns devolve from the complexity of gene interaction. Foods and drugs must be evaluated for safety, and there are debates about where the standard should be set. Even if safe, there will be people who do not want to use products of gene technology. Their reasons range from the "playing God" objection through concerns about risk to feelings of solidarity with small farmers and civil disobedience. Whatever the source of concern, there will be debates over labeling of these products. All the issues of intellectual property rights and the concentration of economic power are there with animal biotechnology as well.[3]

But animal biotechnology does raise a unique set of issues because of its position between biomedical applications of gene technology and the genetic engineering of agricultural crops. Unlike plants, animals have feelings. We can be concerned about their welfare. Animals can be adversely affected by a gene technology, but unlike human patients, animals cannot give or withhold their consent. At the same time, animals have been the object of routine genetic selection through conventional breeding, and millions of animals are slaughtered every year for consumption as food. To be sure, recent years have seen phenomenal growth in ethically-based vegetarianism, especially in response to philosophical work by Peter Singer[4], Tom Regan,[5] and Carol Adams.[6] There are large constituencies for animal protection, and an enormous philosophical literature in support of ending all exploitative use of non-human animals.

There is thus a solid ethical foundation on which to base ethical concern about the impact of gene technology on the animals themselves.

This ethical work oscillates between two poles. On the one hand, there is a perspective that is closely tied to Peter Singer's work on animal ethics: Animal biotechnology matters morally because it could have an impact on the animal's feelings. It could cause pain, distress and suffering, Indeed, some of the mice developed for the study of disease processes are in effect intended to exhibit the debilitating experiences of the disease. On the other hand, the work of Carol Adams and Tom Regan suggests that animals have a "nature" or an integrity to their being that biotechnology fails to respect. Interestingly, this perspective provides a basis for objecting to genetic technologies even if they do not produce any evidence of suffering at all. Hematech's antibody-producing cows, for example, appear perfectly normal, and as they are exceedingly expensive animals that will produce a valuable product throughout their lives, they receive a level of care and husbandry that is unsurpassed. If there is a happy cow anywhere, it would certainly be in Hematech's barn. Yet one might argue that the transformation of even these animals fails to respect their integrity as animals with a nature and an evolutionary history of their own.

If one runs through the list of animal biotechnologies discussed above, any of them, including even animal drugs, might run afoul of the "animal integrity" argument against animal biotechnology. However, in truth, it is difficult to discern what types of livestock production could meet the test of respecting animal integrity. Both Regan and Adams have clearly stated that their positions prohibit any instrumental use of animals that is contrary to their interests—and this excludes *all* forms of food production. However, other ethically-oriented critics of biotechnology suggest that traditional forms of livestock farming set the standard for animal integrity, and given this standard it is highly unclear as to whether Hematech's cows would be viewed as a problem. As such, we must conclude that while the animal integrity argument is represented by some as a very strong case against genetic engineering, it is also a somewhat idiosyncratic standard.

More generally, it seems that for animal drugs or genetic engineering, the specific modification will determine whether animal welfare is an ethical issue. Some transformations do not seem to affect the health or well-being of the animal, and indeed some could eventually be developed to improve it. If the medical community begins to undertake genetic alteration to eliminate genetic disease, it will certainly be attempted first in veterinary medicine. Of the applications discussed above, none shows behavior or veterinary evidence of adverse effects on the animals' well-being. However, one of the first published studies of an experiment with genetic modification on pigs—a transformation using growth hormone genes—resulted in horribly disfigured animals with a joint and limb development that would have certainly been painful had the animals not been quickly euthanized. There thus seems to be no doubt that genetic engineering *can* create significant welfare issues, but there is no evidence to suggest that it inevitably does so.

In addition to these core ethical issues of welfare and integrity in animal biotechnology, there are also some issues that arise in connection to the way that biotechnology intersects with other practices in animal industries. Transformations that would address welfare in industrial production settings are opposed by animal advocates, and the argument that is given is that the industry should respond to these problems by improving animal living conditions, rather than trying to produce an animal that is not adversely affected by existing conditions. At the same time, animal advocates have actually been quite supportive of the most extreme version of this strategy, namely Mark Post's attempt to produce synthetic meat. There seems to be little doubt that Post's lump of cells in a petri dish exhibits all the features of the insentient creature that has been called "chilling" and "horrific" by opponents of animal biotechnology, yet the project of synthetic meat is also extolled as the final solution to the problem of animal welfare—a kind of meat that even ethical vegetarians should be happy to eat.

Issues in Animal Biotechnology: Regulation and Governance

As other chapters in this book document at some length, the overall governance framework for gene technology is far from simple. The Asilomar Conference in 1975 was notable as an event where the research community itself issued a consensus statement on the need for oversight of recombinant methods for altering the genome of any organism capable of reproduction. But this consensus was, in important respects, short-lived, as many researchers came to believe that governance mechanisms developed at government research agencies (such as the US National Institutes of Health) and by state or local authorities were intrusive. Debates over the need for regulatory reform and the recalcitrance of the biotechnology industry have been a bone of ethical contention ever since. In addition to this general debate over the need for a governance, on the one hand, and its stifling impact on innovation, on the other, there are a number of respects in which genetic transformation of animals have been regarded as something of a special case.

As noted already, genetic technologies have the potential to impose suffering on the animals that are created through these means. The governance framework for animal research should theoretically take this suffering into account when transgenic animals are used in biomedical or other scientific studies. In the United States, Europe and most industrialized countries, this legal framework requires review by a board constituted at the level of the institution—a university or government laboratory—performing this research. There are numerous loopholes in this framework. Private entities such as corporations are not covered, for example. Nevertheless, informal soft-law administered through organizations such as the Association for the Assessment and Accreditation of Laboratory Animal Care International (AAALAC) assures that the vast majority of experimental work utilizing animals is subject to institutional review. The basic ethical principles that guide these reviews enjoin researchers to minimize both the number of animals used and the pain or suffering that animals endure through the research process.[7]

There has been much less clarity about the governance framework for the commercial applications of animal biotechnology. The fact that genetically engineered or cloned animals are covered by institutional review boards during research, says nothing about animals that are produced for practical applications. Animals such as those developed for art projects or as pets (such as the Glofish) may not necessarily be subjected to any regulatory review at all. The first several decades of debate over biotechnology was marked by observation of numerous gaps in a regulatory regime that appeared to leave many animal genetic technologies in a totally unregulated status.[8] However, in the United States, the Food and Drug Administration (FDA) issued a finding in 2009 holding that it would regard any genetic engineering of vertebrate animals as a modification of the animal's structure and function. This finding gives FDA's Center for Veterinary Medicine very broad authority to oversee animal biotechnology under its statutory authority to regulate all animal drugs. There is no evidence that this finding has resulted in retroactive review of products commercialized before 2009, such as laboratory mice and rats or the Glofish, and of course it remains to be seen as to whether this authority will be assiduously applied.

Issues in Animal Biotechnology: Public Outrage and Public Debate

From the earliest days of genetic technology, the transformation and genetic exploitation of animals has been extremely controversial. Early public opinion research revealed that while many people found *all* forms of genetic engineering to be ethically problematic, the number of people who felt that genetic engineering of animals was morally wrong was larger than the number who opposed genetic engineering of human beings on moral grounds. This reaction seems to be based on a gut feeling, rather than any ascertainable reasoning. Some applications of animal biotechnology have barely raised a whimper of public outrage, while others have been the target of extended (and arguably quite successful) campaigns of public protest. Transgenic rats and mice were widely publicized during the early years of animal biotechnology, with sensational images published in national media. Yet the biomedical research field can now expect to reliably utilize transgenic lab animals with barely a raised eyebrow from the broader public.

Similarly, unusual applications of animal biotechnology seem to be guaranteed a certain amount of play in the media. The spider silk goats and the Enviro-pig have generated multiple episodes of coverage in mass media outlets—hardly the norm for the results of a scientific experiment. Artists such as Kac generate far more attention from animal projects than from genetically transformed plants. Efforts to solicit public reaction always include a few respondents who find the projects highly objectionable, as well as some feeling that artistic projects trivialize the technology and exploit the animals used. But as the shock value fades, these applications do not seem to generate very potent moral responses from the public at large. An art project that blasphemes a revered religious figure will almost certainly result in more intense and sustained opposition.

But there are animal biotechnologies that have been, to all appearances, successfully opposed. As noted above, AquaBounty's AquaAdvantage salmon have appeared to be on the verge of regulatory approval for an exceedingly long time. They have been opposed—primarily on environmental grounds—throughout their history by a long list of activists and environmental groups. There is every indication that the commercial salmon industry views them as anathema, a product that will only confuse customers and tarnish the image of all commercial fish production. It is difficult to avoid the impression that their long path to regulatory approval has been lengthened considerably by this public outrage. Furthermore, *if* they are eventually approved, it seems clear that the FDA will find some moment to announce the decision at a time when everyone's attention will be directed elsewhere.

The ethical significance of public outrage over animal biotechnology resides in the question of whether, and how, scientific activity and the commercialization of scientific discoveries should be responsible to public morality. Industrialized democracies have tended to approach this question through an application of political liberalism. That is, government agencies have mandates to intervene in the activity of private persons or private business ventures when the activity threatens to harm third parties. Critics have long contended that government has been insufficiently aggressive in protecting public health, and it is clear that at least some of the debates over the governance of genetic technology are motivated by contradictory opinions of whether regulatory agencies are fulfilling a fiduciary responsibility to protect the public. However, the case of animal biotechnology raises a further issue in that a significant portion (perhaps a majority) of the public finds many applications to be morally objectionable on grounds that relate to animal cruelty or exploitation, rather than public health. Regulatory agencies have been reluctant to interpret their legal mandates broadly enough to provoke action on such grounds, and it is likely that gene technologies that appear to involve the exploitative use of animals will continue to provoke controversy for the foreseeable future.

Issues in Animal Biotechnology: The Hidden Dimension

Thus far the focus has been on articulating the basic ethical questions that have been raised in connection with animal biotechnology, and the preceding two sections have identified well-publicized issues involving animal biotechnology. In this final section, I will review two areas where actors (sometimes including government agencies) have been less than fully forthcoming in their willingness to share information about potential issues.

As noted above, all animal research is reviewed by institutional review boards that are charged with ensuring that the use of animals conforms to ethical parameters intended to protect the research animal's interest. In addition to reviewing the basic housing and daily treatment of research animals, institutional review boards must certify that any suffering the animals endure is strictly needed to obtain the research results being sought. Different organizations vary considerably in how assiduously they pursue this objective, and one upshot of the transgenic rat and mouse industry has been to allow those boards inclined to shirk their ethical duties new opportunities to do so. For example, some labs request and receive blanket authorization to use transgenic rats and mice, without even specifying the specific transformation being made. Although review boards would still have some idea of what the research was intended to achieve, they are blinded with respect to key details in the protocol and simply cannot perform an informed evaluation of the protocol's ability to achieve a valid finding while minimizing suffering, These transgenic animals have effectively become laboratory supplies, and the nature of the transformation or its impact on the animal's well-being is no longer the concern of the chief institutional body charged with protecting the interests of the animal subjects of biomedical research.

Besides genetically engineered fish, cloned animals were also the target of significant protest. Cloned livestock are a second area where the public may be deceived. The FDA approved cloning for use in commercial livestock production in 2008, but encouraged the companies offering this service to ensure that cloned animals would not appear in the food chain. The procedure remains highly controversial in Europe, where proposals to ban livestock cloning are still actively discussed. A consortium of American cloning companies did indeed announce a moratorium on the sale of cloned animals for food purposes at the time of the FDA's approval, stating that the animals would be used strictly for breeding purposes. However, there is no discernible follow up to this announcement. Government agencies lack the authority (as well as the budget) to certify the activity of the private sector in this manner, and the reportage in the world press appears to have been thoroughly unresponsive to coverage of how or even whether industry has implemented its pledge in the intervening years.

While some animal activists remain concerned about these issues, any special issues associated with transgenic animals tend to be swamped by the animal activist community's opposition to any form of animal research at all. They are too busy protesting routine animal research to formulate investigations or analyses specific to transgenic rats and mice. As such, the impact of genetic engineering on laboratory research animals has failed to receive a great deal of attention from anyone. It is largely a black box about which little is known and no one seems to care. Yet it is worth stressing that in terms of the number of animals affected, biomedical research is far and away the most significant application for animal biotechnology in general, and for transgenic mammals, in particular.

In conclusion, it is difficult to avoid the conclusion that public sentiment has put a damper on all research that would develop products of animal biotechnology for food and agricultural use. At the very least, this research has gone deeply underground. Virtually all of the university and government websites developed to promulgate information about ongoing animal biotechnology research activities have not been updated in this decade. Animal science departments continue to maintain faculty with vitae indicating expertise in animal biotechnology, but any actual effort to deploy these tools in commercial applications is now thoroughly hidden from public view.

Conclusion: The Future of GM Food

SHELDON KRIMSKY

The essays in this volume were written by more than fifty scientists and public policy experts, whose analyses of GMOs represent many disciplines and public interest perspectives. If we add to their voices the viewpoints of the references they cite, we have literally hundreds of commentaries that bear witness to the deceptions associated with the promoters of GMOs. The real and potentially adverse effects of GMOs have been understated or negated by government and corporations, neglected by the press, and ignored by many in the scientific community who accept uncritically a corporate-crafted message. A fair-minded and unbiased individual looking at all the evidence must reach the conclusion that there is a great deal we do not know and what we do know impels us to be both cautious and concerned, skeptical of an early manufactured consensus, and critical of a framing that fails to recognize the diversity of public objections to GMOS. What follows are the key findings in this volume that support the premise behind "The GMO Deception."

Imprecision

Notwithstanding the claims of the biotechnology industry, plant and animal genetic engineering is not a precise science. Indeed, claims of the precision science of gene splicing represent an outdated view that has long been discredited by credible scientists of human biology and yet continues to be stubbornly advanced by special interests.[1] The genome of plant seeds and animals are not like a set of Legos where biotechnicians can plug in or delete genetic components with great precision. These genomes are more likened to ecosystems where one change in a gene can induce unpredictable changes in the other parts of the system. The only way to know what properties have been changed in the organism is to test each and every product created for a variety of traits. Anyone who claims, "Trust us, we know exactly how the organism will be changing" is deceiving the public. As Dr. Fagan notes in his primer on DNA:

> the GMO gene must enter the nucleus of the cell. Then, once it's in the nucleus, at some low frequency, it becomes inserted into the cell's own DNA. Scientists do not understand the mechanism by which the DNA insertion process occurs, and they have no control over it. This is the big problem because these mutations can give rise to unintended, unexpected damage to the functioning of the organism.

The imprecision and unintended effects inherent to GMO technology necessitate testing GMO products for health and environmental effects before they are introduced into the food supply. Otherwise the burden of assessing risk is shifted from corporations and government regulators to consumers.

Evidence of Potential Harm to Consumers

There is *prima facie* evidence from animal studies of potential harm from some GMO products. This evidence appears in peer-reviewed publications by independent investigators who stand to make no financial gain from reaching these conclusions. The experiments that show adverse effects on animals include those published by Árpád Pusztai on genetically modified potatoes,[2] Gilles-Eric Séralini on genetically modified maize[3], and Malatesta et al on genetically modified soybean.[4] In 2008, the latter group found that GM soybean intake can influence some liver functions during ageing and that senescence pathways are significantly activated in GM-fed mice. The authors emphasize the importance of investigating the long-term consequences of GM-diets and the potential synergistic effects with ageing.[5]

In 2009 de Vendomois et al. fed rats three commercialized GM maize varieties and found newly observed side effects with the kidney and liver and other effects observed in the heart, adrenal glands, spleen, and blood (hematopoietic) system.[6] In the same year another group of researchers fed transgenic and organic soy to rats. The GMO soy altered the ovulation cycle of the rat compared to organic soy or non-soy diets. They also observed increased cell growth in the uterus for rats fed with GMO soy.[7] Rabbits fed GM soya-bean meal exhibited significant differences in enzyme levels in three organs from GM-fed rabbits.[8]

There have been dozens of animal feeding studies of GMOs. Most of them have not shown adverse effects.[9] But without a systematic testing program overseen by a federal agency, the public must take seriously a dozen or so animal studies that do show adverse effects. In the field of science, negative findings to a widely held view that a substance is safe are more important than studies that show no adverse effects. These negative findings must be pursued to ascertain if they stand up to replication and if those findings on animal models can be extrapolated to humans. Additional long-term studies are also required. Assurances that GMOs are safe cannot be considered compelling in the absence of further study. Instead, what we have seen are ad hominem corporate-sponsored attacks on those few scientists who have engaged in such studies and reported concerns.

Ecological Harm

A number of actual and potential ecological impacts of GMOs have been identified. Evidence for some of these impacts has been published. Other potential impacts remain possibilities and have not been adequately tested in field studies. We will focus on those impacts cited in this volume for which there is evidence of GMO-induced ecological harm.

GMO seeds that are resistant to the herbicide glyphosate are pervasively being used, displacing traditional varieties of non-GMO seeds and exposing greater acreages in the United States to the herbicide glyphosate.[10] According to the widely cited study by Charles Benbrook, "Herbicide-resistant crop technology has led to a 239 million kilogram (527 million pound) increase in herbicide use in the United States between 1996 and 2011, while Bt crops have reduced insecticide applications by 56 million kilograms (123 million pounds). Overall, pesticide use increased by an estimated 183 million kgs (404 million pounds), or about 7 percent."[11] The expanded use of the herbicide has reduced the amount of milkweed plants. Milkweed is the one food source for Monarch butterflies. The decline of Monarchs has been, in part, caused by the introduction of GMO glyphosate resistant seeds.[12]

Through pollen drift, GMOs have been shown to have contaminated organic farms. Once an organic farm has been contaminated by GMO pollen it can take years for the farm to be re-accredited as organic. Not only have companies like Monsanto been able to avoid paying damages for such contamination by its seeds, they have actually sued the farmers/victims for patent infringement.

Although widely touted as "green technology" the most credible evidence to date is that GMOs are more chemically dependent than conventional seeds. That's what "Round-up Ready" means—ready for the herbicide. Indeed, GMOs are actually falling behind productivity improvements in more traditional breeding and crop production methods while continuing to reinforce the spread of industrial monoculture. As noted by Gurian-Sherman in this volume, "Monocultures are contrary to agro-ecologically sound farming systems based on crop and ecosystem diversity."

Impact on Small Farmers

GMO technologies have introduced a new form of dependency of small farmers to seed manufacturers. They have become transformed into serfs—workers who must buy their seeds under contract and who are restrained under patent protections from exchanging seeds with other farmers, submitting the seeds for scientific testing, or saving the seeds for the next season. Indian farmers opposed the introduction of Bt brinjal because of its disruption of locally controlled seed trading and because of fears that the GMO product would alter the metabolism of the plant, which has been widely used in alternative medicine for its medicinal properties. Also in India, the use of herbicide tolerant crops has displaced workers who do the weeding, which is an important income source in rural areas. Transnational companies like Monsanto have no concerns about rural labor in developing countries. Often, rural farmers in a developing world are part of an agro-political economy that includes government programs, banks, and international organizations that force them to adopt the latest GMO seed, while making conventional varieties unavailable.

In this volume Dr. Mira Shiva reports that for Indian farmers the price of cotton seeds rose astronomically, five hundred times what they used to pay for conventional seed, half of which was in royalty payments. This resulted in indebtedness and, according to some observers, drove some farmers to suicide. Too often, top-down agricultural policies are specifically designed to maximize corporate profits at the expense of the actual needs of local populations. To succeed, any solution must empower citizens to define their own agricultural management system unrestricted by intellectual property rights and GM patents.

Alteration of the Allergenicity and Immunigenicity of Plants

It is already known from laboratory experiments that transgenes can alter the allergenicity and immunogenicity of a plant in unexpected ways.[13] When a pea plant was genetically modified with genes from the common bean to protect it from insects, the GMO peas caused an immune response in mice that could elicit inflammatory reactions.[14] Unless every GMO plant that is grown can be shown to be free of allergenic and immunogenic effects and their genomes are stable, the laboratory evidence is suggestive of increasing effects on human allergenicity. Food allergies are a growing food safety and public health concern that affect an estimated 4 to 6 percent of children in the United States.[15] Allergic reactions to foods can be life threatening and have far-reaching effects on children and their families. Without systematic testing, we will never know until it is too late.

Rise of Insect and Weed Resistance

Every responsible ecologist knows that if you overuse an insect toxin or an herbicide the result will be an increase in insect and herbicide resistance. We have faced a similar effect in the overuse of antibiotics. The same antibiotics used to offset human infection are used in animals to prevent infection. As a consequence, we have seen the rise of antibiotic resistance genes spread throughout human societies. The purveyors of GMOs are not supporting Integrated Pest Management strategies that would reduce the rise of resistant weeds and insects.

Addressing World Hunger

One of the monumental and continuous deceptions of GMOs is that it represents a miracle technology that will reduce world hunger and increase food security across the poorest nations. In reality, social, political, and economic factors must first be addressed in order to ensure food access and appropriate development. A process that was intended to provide a vision for how agriculture will meet the needs of the world's 850 million poorest over the next fifty years was initiated through the International Assessment of Agricultural Science and Technology for Development (IAASTD). Launched in 2005 under the auspices of five United Nations agencies, the World Bank, and the World Health Organization, the IAAST 2007 report could find little if any major contribution to food security and world hunger from GMOs and little potential to do so.

Forbidden Labels

We have labels for all sorts of food choices including additives, fishing methods (dolphin-free tuna), types of fat, foreign proteins (except when they are incorporated into the plant genome), animal care (free range), or types of agriculture (organically grown). Food is even labeled by where it is grown. One of the great American corporate deceptions is that the public does not have a reason to require the labeling of genetically adulterated food. Every person has a right to be the first, last, or non-user of a new technology. Yet with GAUF—genetically adulterated unlabeled food—this is a right that has been kept from the American consumer. The government argument is that if the food looks, feels, and tastes like its non-GMO counterpart, then it is "substantially equivalent" and should not be labeled. This rationale has no scientific basis and was imposed on consumers by an industry that did not wish consumers to know how their food was created.

Climate Change and Sustainable Agriculture

There is no credible evidence that GMOs have contributed to a sustainable agriculture or to reducing the carbon footprint of food production. The general consensus of the authors in this volume is that GMOs are responsible for increasing chemical inputs since they are tied to the seeds. Most of the gains in drought resistance have come through traditional breeding, agronomy, and ecological agricultural practices, and not GMOs. Dozens of articles have been published about the effect of glyphosate on animals and plants. With GMOs, glyphosate use has become ubiquitous. Findings in the scientific literature are disturbing: glyphosate-based herbicides are toxic and endocrine disruptors in human cell lines; glyphosate induces carcinogenicity in mouse skin, it provokes cell division dysfunction, causes teratogenic effects on vertebrates, and produces adverse effects on human placental cells. In the absence of evidence that genetically modified foods are cheaper, produce greater yields, or even work particularly well lies one widely recognized conclusion: GMO foods provide no added nutritional or cost benefit to the consumer.

This book is trying to set the record straight on GMOs. There is no scientific consensus on the safety and agricultural value of GM crops. We cannot accept the argument that because millions of people have eaten GMOs on their dinner plates then they must be safe. Sometimes it takes years and focused research programs to uncover the hazards of products that have become commonplace. DDT, PCBs, asbestos, tobacco, lead, and benzene are just a few examples of products that were marketed for decades before their danger to human health was documented. As more and more crops and processed foods containing GMOs find their way onto supermarket shelves, the public is justified in demanding a full and transparent investigation. Without one, the GMO Deception will continue unabated.

Resources:
What You Can Do about GMOs

Jeremy Gruber

The presence of risk and the absence of reward have left many consumers wary of GMOs. These consumers are not only supportive of more studies and risk assessments for GMO foods, they are also demanding to know which foods have GMO ingredients before they choose what to feed themselves and their families.

Such consumers have many options at their disposal. In the following pages you will find our "Seven Steps to Take Action on Genetically Modified Foods," a list of organizational resources and suggested readings to learn even more about the subject.

Seven Steps to Take Action on Genetically Modified Foods

1. **Eat fresh and organic ingredients or processed foods that have been identified as non-GMO.** Though a few items of fresh produce may be genetically modified, most GMO ingredients are found in non-organic processed food, particularly ones containing corn or soy. Look for the USDA Organic seal and buy organic. The National Organic Program Standards prohibit use of genetically engineered organisms (GMOs), defined in the rules as "excluded methods." You can also identify products without GMO ingredients through the Center for Food Safety's True Food Shopper's Guide (http://www.centerforfoodsafety.org/fact-sheets/1974/true-food-shoppers-guide-to-avoiding-gmos) and the NON-GMO Project's certification system (http://www.nongmoproject.org/find-non-gmo/search-participating-products/). Some milk producers stipulate on the packaging that they are BGH-free. BGH (or bovine growth hormone) is a protein made with genetic engineering techniques and injected into cows to increase their milk production. Careful consumers can avoid BGH milk by buying organic or reading the labels for BGH-free milk.
2. **Call the manufacturer of your favorite foods.** Ask if they contain GMOs, and let them know that the answer will determine your food shopping choices.

3. **Tell your member of Congress and the USDA to stop all open-air field trials of GMO crops.** GMO crops have been found to contaminate non-GMO crops including organic crops. Stronger regulation is required to ensure that such contamination is investigated to determine its prevalence and to make sure it does not continue. Until then, the USDA at the very least should immediately place a moratorium on open-air field testing of genetically engineered crops.
4. **Tell the FDA to require labeling of GMO foods.** More than sixty countries have already enacted laws banning or mandating the labeling of GMOs. The EU has been labeling GMOs since 1998, and China and Saudi Arabia have been doing so since 2002. And in July 2011, Codex Alimentarius, the intergovernmental food commission, recognized the right of all nations to label GMO foods. The Center for Food Safety has filed a formal legal petition with the FDA demanding that the agency require the labeling of GMO foods and is spearheading a drive with the Just Label It Campaign to direct one million comments to the FDA in support of the petition. Send your comments to the FDA and President Obama in support of mandatory labeling of genetically engineered foods to:

   U.S. Food and Drug Administration
   Center for Food Safety and Applied Nutrition
   Outreach and Information Center
   5100 Paint Branch Parkway HFS-009
   College Park, MD 20740-3835
   Toll-Free Information Line:

   1-888-SAFEFOOD
   (1-888-723-3366)
   Email:
   consumer@fda.gov

5. **Support your local state GMO labeling efforts.** To fill the void in the absence of a federal GMO labeling law, groups in thirty-seven states and Washington, D.C. have begun campaigns to mandate GMO labeling in their state. Twenty-five states have introduced legislation and bills requiring GMO labeling (this legislation has only been approved in Connecticut and Maine), but will only come into force if other states, including a neighboring

state, pass labeling requirements. The Right to Know website maintains a map of state campaigns (http://www.righttoknow-gmo.org/states).

6. **Support only non-GMO seeds by participating in the Council for Responsible Genetics' Safe Seed Program.** The Safe Seed Program helps to connect non-GMO seed sellers, distributors, and traders to the growing market of concerned gardeners and agricultural consumers. The Safe Seed Pledge allows businesses and individuals to declare that they "do not knowingly buy, sell or trade genetically engineered seeds," thus assuring consumers of their commitment. CRG formally recognizes commercial vendors through the Safe Seed Resource List at http://www.councilforresponsiblegenetics.org/ViewPage.aspx?pageId=261. Sellers are encouraged to advertise the Pledge to consumers through seed catalogs and package labels. So far, more than a hundred commercial seed sellers have joined this growing movement for agricultural sustainability. You can also save seeds yourself and participate in seed swapping through the *The GE-Free Seed Trader* which allows commercial and non-commercial seed savers, growers, traders, buyers and sellers alike to share safe, genetically engineered-free seeds at http://gefreeseedtrader.com/.
7. **Join the Campaign to Stop GMO Fish.** The FDA is very close to approving genetically-engineered salmon, the first genetically-engineered animal that would be allowed into the food supply. The agency has stated it will not require such salmon to be labeled as such, making it indistinguishable from non-GE salmon in the marketplace. The Center for Food Safety is leading a campaign against this approval which can be found at: http://www.centerforfoodsafety.org/issues/309/ge-fish/join-the-campaign-to-stop-ge-fish

Organizations

There are a number of organizations working on GMO food issues from a variety of angles. All offer a wealth of educational materials, opportunities to get involved, and other resources to help consumers understand the health, ecological, and agricultural issues surrounding GMOs.

1. **Agra Watch**
   AgraWatch is a grassroots, membership-based organization in Seattle that works for a just local and global economy. CAGJ has three programs: Food Justice Project, AGRA Watch, and Trade Justice

Email: agrawatch@seattleglobaljustice.org
Website: http://www.seattleglobaljustice.org/agra-watch/

2. **Center for Food Safety**
Center for Food Safety (CFS) is a national non-profit public interest and environmental advocacy organization working to protect human health and the environment by curbing the use of harmful food production technologies and by promoting organic and other forms of sustainable agriculture.

660 Pennsylvania Avenue, SE, #302
Washington DC 20003
Phone: 202-547-9359
Email: office@centerforfoodsafety.org
Website: http://www.centerforfoodsafety.org/

3. **Consumers Union**
Consumers Union is the policy and action division of Consumer Reports. They work with activists to pass consumer protection laws in states and in Congress. They criticize corporations that do wrong by their customers, and encourage companies that are heading in the right direction.

101 Truman Avenue
Yonkers, NY 10703-1057
Phone: 914-378-2000
Website: http://consumersunion.org/

4. **Council for Responsible Genetics**
The Council for Responsible Genetics (CRG) serves the public interest and fosters public debate about the social, ethical, and environmental implications of genetic technologies.
5 Upland Road, Suite 3
Cambridge, MA 02140
Phone: 617-868-0870

New York Office:
30 Broad Street, 30th Fl.

New York, NY 10004
Phone: 212-361-6360
Email: crg@gene-watch.org
Website: http://www.councilforresponsiblegenetics.org/

5. **Earth Open Source**
Earth Open Source is a not-for-profit organization dedicated to assuring the sustainability, security, and safety of the global food system.

2nd Floor 145-157, St John Street
London EC1V 4PY, United Kingdom
Phone: +44 203 286 7156
Website: http://www.earthopensource.org/

6. **European Network of Scientists for Social and Environmental Responsibility**
The European Network of Scientists for Social and Environmental Responsibility (ENSSER) brings together independent scientific expertise to develop public-good knowledge for the critical assessment of existing and emerging technologies.

Marienstrasse 19/20
D-10017 Berlin
Germany
Phone: +49 (0)30-21234056
Email: office@ensser.org
Website: http://www.ensser.org/

7. Food and Agriculture Organization of the United Nations
Achieving food security for all is at the heart of FAO's efforts to make sure people have regular access to enough high-quality food to lead active, healthy lives. Their mandate is to improve nutrition, increase agricultural productivity, raise the standard of living in rural populations, and contribute to global economic growth.
Viale delle Terme di Caracalla
00153 Rome, Italy
Phone:(+39) 06 57051

Email: FAO-HQ@fao.org
Website: http://www.fao.org/home/en/

8. Food Democracy Now!

Food Democracy Now! is a grassroots movement of more than 650,000 farmers and citizens dedicated to building a sustainable food system that protects the natural environment, sustains farmers, and nourishes families.

Email: info@fooddemocracynow.org
Website: http://www.fooddemocracynow.org/

9. Food Policy Research Center

The Food Policy Research Center (FPRC) examines the impact of the political, technical, environmental, economic, and cultural forces that have an impact on what is eaten, illuminating the science behind food issues and policies from an interdisciplinary perspective.

Their goal is to arm lawmakers, consumers, and industry representatives with scientifically sound information about how we grow, process, package, distribute, and prepare what we eat.

6004A Campus Delivery Code
1354 Eckles Avenue
St. Paul, MN 55108
Phone: 612-624-6772
Email: fprc@umn.edu
Website: http://www.foodpolicy.umn.edu/

10. Food & Water Watch

Food & Water Watch works to ensure the food, water, and fish we consume is safe, accessible, and sustainably produced, and educates about the importance of keeping the global commons under public control.
1616 P Street NW, Suite 300
Washington, DC 20036
Phone: 202-683-2500
Website: http://www.foodandwaterwatch.org/

11. GeneWatch UK

GeneWatch UK is a not-for-profit policy research and public interest group. They investigate how genetic science and technologies will impact our food, health, agriculture, environment, and society.

60 Lightwood Road Buxton Derbyshire
SK17 7BB
Phone: +44 (0)1298 24300
Email: mail@genewatch.org
Website: http://www.genewatch.org/

12. GMO Free USA

GMO Free USA's mission is to harness education, advocacy, and bold action to foster consumer rejection of genetically modified organisms until they are proven safe.

Website: http://gmofreeusa.org/

13. GM Watch

GMWatch is an independent organization that seeks to counter the enormous corporate political power and propaganda of the biotech industry and its supporters.
26 Pottergate
Norwich Norfolk
NR2 1DX UK
Website: http://www.gmwatch.org/

14. Institute for Responsible Technology

The Institute for Responsible Technology educates policy makers and the public about genetically modified (GM) foods and crops. They investigate and report their risks and impact on health, environment, the economy, and agriculture, as well as the problems associated with current research, regulation, corporate practices, and reporting.

PO Box 469
Fairfield, IA 52556

Phone: 641-209-1765
Email: info@responsibletechnology.org
Website: http://www.responsibletechnology.org/

15. **Just Label It!**
The Just Label It campaign was created to advocate for the labeling of GE foods.

1436 U Street NW, Suite 205
Washington, D.C. 20009
Phone: 202-688-5834
Email: info@justlabelit.org
Website: http://justlabelit.org/

16. **Non-GMO Project**
The Non-GMO Project is a non-profit organization committed to preserving and building sources of non-GMO products, educating consumers, and providing verified non-GMO choices.

1200 Harris Avenue, Suite #305
Bellingham, WA 98225
Phone: 877-358-9240
Email: info@nongmoproject.org
Website: http://www.nongmoproject.org

17. **Oakland Institute**
The Oakland Institute's mission is to increase public participation and promote fair debate on critical social, economic and environmental issues in both national and international forums.

4173 MacArthur Boulevard, Suite 225
Oakland, CA 94619
Phone: 510-474-5251
Email: info@oaklandinstitute.org
Website: http://www.oaklandinstitute.org/

18. Organic Consumers Association

The Organic Consumers Association (OCA) is an online and grassroots non-profit, public interest organization campaigning for health, justice, and sustainability. The OCA deals with crucial issues of food safety, industrial agriculture, genetic engineering, children's health, corporate accountability, Fair Trade, environmental sustainability, and other key topics.

6771 South Silver Hill Drive
Finland MN 55603
Phone: 218-226-4164
Website: http://www.organicconsumers.org/

19. The Organic and Non-GMO Report

The Organic and Non-GMO Report is the only news magazine exclusively dedicated to information you need to respond to the challenges of genetically modified (GM) foods.

PO Box 436
Fairfield, IA 52556
Phone: 641-209-3426
Email: ken@non-gmoreport.com
Website: http://www.non-gmoreport.com/

20. Right to Know GMO

The Right to Know GMO, A Coalition of States is a broad coalition of state leaders, nonprofit organizations, and organic companies that have a shared goal of winning mandatory labeling of genetically-engineered foods in the U.S.

Website: http://www.righttoknow-gmo.org/

21. Truth in Labeling Coalition

The Truth in Labeling Coalition works for the right of American families to an informed choice about the food we eat.
info@truthinlabelingcoalition.org

Website: http://truthinlabelingcoalition.org/

22. **Union of Concerned Scientists**

    The Union of Concerned Scientists is an alliance of more than 400,000 citizens and scientists that puts rigorous, independent science to work to solve our planet's most pressing problems. They combine technical analysis and effective advocacy to create innovative, practical solutions for a healthy, safe, and sustainable future.

    Two Brattle Square
    Cambridge, MA 02138-3780
    Phone: 617-547-5552
    Website: http://www.ucsusa.org/

23. **World Health Organization Department of Food Safety**

    The Department of Food Safety and Zoonoses (FOS), provides leadership in global efforts to lower the burden of diseases from food and animals.

    Avenue Appia 20
    CH-1211 Geneva 27, Switzerland
    Email: foodsafety@who.int
    Website: http://www.who.int/foodsafety/en/

Books

There are several other excellent books on GMOS including the following:

Agriculture, Biotechnology and the Environment
Sheldon Krimsky and Roger Wrubel
University of Illinois Press, 1996

Animal, Vegetable, Miracle: A Year of Food Life
Barbara Kingsolver
Harper Perennial, first edition, 2008

Engineering the Farm: The Social and Ethical Aspects of Agricultural Biotechnology
Edited by Mark Lappe and Britt Bailey
Island Press, 2002

Fast Food Nation: The Dark Side of the All-American Meal
Eric Schlosser
Houghton Mifflin Company, first edition, 2001

Food, Inc.: Mendel to Monsanto—The Promises and Perils of the Biotech Harvest
Peter Pringle
Simon and Schuster, 2005

Food Politics: How the Food Industry Influences Nutrition, and Health
Marion Nestle
University of California Press, revised and expanded edition, 2007

Intervention: Confronting the Real Risks of Genetic Engineering and Life on a Biotech Planet
Denise Caruso
The Hybrid Vigor Institute, 2006

The Omnivore's Dilemma: A Natural History of Four Meals
Michael Pollan
Penguin, 2007

Safe Food: The Politics of Food Safety
Marion Nestle
University of California Press, updated and expanded edition, 2010

Uncertain Peril: Genetic Engineering and the Future of Seeds
Claire Cummings
Beacon Press, first edition, 2008

EPILOGUE

Two years have passed since the first edition of *The GMO Deception*. This epilogue is intended to bring the reader up to date on some key issues, including the potential human health impacts of herbicides and herbicide-tolerant plants, GMO labeling initiatives, new European Union policies, ecological effects, and scientists' concerns over GMOs.

Glyphosate

There are several essays in this volume that discuss the most widely used herbicide in the world—glyphosate. Monsanto's version of this, its signature herbicide, is known under the trade name Roundup. In the book's concluding chapter we summarized what was known at the time of the first printing about the toxicity of glyphosate (p. 345). The information was mostly gleaned through animal studies.

On March 20, 2015, the International Agency for Research on Cancer (IARC), an agency of the World Health Organization, released a report from its meeting in Lyon, France. The report revisited the carcinogenicity of five organophosphate insecticides and herbicides, including glyphosate. The agency released its reassessment in preparation for the publication of its report entitled "Some Organophosphate Insecticides and Herbicides: Diazinon, Glyphosate, Malathion, Parathion, and Tetrachlorvinphos."

The second report was the culmination of work produced by a working group organized by the IARC, which consisted of seventeen experts from eleven countries who met in person March 3–10, 2015. The weeklong meeting followed nearly a year of preparation by the working group and the IARC secretariat. During this period the working group had considered reports that have been published or accepted for publication in open scientific literature and data that are part of publicly available governmental documents. No one in the toxicology community doubts that this was a serious review. Based on the latest animal and human studies related to glyphosate and glyphosate formulations, especially studies of agricultural workers, and acknowledging the limited amount of human data, the report of the IARC working group classified glyphosate as "probably carcinogenic to humans," one category below the top category. (IARC's five categories are: Group 1: Carcinogenic to humans; Group 2A: Probably carcinogenic to humans; Group 2B: Possibly carcinogenic to humans; Group 3: Not classifiable as to its carcinogenicity to humans; Group 4: Probably not carcinogenic to humans.)

Shortly after the meeting and the reports, on March 25, 2015, the *New York Times* ran an op-ed by Mark Bittman, its food columnist, entitled "Stop Making Us Guinea Pigs." The article raised questions about eating food containing residues from a substance that is a probable human carcinogen: Roundup. In response, Monsanto sent in a letter to the *Times* (published on April 8,2015) that disputed the IARC finding. In it he wrote, "Overwhelming evidence regarding glyphosate supports a conclusion of no cancer risk." The "evidence" resulted from studies funded by Monsanto.

In 1996, we wrote in a book entitled *Agricultural Biotechnology and the Environment*: "The factor that most threatens the success and agronomic usefulness of HRCs [herbicide resistant crops] is the potential for weeds to develop resistance to the associated herbicides. The extensive and continuous use of herbicides since the 1950s has resulted in more than a hundred weed species resistant to one or more herbicides." The National Academy of Sciences released a report in April 2010 warning that weeds are becoming increasingly resistant to glyphosate. Five years later, in April 2015, the journal *Nature Biotechnology* reported that weeds throughout the world have become resistant to glyphosate. To combat this, Monsanto added another herbicide-tolerant property to its GMOs (stacked properties) so farmers can either use glyphosate or dicamba. If it's approved for the two to be combined, stacked weed-resistant GMOs will witness the explosion of the herbicide dicamba just as they had glyphosate. There is already evidence that dicamba is an endocrine disrupter in fish. Undoubtedly, we will hear more about its toxicological effects as its use increases in GMOs.

Labeling

Several grassroots initiatives to pass legislation for labeling GMOs in food products have failed when anti-labeling advertising by agribusiness frightened the public about higher food prices; the industry claimed it would have to pass on the costs of labeling to consumers. GMO corporate interests have lobbied Congress to introduce a labeling preemption bill that would not allow any state to mandate GMO labeling. Meanwhile, the State of Vermont passed the first US law requiring the labeling of food produced from genetic engineering. The landmark legislation was signed by Governor Peter Shumlin on May 8, 2014. The Grocery Manufacturers Association, a trade organization for agri-industry, filed suit against the State of Vermont asking the courts to overturn Act 120, which is what requires the labeling of GMO foods. A US District Court Judge decided in April 2015 to let Vermont proceed with its plans to require the labeling of food containing genetically modified ingredients starting July 1, 2016, but also to allow the Grocery Manufacturers of America (GMA) to continue its appeal against labeling.

On January 8, 2016, the Cambell's Soup Company, breaking away from the GMA, announced its support for mandatory national labeling of products that may contain genetically modified organisms and proposed that the federal government provide a national standard for non-GMO claims made on food packaging. By this move, Campbell's, which owns Pepperidge Farm, Plum Organics, V8, and Prego, has declared full transparency on how its foods are manufactured and processed.

European Union Policies on GMOs

The European Union established from the outset, through Directive 2001/18/EC, that for a GMO to be placed in the market of member states it must meet pre-market authorization and post-market environmental monitoring. Both the member states and the European Commission are responsible for authorizing the acceptance of GMOs, which includes a risk assessment for any GMO crop approved for cultivation. Then on March 11, 2015, Directive (EU) 2015/412 of the European Parliament and the Council was adopted, amending Directive 2001/18/EC. It gave to any member states the opportunity to restrict or prohibit the cultivation of genetically modified organisms in its territory. After GMO products have been authorized for cultivation by the EU, individual states may decide to restrict or prohibit GMOs on the basis of broad criteria including: environmental policy objectives; town and country planning; land use; socioeconomic impacts; avoidance of GMO presence in other products; and agricultural or public policy objectives (Directive [EU] 2015/412, European Parliament and Council of the European Union, March 11, 2015).

A number of European countries and regions have banned or restricted the cultivation of GMOs, including Russia, Austria, Germany, Hungary, Luxembourg, Greece, Bulgaria, Poland, and thirty Italian regions, among them Tuscany, Rome, Milan, and Genoa.

United States Department of Agriculture Plans for Certifying Non-GMO Foods

In 2015, Chipotle was the first national food chain to announce its plans to eliminate all GMO ingredients from its foods. Unlike the EU, which requires the labeling of GMO foods unless their presence is less than 0.9 percent of the the animal's food of feed, the US has the organic standard which precludes GMOs. After years of remaining unresponsive to public polls supporting GMO labeling, the USDA announced the creation of a new government certification program for labeling GMO-free foods. In October 2014 it developed procedures for including a non-GMO statement on the labeling of meat and poultry products under the organic standard. The 2015 USDA initiative applies to all food products whether or not they are under the organic standard. The certification would be voluntary and companies would have to pay for it. The label would state "USDA Process Verified" with a claim that the product is free of GMOs. This is a sign that US public opinion in favor of GMO labeling is gaining traction.

No Scientific Consensus on the Safety of GMOs

The claim of a scientific consensus on the health and safety of GMOs is frequently cited in the media, but the evidence suggests otherwise. Much of the public debate has focused on the safety of consuming GMOs while relatively little attention has been devoted to potential environmental harms. Preeminent scientific bodies such as the National Research Council, the Royal Society of Canada, and the British Medical Journal have recognized that some engineered foods could pose considerable risk depending on the particular engineered crop and the foreign gene introduced. A statement in the journal *Environmental Sciences Europe* first published online on January 24, 2015, and later in print, was developed and signed by more than three hundred scientists, physicians, and scholars, and asserted that there is no scientific consensus on the safety of GMOs. The statement concluded that "the scarcity and contradictory nature of the scientific evidence published to date prevents conclusive claims of safety, or of lack of safety, of GMOs. Claims of consensus on the safety of GMOs are not supported by an objective analysis of the refereed literature."

GMOs and Monarch Butterflies

In the late 1990s, Monarch butterfly populations began to drop at about the same time that Roundup-ready corn and soybeans entered the market. Since then, instead of tilling weeds, farmers spray entire fields with Roundup. And as insect resistance rises, they spray more and more. Not surprisingly, this spraying kills every plant except Roundup-ready crops, including milkweed which is the main food source of the Monarch butterfly. Additional harms from ingestion may include a compromised immune system of some species. While there is not yet certain proof that the expansion of GMO crops is contributing to the milkweed decline, there is a strong correlation among the increased use of herbicide-tolerant crops, fewer milkweed plants, and the declining Monarch population that has prompted many environmental activists to cry foul. They point to a 2012 study in the journal *Insect Conservation and Diversity* that estimated that between 1990 and 2010, milkweed prevalence declined 58 percent in Midwestern agricultural areas. Over the same time frame, the Monarch population declined 81 percent. On March 13, 2015, fifty-two members of Congress sent President Barack Obama a letter expressing concern over the proliferation of GMOs and the decline of Monarchs.

Transatlantic Trade and Investment Partnership

At the time of this writing a major trade agreement is being negotiated between Europe, the United States, and Canada with the active participation of multinational companies. Known as the Transatlantic Trade and Investment Partnership (TTIP), it has come under criticism from a number of democracy campaigners for its lack of transparency. A recent European Commission decision to let member states have "opt-outs" on imports of genetically modified food and feed is evoking a particularly strong reaction from the United States. Strong pressure is being brought to bear from

US industries to allow GMO products and other foods into EU markets that would violate the EU's current standards, in the name of free trade. Campaigners, concerned that the EU might reverse itself, point to a new EU deal with Canada—one outside of the TTIP's initiatives but related to them—wherein the two have agreed to have a "shared objective" of minimizing the disruption to trade from their different GMO rules.

—Sheldon Krimsky, 2016

ENDNOTES

Foreword by Ralph Nader

1. See Sheldon Krimsky, *Genetic Alchemy: The Social History of the Recombinant DNA Controversy.* Cambridge, MA: The MIT Press, 1982.
2. Michael Pollan, "The Great Yellow Hope." *New York Times*, March 4, 2001.
3. Jonathan Latham, "Fakethrough! GMOs and the Capitulation of Science Journalism." *Independent Science News*, January 7, 2014.
4. The Editors, "Do Seed Companies Control GM Crop Research?" *Scientific American*, July 20, 2009.
5. Ellen Barry, "After Farmers Commit Suicide, Debts Fall on Families in India." *New York Times*, February 22, 2014.
6. David Michaels, *Doubt is Their Product: How Industry's Assault on Science Threatens Your Health.* Oxford, UK: Oxford University Press, 2008.

Introduction: *The Science and Regulation behind the GMO Deception* by Sheldon Krimsky and Jeremy Gruber

1. Ania Wieczorek and Mark Wright, History of Agricultural Bioetechnology: How Crop Development has Evolved. *Nature Education Knowledge* 3(3)9-15 (2012).
2. Indra K. Vasil. A History of Plant Biotechnology: From the Cell Theory of Schleiden and Schwann to Biotech Crops. *Plant Cell Rep* 27:1423–1440 (2008).
3. Francis Bacon, *The New Atlantis*.
4. Quote taken from a blog on the American Council of Science and Health website. October 24, 2013
5. Doug Gurian-Sherman, High and Dry: Why Genetic Engineering is Not Solving Agriculture's Drought Problem in a Thirsty World. Union of Concerned Scientists Report, June 2012. http://www.ucsusa.org/assets/documents/food_and_agriculture/high-and-dry-report.pdf
6. Chris Parker, The Monsanto Menace, *Village Voice* July 24-30, 2013.
7. http://www.isaaa.org/resources/publications/briefs/44/executivesummary/ Accessed November 13, 2013.

8. Wieczorek and Wright, 2012.
9. M. Antoniou, C. Robinson, and J. Fagan. *GMO Myths and Truths*, June 2012. Earth Open Source Publishers. http://earthopensource.org/files/pdfs/GMO_Myths_and_Truths/GMO_Myths_and_Truths_1.3b.pdf. Accessed November 1, 2013.
10. Alessandro Nicolia, Alberto Manzo, Fabio Veronesi, and Daniele Rosellini, An overview of the last 10 years of genetically engineered crop safety research. *Critical Reviews in Biotechnology* September 16, 2013. Early online. (doi:10.3109/07388551.2013.823595).
11. http://www.isaaa.org/resources/publications/briefs/44/executivesummary/ Accessed November 13, 2013.
12. Statement of the European Network of Scientists for Social and Environmental Responsibility (ENSSER), October 21, 2013. www.ensser.org. Accessed November 12, 2013.
13. US FDA website. http://www.fda.gov/Food/GuidanceRegulation/GuidanceDocumentsRegulatoryInformation/Biotechnology/ucm096156.htm Accessed November 2, 2013
14. US FDA Website. http://www.fda.gov/Food/GuidanceRegulation/GuidanceDocumentsRegulatoryInformation/LabelingNutrition/ucm059098.htm Accessed November 2, 2013.
15. http://www.epa.gov/oppbppd1/biopesticides/pips/pip_list.htm. Accessed November 3, 2013
16. E. Millstone, E. Brumer and S. Mayer. Beyond Substantial Equivalence. *Nature* 401:525-526 (1999).
17. Marc Lappé, Ch. 10, A perspective on anti-biotechnology convictions In: *Engineering the Farm,* Britt Bailey and Marc Lappé. Washington: Island Press, 2002, p. 135
18. Árpád Pusztai and Susan Bardocz, *Potential Health Effects of Foods Derived from Genetically Modified Plants: What are the Issues?* Ch. 5, Problems and Perspectives. Penang, Malaysia: Third World Network, 2011.
19. T.G. Neltner, H.M. Alger, J.T. Oreilly, S. Krimsky, L.A. Bero, and M.V. Maffini, Conflicts of interest in approvals of additives to food determined to be generally recognized as safe: Out of balance. *JAMA Internal Medicine* August 7, 2013. doi:10.1001/jamainternmed.2013.10559.

What Is Genetic Engineering? An Introduction to the Science **by John Fagan, Michael Antoniou, and Claire Robinson**

1. Antoniou, Michael, Fagan, John, Robinson, Claire GMO Myths and Truths An evidence-based examination of the of the claims made for the safety and efficacy of genetically modified crops, June 2012, earthopensource http://earthopensource.org/files/pdfs/GMO_Myths_and_Truths/GMO_Myths_and_Truths_1.3b.pdf

Part 1: Safety Studies: Human and Environmental Health

Chapter 1: *The State of the Science* by Stuart Newman

1. Weinstein, I. B., et al. 1984. Cellular targets and host genes in multistage carcinogenesis. Fed Proc. 43: 2287–2294.
2. Hsieh-Li, H. M., et al. 1995. Hoxa 11 structure, extensive antisense transcription, and function in male and female fertility. Development. 121: 1373–1385.
3. Potrykus, I. 2010. Regulation must be revolutionized. Nature. 466: 561.
4. Zhang, W. & F. Shi. 2010. Do genetically modified crops affect animal reproduction? A review of the ongoing debate. Animal. 5: 1048–1059.
5. de Vendomois, J. S., et al. 2010. Debate on GMOs health risks after statistical findings in regulatory tests. Int J Biol Sci. 6: 590–598.
6. Krzyzowska, M., et al. 2010. The effect of multigenerational diet containing genetically modified triticale on immune system in mice. Pol J Vet Sci. 13: 423–430.
7. Domingo, J. L. 2007. Toxicity studies of genetically modified plants: a review of the published literature. Crit Rev Food Sci Nutr. 47: 721–733.
8. Domingo, J. L. & J. Gine Bordonaba. 2011. A literature review on the safety assessment of genetically modified plants. Environ Int. 37: 734–742.
9. Redenbaugh, K. 1992. Safety assessment of genetically engineered fruits and vegetables: a case study of the FLAVR SAVR tomato. CRC Press. Boca Raton, Fla.
10. Newman, S. A. 2009. Genetically modified foods and the attack on nature. Capitalism Nature Socialism. 20: 22–31.
11. Silver, L. M. 2006. Why GM Is good for us: genetically modified foods may be greener than organic ones. In Newsweek International, March 20: 57–58. http://128.112.44.57/CNmedia/articles/06newsweekpig1s1.pdf

12. Shermer, M. 2013. The liberals' war on science. ScientficAmerican.com, January 21. http://www.scientificamerican.com/article.cfm?id=the- liberals-war-on-science
13. Vaughan, A. 2012. Prop 37: Californian voters reject GM food labelling. Guardian.co.uk, November 7. http://www.guardian.co.uk/environment/2012/nov/07/prop-37-californian-gm-labelling
14. Haskell, M. J. 2012. The challenge to reach nutritional adequacy for vitamin A: beta-carotene bioavailability and conversion--evidence in humans. Am J Clin Nutr. 96: 1193S–1203S.

Chapter 2: *Antibiotics in Your Corn* by Sheldon Krimsky

1. Ma JKC, Chikwamba R, Sparrow P, Fischer R, Mahoney R, Twyman RM. 2005. Plant-derived pharmaceuticals—the road forward. Trends Plant Sci. 10;12:580-5. http://dx.doi.org/10.1016/j.tplants.2005.10.009.
2. Bauer A. 2006. Pharma crops. State of field trials worldwide. Munich: Umweltinstitut München e.V. 37 pp. http://www.umweltinstitut.org, http://www.organicconsumers.org/articles/article_1419.cfm;
   Spök A. 2007. Molecular farming on the rise—GMO regulators still walking a tightrope. Trends Biotechnol. 25;2:74-82. http://dx.doi.org/10.1016/j.tibtech.2006.12.003;
   APHIS. 2008. Release Permits for Pharmaceuticals, Industrials, Value Added Proteins for Human Consumption, or for Phytoremediation Granted or Pending by APHIS as of August 19, 2008. http://www.aphis.usda.gov/brs/ph_permits.html;
   JRC. 2008. Deliberate releases and placing on the EU market of Genetically Modified Organisms (GMOs). Ispra, Italy: EU Joint Research Center. http://gmoinfo.jrc.it.
3. FDA and USDA, 2002. A look at the benefits and risks of bioengineering plants to produce pharmaceuticals. Pew Initiative on Food and Biotechnology, Proc, 'Pharming the Field' Workshop. http://pewagbiotech.org/events/0717/ConferenceReport.pdf;
   Spök A. 2006. From Farming to "Pharming"- Risks and Policy Challenges of Third Generation GM Crops. ITA-06-06. Vienna: Institute for Technology Assessment. http://hw.oeaw.ac.at:8000/ita/ita-manuscript/ita_06_06.pdf.

4. Commandeur U, Twyman RM, Fischer R. 2003. The biosafety of molecular farming in plants. AgBiotechNet 5 ABN 110:1-9. http://www.agbiotechnet.com/Reviews.asp?action=display&openMenu=relatedItems&ReviewID=820.
5. Spök 2007.
6. Stein KE, Webber KO. 2001. The regulation of biologic products derived from bioengineered plants. Curr. Opin. Biotechnol. 12;3:308-11. http://dx.doi.org/10.1016/S0958-1669(00)00217-2.
7. NFPA. 2003. News release, "No use of food or feed crops for plant-made pharmaceutical production without a '100percent guarantee' against any contamination, says NFPA," Washington, DC: National Food Processors Association, Feb 6.
8. Freese B. 2002. Manufacturing drugs and chemicals in crops: Biopharming poses new threats to consumers,farmers, food companies and the environment. Report for Genetically Engineered Food Alert. p. 97. www.gefoodalert.org, www.foe.org/camps/comm/safefood/biopharm/BIOPHARM_REPORT.pdf.
9. Marvier M, Van Acker RC. 2005. Can crop transgenes be kept on a leash? Frontiers Ecol. Environ. 3;2:93-100. http://www.scu.edu/cas/environmentalstudies/upload/Marvierpercent20&percent20VanAcker.pdf.
10. Castle D. 2008. The Future of Plant-Derived Vaccines. Undated web article, Bioscienceworld. http://www.bioscienceworld.ca/TheFutureofPlant-DerivedVaccines (last visited Aug 26th 2008).
11. Sauter A, Hüsing B. 2006. Grüne Gentechnik—Transgene Pflanzen der 2 und 3 Generation. Berlin: Office of Technology Assessment at the German Parliament, TAB. 304 pp. http://www.tab.fzk.de/de/projekt/zusammenfassung/ab104.pdf.
12. Ellstrand NC. 2003. Going to 'great lengths' to prevent the escape of genes that produce specialty chemicals. Plant Physiol. 132: 1770-4. http://www.plantphysiol.org/cgi/reprint/132/4/1770.pdf.
13. Spök 2007; see also Levidow L. 2007. Making Europe Unsafe for Agribiotech. Manuscript for Handbook of Genetics & Society, 16/10/2007. 13 pp.
14. Etty TFM. 2007. Coexistence—The Missing Link in the EU Legislative Framework. Proc. 3rd Int Conf 'GMO Free Regions, Biodiversity and

Rural Development', Brussels, April 19-20, 2007. 5 pp. http://www.gmo-free-regions.org.

15. EPC. 2005. Emerging biotechnology applications: EU, US and global regulatory perspectives. Report of Meeting, 4-5 Dec 2005, Lille. European Policy Centre. www.epc.eu/ER/pdf/119600569_Lillepercent20Conferencepercent20Reportpercent204-5.12.2005.pdf.
16. Freese, 2002; APHIS 2004. Environmental impact statement; introduction of genetically engineered organisms. Fed Reg. 69:3271-72. http://www.epa.gov/EPA-IMPACT/2004/January/Day-23/i1411.htm.

Chapter 3: *A Conversation with Dr. Árpád Pusztai* by Samuel W. Anderson

1. Prescott, Vanessa, et al. "Transgenic Expression of Bean r-Amylase Inhibitor in Peas Results in Altered Structure and Immunogenicity." Journal of Agricultural and Food Chemistry. No. 53, 9023–9030 (2005).

Chapter 4: *Glypho-Gate* by Gilles-Eric Séralini

1. Gilles-Eric Séralini, Emile Clair, Robin Mesnage, Steeve Gress, Nicolas Defarge, Manuela Malatesta, Didier Hennequin, Jo?l Spiroux de Vendômors. Longterm toxicity of a Roundup herbicide and a Roundup-tolerant genetically modified maize. Food and Chemical Toxicology 50(2012):4221–4231.
2. Gilles-Eric Séralini, et. al. Answers to critics: Why there is a long term toxicity due to a Roundup-tolerant genetically modified maize and to a Roundup herbicide. Food and Chemical Toxicology 53(2013):476–483.

Chapter 7: *Busting the Big GMO Myths* by John Fagan, Michael Antoniou, and Claire Robinson

1. Antoniou, M, Fagan, J, Robinson, C. GMO Myths and Truths An evidence-based examination of the of the claims made for the safety and efficacy of genetically modified crops, June 2012, Earth Open Source. http://earthopensource.org/files/pdfs/GMO_Myths_and_Truths/GMO_Myths_and_Truths_1.3b.pdf
2. "Safety and health in agriculture". International Labour Organization. 21 March 2011.

3. Antoniou, M, Fagan, J, Robinson, C. GMO Myths and Truths An evidence-based examination of the of the claims made for the safety and efficacy of genetically modified crops, June 2012, Earth Open Source. http://earthopensource.org/files/pdfs/GMO_Myths_and_Truths/GMO_Myths_and_Truths_1.3b.pdf
4. Séralini, G-E, et al. Long-term toxicity of a Roundup herbicide and a Roundup-tolerant genetically modified maize. Food and Chemical Toxicology. 2012. 4221–4231.
5. David Schubert U-T San Diego, 8 Jan 2014. http://www.utsandiego.com/news/2014/jan/08/science-food-health/. European Network of Scientists for Social and Environmental Responsibility (ENSSER) Comments on the Retraction of the Séralini et al. 2012 Study http://www.ensser.org/democratising-science-decision-making/ensser-comments-on-the-retraction-of-the-seralini-et-al-2012-study/
6. Comision Provincial de Investigación de Contaminantes del Agua (2010). Primer informe [First report]. Resistencia, Chaco, Argentina. http://www.gmwatch.eu/files/Chaco_Government_Report_Spanish.pdf; http://www.gmwatch.eu/files/Chaco_Government_Report_English.pdf
7. Comision Provincial de Investigación de Contaminantes del Agua (2010). Primer informe [First report]. Resistencia, Chaco, Argentina. http://www.gmwatch.eu/files/Chaco_Government_Report_Spanish.pdf ; http://www.gmwatch.eu/files/Chaco_Government_Report_English.pdf
8. Sahai, S. (2003). The Bt Cotton Story: The ethics of science and its reportage. Current Science. 84: 974–975.
9. Gurian-Sherman D. Failure to yield: Evaluating the performance of genetically engineered crops. Union of Concerned Scientists. 2009. http://www.ucsusa.org/assets/documents/food_and_agriculture/failure-to-yield.pdf
10. Hattori, Y., et al. (2009). "The ethylene response factors SNORKEL1 and SNORKEL2 allow rice to adapt to deep water." NATURE 460: 1026–1030.
11. IAASTD website: http://www.unep.org/dewa/Assessments/Ecosystems/IAASTD/tabid/105853/Default.aspx
12. Hine R, Pretty J, Twarog S. Organic agriculture and food security in Africa. New York and Geneva. UNEP-UNCTAD Capacity- Building Task Force on Trade, Environment and Development. 2008. http://bit.ly/KBCgY0

13. Leahy S. Africa: Save climate and double food production with eco-farming. IPS News. 8 March 2011. http://allafrica.com/stories/201103090055.html
14. Id.

Part 2: Labeling and Consumer Activism

Chapter 9: *Consumers Call on FDA to Label GMO Foods* by Colin O'Neil

1. http://gefoodlabels.org/gmopercent20labeling/polls-on-gmo-labeling/
2. See Aris A., Leblanc S., "Maternal and fetal exposure to pesticides associated to genetically modified foods in Eastern Townships of Quebec, Canada," (Feb. 18, 2011) available at http://www.sciencedirect.com/science/article/pii/S0890623811000566 (last visited May 25, 2011). In approving Bt corn, FDA had previously relied on the industry's assurances that the Bt toxin would be broken down during digestion.
3. http://www.greenerchoices.org/foodpoll2008/

Chapter 10: *Genetically Engineered Foods: A Right to Know What You Eat* by Phil Bereano

1. John B. Fagan, Assessing the Safety and Nutritional Quality of Genetically Engineered Foods, PSRAST (July 1996)

Part 3: GMOs in the Developing World

Chapter 15: *The Agrarian Crisis in India* by Indrani Barpujari and Birenda Biru

1. Patnaik, P., "The Crisis in India's Countryside", http://ccc.uchicago.edu/docs/India/patnaik.pdf , accessed on May 5, 2007.
2. Sahai, S., "Are Genetically Engineered Crops the Answer to India's Agrarian Crisis".
3. Mishra, D.K., (undated), "Behind Agrarian Distress: Interlinked Transactions as Exploitative Mechanisms", Epov (a newsletter of the Centre for Science and Development)
4. Patnaik, U., "It is Time for Kumbhakarna to Wake up", The Hindu, August 5, 2005
5. Sahai, S., "Are Genetically Engineered Crops the Answer to India's Agrarian Crisis?"
6. Shukla, S.P., "An Initiative for Agrarian Analysis and Action", ATIS (Agricultural Trade Initiative from the South", September 30, 2005.

7. Sahai, S. "Are Genetically Engineered Crops the Answer to India's Agrarian Crisis?"
8. "Performance of Bt .cotton", Economic & Political Weekly, July 26- Aug 2003, Vol. XXXV111, no. 30; pp.3139–3141.
9. Tata Institute of Social Sciences, "Causes of Farmer Suicides in Maharashtra: An Enquiry", March 15, 2005. 10. "Science Finds against Bt Cotton", Genet, February 5, 2006.
10. "Most Farmers who Committed Suicide were Bt Cotton Growers: VJAS", Genet 20/04/06
11. "Farm bodies seek ban on Bt cotton cultivation", Genet 26/09/2005.
12. "A Disaster Called Bt Cotton", Genet 05/12/05.
13. "Bt Cotton and Farmers' Health", Genet 24/03/2006.
14. "1600 sheep die after grazing in Bt cotton field", Genet 04/05/06.
15. " Goats/ sheep Mortality after Grazing on Bt Cotton", Genet 10/02/07

Chapter 16: *Bill Gates's Excellent African Adventure: A Tale of Technocratic AgroIndustrial Philanthrocapitalism* by Phil Bereano

1. http://www.ucsusa.org/food_and_agriculture/our-failing-food-system/genetic-engineering/failure-to-yield.html
2. http://nature.berkeley.edu/~miguel-alt/what_is_agroecology.html
3. http://www.conservation.org/newsroom/pressreleases/Pages/Global_Tool_to_Gauge_Earths_and_Humanitys

Chapter 18: *Hearts of Darkness* by Doreen Stabinsky

1. Benbrook, Charles. "Comments to the Zambian delegation." September 13, 2002, www.biotech-info.net; see "Better Dead Than GM Fed," *Seedling*, October 2002, p. 15.

Chapter 19: *Rooted Resistance: Indian Farmers Stand against Monsanto* by Mira Shiva

1. http://www.i-sis.org.uk/MDSGBTC.php
2. http://www.thehindu.com/opinion/columns/sainath/article995824.ece?homepage=true
3. http://articles.timesofindia.indiatimes.com/2010-02-06/india/28128712_1_bt-brinjal-public-consultations-food-crop/2

Chapter 20: *Why GM Crops Will Not Feed the World* by Bill Freese

1. "Food crisis threatens security, says UN chief," The Guardian, April 21, 2008. http://www.guardian.co.uk/environment/2008/apr/21/food.unitednations; for tortilla prices, see: "A culinary and cultural staple in crisis," The Washington Post, January 27, 20007. http://www.washingtonpost.com/wp-dyn/content/article/2007/01/26/AR2007012601896.html.
2. Rising food prices: Policy options and World Bank response," World Bank, April 2008. http://siteresources.worldbank.org/NEWS/Resources/risingfoodprices_backgroundnote_apr08.pdf.
3. Runge, C.F. & Senauer, B. (2007). "How Biofuels Could Starve the Poor," Foreign Affairs, May/June 2007. http://www.foreignaffairs.org/20070501faessay86305/c-ford-runge-benjamin-senauer/how-biofuels-could-starve-the-poor.html
4. For one of many examples, see: "Biotech crops seen helping to feed hungry world," Reuters, June 18, 2008. http://uk.reuters.com/article/rbssIndustryMaterialsUtilitiesNews/idUKN1841870420080618.
5. Sullivan, D. (2008). "Groundbreaking report offers holistic remedies for famine relief and environmental protection in developing countries," The Rodale Institute, April 18, 2008. http://www.rodaleinstitute.org/20080418/fp1; for report and commentaries, see: www.agassessment.org.
6. "Deserting the Hungry? Monsanto and Syngenta are wrong to withdraw from an international assessment on agriculture," Nature 451:223-224 (January 17, 2008).
7. ISAAA (2007). "Global Status of Commercialized Biotech/GM Crops: 2007," International Service for Acquisition of Agri-biotech Applications, ISAAA Brief 37-2007.
8. FoEI-CFS (2008). "Who Benefits from GM Crops? The Rise in Pesticide Use," Friends of the Earth International-Center for Food Safety, January 2008, Figure 1 & Table 1.
9. ISAAA (2007), op. cit.
10. Benbrook, C. (2005). "Rust, resistance, run down soils, and rising costs: problems facing soybean producers in Argentina," AgBioTech InfoNet, Technical Paper No. 8, Jan. 2005, p. 26. http://www.aidenvironment.org/soy/08_rust_resistance_run_down_soils.pdf.

11. FoEI-CFS (2008), op. cit., pp. 23-24.
12. ISAAA (2007), op. cit. On an acreage basis, 63 percent of biotech crops have the herbicide-tolerance trait alone, while another 19 percent are both herbicide-tolerant and insect-resistant.
13. Benbrook (2005), op. cit., pp. 11, 27.
14. Benbrook, C. (2004). "Genetically Engineered Crops and Pesticide Use in the United States:
    The First Nine Years," AgBioTech InfoNet, Technical Paper No. 7, October 2004.
    http://www.biotech-info.net/Full_version_first_nine.pdf
15. Roberson, R (2006). "Herbicide resistance goes global," Southeast Farm Press, 12/1/08. http://southeastfarmpress.com/mag/farming_herbicide_resistance_goes/; Boerboom, C. et al (2004). "Selection of Glyphosate-Resistant Weeds," available at http://www.extension.umn.edu/cropenews/2004/04MNCN43.htm.
16. FoEI-CFS (2008), op. cit., pp. 7–12; 18–22.
17. Benbrook, C. (2001). "Troubled Times Amid Commercial Success for Roundup Ready Soybeans: Glyphosate Efficacy is Slipping and Unstable Transgene Expression Erodes Plant Defences and Yields," AgBioTech InfoNet Technical Paper No. 4, May 2001, p. 3. http://www.biotech-info.net/troubledtimes.html; Elmore et al (2001). "Glyphosate-Resistant Soybean Cultivar Yields Compared with Sister Lines," Agron J. 93: 408-412; quote from press release announcing study at: http://ianrnews.unl.edu/static/0005161.shtml.
18. Braidotti, G. (2008). "Scientists share keys to drought tolerance," Australian Government Grains Research & Development Corporation, Ground Cover, Issue 72, Jan.–Feb. 2008.
19. FoEI-CFS (2008), op. cit., Chapter 4; FoEI-CFS (2007). "Who Benefits from GM Crops," Friends of the Earth International & Center for Food Safety, January 2007, Chapter 4.
20. ETC Group (2007), "The World's Top 10 Seed Companies—2006." http://www.etcgroup.org/en/materials/publications.html?pub_id=656.
21. CFS (2005). "Monsanto vs. US Farmers," Center for Food Safety, 2005. See also 2007 update. Both available at: http://www.centerforfoodsafety.org/Monsantovsusfarmersreport.cfm.
22. GRAIN (2007). "The end of farm-saved seed?" GRAIN Briefing, Feb. 2007.

23. Data compiled by CFS from USDA website of approved (deregulated) GM crop events at http://www.aphis.usda.gov/brs/not_reg.html. Figure for "big five" includes companies subsequently bought out by these five biotech firms.
24. Based on figures from Monsanto (2008) and ISAAA (2007).
25. FoEI-CFS (2008), op. cit., pp. 7–8, 10.
26. Monsanto (2006). "Delta and Pine Land Acquisition: Investor Conference Call," Power Point presentation, 8/15/06. http://www.monsanto.com/pdf/investors/2006/08-15-06.pdf.
27. Reviewed in: Halweil, B. (2006). "Can Organic Farming Feed Us All?," Worldwatch 19(3), May/June 2006. http://www.organic-center.org/reportfiles/EP193A.Halweil.pdf
28. Gatsby (2005). "The Quiet Revolution: Push-Pull Technology and the African Farmer," The Gatsby Charitable Foundation, April 2005. http://www.push-pull.net/Gatsby_paper.pdf.

Part 4: Corporate Control of Agriculture

Chapter 21: *Patented Seeds vs. Free Inquiry* by Martha L. Crouch

1. 2009 Monsanto Technology/Stewardship Agreement (Limited Use License)
2. Gorman ME, Simmonds J, Ofiesh C, Smith R, and Werhane PH (2001) Monsanto and Intellectual Property, In "Teaching Ethics, Fall 2001", University of Virginia Darden School Foundation, Charlottesville, VA. http://www.uvu.edu/ethics/seac/Monsantopercent20andpercent20Intellectualpercent20Property.pdf; accessed 18 Feb 2013.
3. Center for Food Safety & Save Our Seeds (2013) Gene Giants vs. US Farmers. http://www.centerforfoodsafety.org/wp-content/uploads/2013/02/Seed-Giants_final.pdf; accessed 18 Feb 2013
4. Waltz E (2009) Under wraps. Nature Biotechnology 27 (10): 880–882. http://www.emilywaltz.com/Biotech_crop_research_restrictions_Oct_2009.pdf, accessed 18 Feb 2013
5. Sappington TW, Ostlie KR, DiFonzo C, Hibbard BE, Kurpke CH, Porter P, Pueppke S, Shields EJ and Tollefson JJ (2010) Conducting public-sector research on commercialized transgenic seed. GM Crops 1 (2): 55 - 58. http://www.landesbioscience.com/journals/gmcrops/02SappingtonGMC1-2.pdf, accessed 18 Feb 2013

6. Stutz B (2010) Companies Put Restrictions on Research into GM Crops. Yale Environmnet 360. http://e360.yale.edu/content/print.msp?id=2273, accessed 18 Feb 2013
7. Understandably, the people who told me their experiences did so in confidence.
8. Food & Water Watch (2012) Public Research, Private Gain: Corporate Influence Over University Agricultural Research. http://documents.foodandwaterwatch.org/doc/PublicResearchPrivateGain.pdf, accessed 18 Feb 2013
9. Crouch ML (1991) The very structure of scientific research mitigates against developing products to help the environment, the poor, and the hungry. J. Agricultural and Environmental Ethics 4:151–158

Chapter 23: *Changing Seeds or Seeds of Change?* by Natalie DeGraaf

1. Black, R., Fava, F., Mattei, N., Robert, V., Seal, S., & Verdier, V. (2011). Case studies on the use of biotechnologies and on biosafety provisions in five African countries. Journal of biotechnology. doi:10.1016/j.jbiotec.2011.06.036
2. http://www.theecologist.org/News/news_round_up/522538/agroecological_farming_methods_being_ignored_says_un_expert.html
3. Levitt, T. (2011). Worldwatch report attacks criminalising of seed saving and promotes agroecology—The Ecologist. Ecologist. Retrieved November 4, 2011, from http://www.theecologist.org/News/news_round_up/727000/worldwatch_report_attacks_criminalising_of_seed_saving_and_promotes_agroecology.html
4. Matt Styslinger. (n.d.). Debating the Ethics of Biotechnology: An Interview with Philip Bereano . World Watch Institute. Retrieved November 9, 2011, from http://www.worldwatch.org/node/6522
5. http://www.enn.com/top_stories/article/37741

Chapter 24: *Food, Made from Scratch* by Eric Hoffman

1. Hylton, Wils S. "Craig Venter's Bugs Might Save the World." New York Times 30 May 2012
2. "What Is GMO?" The Non-GMO Project, <http://www.nongmoproject.org/learn-more/what-is-gmo/>.
3. "Global Market for Synthetic Biology to Grow to $10.8 Billion by 2016." BCC Research, Nov. 2011.

4. Fehrenbacher, Katie. "Monsanto Backs Algae Startup Sapphire Energy." Thomson Reuters, 8 Mar. 2011.
5. "Biofuel Crops Breeding and Improvement." Agradis, Web.
6. National Renewable Energy Laboratory. NREL's Multi-Junction Solar Cells Teach Scientists How to Turn Plants into Powerhouses. N.p., 12 May 2011.
7. Synthetic Biology: 10 Key Points for Delegates. ETC Group, Oct. 2012. <http://www.etcgroup.org/sites/www.etcgroup.org/files/synbio_ETC4COP11_4web_0.pdf>.
8. Rodemeyer, Michael. New Life, Old Bottles: Regulating the First-Generation Products of Synthetic Biology. Woodrow Wilson International Center for Scholars, Synthetic Biology Project, 2009.
9. Wade, Nicholas. "Researchers Say They Created a 'Synthetic Cell.'" The New York Times. 20 May 2010.
10. Bittman, Mark. "A Simple Fix for Farming." New York Times, 19 Oct. 2012.
11. "Agroecology and Sustainable Development." Pesticide Action Network North America, Apr. 2009.
12. "Principles for the Oversight of Synthetic Biology." Friends of the Earth, Mar. 2012.

Part 5: Regulation, Policy, and Law

Chapter 27: *AG Biotech Policy: 2012 in Review* by Colin O'Neil

1. For more information, see the Center for Food Safety's Food Safety Review "Going Backwards: Dow's 2,4-D-Resistant Crops and a More Toxic Future." Winter 2012. http://www.centerforfoodsafety.org/wp-content/uploads/2012/02/FSR_24-D.pdf
2. Gillam, Carey. "Dow's controversial new GMO corn delayed amid protests." Reuters. January 18, 2012. Available online at http://www.reuters.com/article/2013/01/18/dow-biotech-idUSL1E9CIBN320130118
3. Darek T. R. Moreau, Corinne Conway, Ian A. Fleming. (2011) "Reproductive performance of alternative male phenotypes of growth hormone transgenic Atlantic salmon (Salmo salar)." Evolutionary Applications, Blackwell Publishing, Ltd.
4. Living Oceans Society, Media Release. "ISA virus confirmed in AquaBounty's genetically-engineered salmon." Reposted on December 20, 2012. Available at: http://www.livingoceans.org/media/releases/salmon-farming/isa-virus-confirmed-aquabounty's-genetically-engin

Part 6: Ecology and Sustainability

Chapter 35: *The Role of GMOs in Sustainable Agriculture* by Doug Gurian-Sherman

1. Rockström, J., et al. 2009. A safe operating space for humanity. *Nature* 461:472–475.
2. IAASTD. 2009. International Assessment of Agricultural Knowledge, Science and Technology for Development: Global report, McIntyre, B., et a., eds. Island Press, 1718 Connecticut Avenue, NW, Suite 300, Washington, DC 20009.
3. Benbrook, C.M. (2012). Impacts of genetically engineered crops on pesticide use in the US—The first sixteen years. *Environmental Sciences Europe* 24:24.
4. National Research Council. 2010. The impact of genetically engineered crops on farm sustainability in the United States. The National Academies Press, Washington, DC.
5. eg. Meehan, T.D., B.P. Werling, D.A. Landis, and C. Gratton. 2011. Agricultural landscape
simplification and insecticide use in the midwestern United States. Proceedings of the National Academy of Sciences 108(28):11500–11505.
6. Blesh, J., & Drinkwater, L.E. (2013). The impact of nitrogen source and crop rotation on nitrogen mass balances in the Mississippi River Basin. *Ecological Applications*. In press. Online at http://dx.doi.org/10.1890/12-0132.1.
7. Davis, A.S., Hill, J.D., Chase, C.A., Johanns, A.M., & Liebman, M. 2012. Increasing cropping system diversity balances productivity, profitability and environmental health. *PLOS ONE* 7(10):e47149. doi:10.1371/journal.pone.0047149.RCS
8. Over 99 percent of planted acreage consists of a few patented herbicide and insect resistance traits, owned by a few large transnational companies.
9. Gurian-Sherman, D. (2009). Failure to Yield: Evaluating the performance of Genetically engineered crops. Union of Concerned Scientists, Cambridge, MA.
10. Hilbeck, A., Lebrecht, T., Vogel, R., Heinemann, J.A., & Binimelis, R. (2013). Farmer's choice of seeds in four EU countries under different levels of GM crop adoption. *Environmental Sciences Europe*. 25:12 http://www.enveurope.com/content/25/1/12

11. Cook,S.M., Khan,Z.R. & Pickett, J.A. (2007). Use of push-pull strategies in integrated pest management. *Annual Review of Entomology*. 52:375–400
12. Gouse, M. et al. 2009. Assessing the performance of GM maize amongst smallholders in KwaZulu-Natal, South Africa. *AgBioForum* 12(1): 78–89.
13. McCouch, S. et al. 2013. Agriculture: Feeding the future. *Nature* 499: 23–24.
14. Gurian-Sherman, D. & Gurwick, N. (2009). No sure fix: Prospects for reducing nitrogen pollution through genetic engineering. Union of Concerned Scientists, Cambridge, MA.
15. Blesh, J., & Drinkwater, L.E. (2013). The impact of nitrogen source and crop rotation on nitrogen mass balances in the Mississippi River Basin. *Ecological Applications*. In press. Online at http://dx.doi.org/10.1890/12-0132.1.
16. Gurian-Sherman, D. (2012). High and dry: Why genetic engineering is not solving agriculture's drought problem in a thirsty world. Union of Concerned Scientists, Cambridge, MA.
17. Gurian-Sherman, D. (2012). High and dry: Why genetic engineering is not solving agriculture's drought problem in a thirsty world. Union of Concerned Scientists, Cambridge, MA.
18. Goodman, M.M., and M.L. Carson. 2000. Reality vs. myth: Corn breeding, exotics, and genetic engineering. *Annual Corn Sorghum Research Conference Proceedings* 55:149–172.
19. Phillips McDougall. (2011). The cost and time involved in the discovery, development
20. Goodman, M.M., and M.L. Carson. 2000. Reality vs. myth: Corn breeding, exotics, and genetic engineering. *Annual Corn Sorghum Research Conference Proceedings* 55:149–172.
21. Gurian-Sherman, D. (2012). High and dry: Why genetic engineering is not solving agriculture's drought problem in a thirsty world. Union of Concerned Scientists, Cambridge, MA.
22. Blesh, J., & Drinkwater, L.E. (2013). The impact of nitrogen source and crop rotation on nitrogen mass balances in the Mississippi River Basin. *Ecological Applications*. In press. Online at http://dx.doi.org/10.1890/12-0132.1.
23. Davis, A.S., Hill, J.D., Chase, C.A., Johanns, A.M., & Liebman, M. 2012. Increasing cropping system diversity balances productivity, profitability and environmental health. *PLOS ONE* 7(10):e47149. doi:10.1371/journal.pone.0047149.RCS

24. Mortensen, D.A., J.F. Egan, B.D. Maxwell, R.M. Ryan, and R.D. Smith. 2012. Navigating a critical juncture for sustainable weed management. *BioScience* 6(1):75–84.

Chapter 36: *Genetically Modified Crops and the Intensification of Agriculture* by Bill Freese

1. FAO (2012). The State of Food Insecurity in the World. Food and Agriculture Organization of the United Nations, Rome 2012, Figure A2.2, p. 55. Different estimates based on dietary energy required for sedentary lifestyle (852 million) and for normal activity levels (1.52 billion).
2. WHEF (2013). 2013 World Hunger and Poverty Facts and Statistics, World Hunger Education Service. http://www.worldhunger.org/articles/Learn/world%20hunger%20facts%202002.htm.
3. "DuPont, Monsanto Others Plan $50 Million Ad Effort on Biotech Foods," Bloomberg, March 10, 2000. http://www.organicconsumers.org/ge/giant-adcamp.cfm.
4. Gillam, C. (2013). GMO companies launch website to fight anti-biotech movement," Reuters, July 29, 2013. http://www.reuters.com/article/2013/07/29/gmo-campaign-idUSL1N0FZ0RE20130729.
5. ISAAA (2011). "Global Status of Commercialized Biotech/GM Crops: 2011," Executive Summary, ISAAA Brief 43-2011, International Service for Acquisition of Agri-biotech Applications
6. FoEI-CFS (2008). "Who Benefits from GM Crops? The Rise in Pesticide Use," Friends of the Earth International-Center for Food Safety, January 2008, Figure 1 & Table 1. http://www.centerforfoodsafety.org/WhoBenefitsPR2_13_08.cfm.
7. ISAAA (2011), op. cit. Most of the remaining 0.5% is GM alfalfa and sugar beets.
8. ISAAA (2011), op. cit. 85% of GM crops incorporate herbicide-resistance (HR): 59% HR alone; 26% combine both HR and insect resistance; while the remaining 15% of GM crop acres are insect-resistant alone. GM crops with a few other traits (e.g. virus-resistant papaya and squash) are planted on just fractions of a percent of overall biotech crop acreage.
9. Benbrook, C (2012). Impacts of genetically engineered crops on pesticide use in the U.S. – the first sixteen years. Environmental Sciences Europe 24:24. http://www.enveurope.com/content/24/1/24.

10. Freese, W. (2010). Congressional Testimony before the House Oversight and Government Reform Committee, Sept. 30, 2010, pp. 1-9. http://www.centerforfoodsafety.org/files/oversight-hearing--freese-response-to-questions-corrected_59304.pdf.
11. Fraser, K (2013). Glyphosate-resistant weeds – intensifying. Stratus Agri-Marketing, Inc., Jan. 25, 2013. http://www.stratusresearch.com/blog07.htm.
12. Eight of 13 GM crops awaiting approval by USDA are herbicide-resistant. See top two tables at http://www.aphis.usda.gov/biotechnology/petitions_table_pending.shtml, last visited 1/15/14.
13. US EPA (2011) "Pesticide Industry Sales and Usage: 2006 and 2007 Market Estimates." Environmental Protection Agency, February: Table 3.4. http://www.epa.gov/opp00001/pestsales/07pestsales/market_estimates2007.pdf
14. USDA (2013). Draft Environmental Impact Statement for 2,4-D-Resistant Corn and Soybeans. US Dept. of Agriculture, 2013, p. 134. http://www.aphis.usda.gov/brs/aphisdocs/24d_deis.pdf
15. Mortensen, D.A., Egan, J.F., Maxwell, B.D., Ryan, M.R. and Smith, R.G. (2012). Navigating a critical juncture for sustainable weed management. Bioscience 62(1): 75–84.
16. DuPont (2009). "Novel Glyphosate-N-Acetyltransferase (GAT) Genes," U.S. Patent Application Publication, Pub. No. US 2009/0011938 A1, January 8, 2009, paragraph 33.
17. Brower L.P., Taylor O.R., Williams E.H., Slayback D.A., Zubieta R.R., Ramírez M.I. (2011) Decline of monarch butterflies overwintering in Mexico: is the migratory phenomenon at risk? Insect Conservation and Diversity **5**:95–100. [online] URL: http://doi.wiley.com/10.1111/j.1752-4598.2011.00142.x
18. Pleasants, J.M. (in press), Monarch Butterflies and Agriculture, Ch. 14 in: "Monarchs in a Changing World: Biology and Conservation of an Iconic Insect" Cornell Univ. Press. Pleasants, J.M. and Oberhauser, K.S. (2012). Milkweed loss in agricultural fields because of herbicide use: effect on the monarch butterfly population. Insect Conservation and Diversity 6(2): 135–144.
19. Mortensen et al (2012), op. cit.
20. Peterson, R.K.D. and Hulting, A.G. (2004). A comparative ecological risk assessment for herbicides used on spring wheat: the effect of glyphosate

when used within a glyphosate-tolerant wheat system. Weed Science 52(5): 834–844, Table 4.

21. EPA (2009). "Risks of 2,4-D Use to the Federally Threatened California Red-legged Frog (*Rana aurora draytonii*) and Alameda Whipsnake (*Masticophis lateralis euryxanthus*)," Environmental Protection Agency, Feb. 2009.
22. NMFS (2011). "Biological Opinion: Endangered Species Act Section 7 Consultation with EPA on Registration of 2,4-D, Triclopyr BEE, Diuron, Linuron, Captan and Chlorothalonil," National Marine Fisheries Services, June 30, 2011.
23. USDA (2010). "2007 National Resources Inventory: Soil Erosion on Cropland," USDA NRCS, April 2010. http://www.nrcs.usda.gov/Internet/FSE_DOCUMENTS/ nrcs143_012269.pdf.
24. Coughenour, DM & S. Chamala (2000). Conservation Tillage and Cropping Innovation: Constructing the New Culture of Agriculture, Iowa State University Press, Ames, Iowa, 2000, pp. 258, 286.
25. NRC (2010). "The Impact of Genetically Engineered Crops on Farm Sustainability in the United States," National Research Council, National Academy of Sciences, 2010 (prepublication copy), p. 2–15.
26. USDA (2007). No shortcuts in checking soil health. Agricultural Research Magazine, July 2007. Agricultural Research Service, USDA. http://www.ars.usda.gov/is/AR/archive/jul07/soil0707.htm.
27. Plourde JD, Pijanowski BC and Pekin BK (2013). "Evidence for increased monoculture cropping in the Central United States," Agriculture, Ecosystems and Environment 165: 50–59.
28. Monsanto (2012). Corn-On-Corn Clinics. http://www.genuity.com/corn/Pages/Corn-on-Corn-Clinics.aspx. Last accessed 1/16/14.
29. Philpott, T (2011). Monsanto (still) denies superinsect problem, despite evidence. Mother Jones, December 8, 2011. http://motherjones.com/tom-philpott/2011/12/superinsects-monsanto-corn-epa.
30. Jongeneel (2013). "Expect more soil insecticide used with Bt hybrids," Ag Professional, April 1, 2013. http://www.agprofessional.com/news/Expect-more-soil-insecticide-used- with-Bt-hybrids-200626161.html.
31. 140 lbs/acre on corn. See "Table 10. Nitrogen used on corn, rate per fertilized acre receiving nitrogen, selected States," accessible from: http://www.ers.usda.gov/data- products/fertilizer-use-and-price.aspx#.Up40LqVXaa4.n.

32. Wise, K. and Mueller, D. (2011). Are fungicides no longer just for fungi? An analysis of foliar fungicide use in corn. American Phytopathological Society. http://www.apsnet.org/publications/apsnetfeatures/Pages/fungicide.aspx.
33. Israel, B. (2013). Fungicide use surges, largely unmonitored. Scientific American, February 22, 2013. https://www.scientificamerican.com/article.cfm?id=fungicide-use-surges-largely-unmonitored
34. Benbrook, C. (2005). "Rust, resistance, run down soils, and rising costs: problems facing soybean producers in Argentina," AgBioTech InfoNet, Technical Paper No. 8, Jan. 2005, p. 11, 27. http://www.aidenvironment.org/soy/08_rust_resistance_run_down_soils.pdf.
35. Guerena, A. (2013). The Soy Mirage. Oxfam Research Reports, August 2013. http://www.oxfam.org/sites/www.oxfam.org/files/rr-soy-mirage-corporate-social-responsibility-paraguay-290813-en.pdf.
36. BaseIS (2010). Socio-environmental impacts of soybean in Paraguay – 2010. Base Investigaciones Sociales-NGO Reporter Brazil, August 2010.
37. Commonly used to describe GM soybean monocultures in Latin America. For example, see: http://www.argentinalatente.org/nacionales/272-agricultura-sin-agricultores.
38. As quoted in: Grit (2012). Respect the rotation events motivate farmers. Grit Blog, 2/10/12. http://www.grit.com/farm-and-garden/respect-the-rotation-events-motivate-farmers.aspx/.
39. Pretty, J. (2009). Can Ecologicaol Agriculture Feed Nine Billion People. Monthly Review 61(6), November 2009. http://monthlyreview.org/2009/11/01/can-ecological-agriculture-feed-nine-billion-people.

Chapter 37: *Science Interrupted: Understanding Transgenesis in Its Ecological Context* by Ignacio Chapela

1. Commoner, B. 2002. Unravelling the DNA Myth: The Spurious Foundations of Genetic Engineering. *Harpers* (February 2002).
2. Delborne, J.A. 2008. Transgenes and Transgressions: Scientific Dissent as Heterogeneous Practice. *Social Studies of Science* 38(4): 509-541. doi: 10.1177/0306312708089716.
3. Ellstrand, N.C. 2003. *Dangerous Liaisons? When cultivated Plants Mate with their Wild Relatives*. Johns Hopkins University Press.
4. Jablonka, E. & Lamb, M.J. 2005. *Evolution in Four Dimensions*. MIT Press.

5. Jian Chen, Min Jin, Zhi-Gang Qiu, Cong Guo, Zhao-Li Chen, Zhi-Qiang Shen, Xin-Wei Wang & Jun-Wen Li. 2012. *Environmental Science and Technology* 2012(46): 13448–13454. doi: 13448dx.doi.org/10.1021/es302760s.
6. Ortiz-García, S., Ezcurra, E., Schoel, B., Acevedo, F., Soberón, J. & Snow, A.A. 2005.
7. Absence of detectable transgenes in local landraces of maize in Oaxaca, Mexico (2003–2004). *Proceedings of the National Academy of Sciences (USA)* 102(35): 12338–12343, doi: 10.1073/pnas.0503356102.
8. Marvier, M., Carrière, Y., Ellstrand, N.C., Gepts, P., Kareiva, P., Rosi-Marshall, E., Tabashnik, B.E. & Wolfenbarger, L.L. 2008. Harvesting Data from Genetically Engineered Crops. *Science* 320(5875): 452-453. doi:10.1126/science.1154521.
9. Piñeyro-Nelson, A., VanHeerwaarden, J., Perales, H.R., Serratos-Hernandez, J.A., Rangel, A., Hufford, M.B., Gepts, P., Garay-Arroyo, A., Rivera-Bustamante, R. & Álvarez-Buylla, E. 2008. Transgenes in Mexican maize: molecular evidence and methodological considerations for GMO detection in landrace populations. *Molecular Ecology* 18(4): 750–761. doi: 10.1111/j.1365-294X.2008.03993.x.
10. Quist, D. & Chapela, I.H. 2001. Transgenic DNA introgressed into traditional maize landraces in Oaxaca, Mexico. *Nature* 414: 541-543. doi:10.1038/35107068.

Chapter 38: *Agricultural Technologies for a Warming World* by Lim Li Ching

1. Schlenker, W. and D.B. Lobell (2010). Robust negative impacts of climate change on African agriculture. Environmental Research Letters, 5, doi:10.1088/1748-9326/5/1/014010.
2. Nelson, G.C., M.W. Rosegrant, J. Koo, R. Robertson, T. Sulser, T. Zhu, C. Ringler, S. Msangi, A. Palazzo, M. Batka, M. Magalhaes, R. Valmonte-Santos, M. Ewing and D. Lee (2009). Climate Change: Impact on Agriculture and Costs of Adaptation. IFPRI, Washington, DC.
3. Gurian-Sherman, D. (2012). High and dry: Why genetic engineering is not solving agriculture's drought problem in a thirsty world. Union of Concerned Scientists, Cambridge, MA. Available at: http://www.ucsusa.org/assets/documents/food_and_agriculture/high-and-dry-report.pdf
4. Ibid, p.3.

5. IAASTD (2009). Agriculture at a Crossroads. International Assessment of Agricultural Knowledge, Science and Technology for Development. Island Press, Washington, DC. http://www.agassessment.org
6. UNEP-UNCTAD Capacity-building Task Force on Trade, Environment and Development (2008). Organic Agriculture and Food Security in Africa. United Nations, New York and Geneva.
7. http://www.rodaleinstitute.org/fst30years/yields

Chapter 43: *Why Context Matters* by Craig Holdrege

1. Mattick, J. S. et al (2010). A Global View of Genomic Information—Moving Beyond the Gene and the Master Regulator. *Trends in Genetics* vol. 26, pp. 21–8.
2. Gerstein, M. B. (2007). What is a Gene, Post-ENCODE? History and Updated Definition. *Genome Research* vol. 17, pp. 669–81.
3. Gelbart, William. (1998). Databases in Genomic Research. *Science* vol. 282, pp. 659–61.
4. Talbott, S. (2012). Getting over the Code Delusion. Available online: http://www.natureinstitute.org/txt/st/mqual/genome_4.htm (See also other articles at:http://natureinstitute.org/txt/st/org/)
5. For numerous examples, see: www.nontarget.org
6. Reddy, M. S. S., F. Chen, G. Shadle, L. Jackson et al. (2005). Targeted Down-regulation of Cytochrome P450 Enzymes for Forage Quality Improvement in Alfalfa (*Medicago sativa* L.). *Proceedings of the National Academy of Sciences* vol. 102, pp. 16573-8. Shadle, G., F. Chen, M. S. S. Reddy, L. Jackson et al. (2007). Down-regulation of Hydroxycinnamoyl CoA: Shikimate Hydroxycinnamoyl Transferase in Transgenic Alfalfa Affects Lignification, Development and Forage Quality. *Phytochemistry* vol. 68, pp. 1521–9.
7. Heap, I. (2014). The International Survey of Herbicide Resistant Weeds. January 08; available online: www.weedscience.org
8. Service, R. (2013). What Happens When Weed Killers Stop Killing? *Science* vol. 341, p. 1329.
9. Lovins, A. (2001). Loaves and Fishes. Interview; available online: http://neweconomy.net/publications/other/lovins/amory/loaves-and-fishes
10. Holdrege, C. (1996). *Genetics and the Manipulation of life.* Great Barrington, MA: Lindisfarne Press. Holdrege, C. & Talbott, S. (2010). *Beyond*

Biotechnology: The Barren Promise of Genetic Engineering. Lexington: The University Press of Kentucky.

Part 8: Modifying Animals for Food

Chapter 49: *Ethical Roots to Bioengineering Animals* by Paul Root Wolpe

1. Jonsen, AR. (1986) "Bentham in a box: Technology assessment and health care allocation" Law, Medicine and Health Care 14:172–4.
2. Talwar, SK; Shaohua, X; Hawley, ES; Weiss, SA; Moxon, KA; Chapin, JK. "Behavioural neuroscience: Rat navigation guided by remote control." Nature 417, 37–38.
3. Whitehouse, D. (2002) "Here come the ratbots." BBC News: May 1, 2002 {http://news.bbc.co.uk/2/hi/science/nature/1961798.stm}

Chapter 52: *In the Bullpen: Livestock Cloning* by Jaydee Hanson

1. John Gurdon, won the Nobel Prize for Medicine for his work developing the techniques now used for cloning for his work in frogs in 1962. The Scottish team's achievement was figuring out how to use this technique, previously successful only in amphibians and fish, in mammals.
2. See Center For Food Safety, "Not Ready for Prime Time: FDA's Flawed Approach to Assessing the Safety of Food from Animal Clones," March 2007, available at: http://www.centerforfoodsafety.org/pubs/FINAL_FORMATTEDprimepercent20time.pdf
3. See Jaydee Hanson, Comments to the US Department of Agriculture, National Organic Program on tracking animal clones using pedigrees, September 20, 2011, pgs. 253-259. Available at: http://www.ams.usda.gov/AMSv1.0/getfile?dDocName=STELPRDC5095829
4. See Melissa Del Bosque, Clone on the Range, Texas Observer, September 14, 2011 available at: http://www.texasobserver.org/clone-on-the-range-2/
5. See Daniel Boffey, "El Cardinal, the Opus Dei devotee behind cloning firm", The Daily Mail, UK, August 20, 2010 available at: http://www.dailymail.co.uk/news/article-1301215/Wisconsins-king-copy-cattle-Farmer-sold-cloned-cow-embryos-Britain-claims-fell-sales-patter-promising-prize-animal-live-ever.html#ixzz2LqRAcC3B
6. See BGI Ark Biotechnology Co. LTD Shenzen (BAB) http://www.babgenomics.com/list.aspx?catid=168 and Christine Larson, Inside China's

GenomeFactory,MITTechnologyReview,Feb.11,2013availableathttp://www.technologyreview.com/featuredstory/511051/inside-chinas-genome-factory/

7. See http://www.dw.de/eu-ministers-approve-sale-of-food-from-cloned-animals-offspring/a-4414990
8. See Director General, SANCO, "Measures on animal cloning for food production in the EU" available at http://ec.europa.eu/dgs/health_consumer/dgs_consultations/animal_cloning_consultation_en.htm

Chapter 54: *Food and Drug Amalgamation* by Eric Hoffman

1. See "Fishy Business at the FDA," from Genewatch, Volume 23 Issue 4: http://www.councilforresponsiblegenetics.org/genewatch/GeneWatchPage.aspx?pageId=289
2. Twelve of the nation's largest environmental groups sent an open letter to the FDA asking for an independent and comprehensive Environmental Impact Statement to be completed in place of the less-thorough Environmental Assessment, as mandated by the National Environmental Policy Act: http://foe.org/sites/default/files/Environmentalpercent20Grouppercent20Letterpercent20topercent20FDApercent20-percent20GEpercent20Salmonpercent20Final.pdf
3. This is an abridgment of DRAFT legislative principles for the regulation of genetically engineered animals developed by the Center for Food Safety.

Chapter 57: *Gene Technology in the Animal Kingdom* by Paul B. Thompson

1. Marchant, Gary E., Guy A. Cardineau and Thomas P. Redick. 2011. *Thwarting Consumer Choice: The Case Against Mandatory Labeling for Genetically Modified Foods.* Washington, D.C.: American Enterprise Institute.
2. Rollin, Bernard A. 1995. *The Frankenstein Syndrome: Social and Ethical Issues in the Genetic Engineering of Animals.* New York: Cambridge University Press.
3. Thompson, Paul B., Hannah, W. 2008. Nanotechnology, Risk and the Environment: A Review. *Journal of Environmental Monitoring*, 10, 291–300.
4. Singer, Peter. 2009 [1974]. *Animal Liberation: The Definitive Classic of the Animal Movement* Updated Edition. New York: Harper Collins.
5. Regan, Tom. 2004 [1984]. *The Case for Animal Rights,* 2nd Edition. Berkeley: University of California Press.

6. Adams, Carol J. 2010 [1985]. *The Sexual Politics of Meat: A Feminist-Vegetarian Critical Theory.* 25th Anniversary Edition. New York: Continuum International Publishing.
7. Thompson, Paul B. 2004. Thompson, "Research Ethics for Animal Biotechnology," in *Ethics for Life Scientists*. M. Korthals and R. J. Bogers, Eds. Dordrecht: Springer. pp. 105–120.
8. NRC (National Research Council). 2002. *Issues in Animal Biotechnology: Science Based Concerns.* Washington, D.C.: National Academy Press.

Conclusion: *The Future of GM Food* by Sheldon Krimsky

1. S. Krimsky and J. Gruber, eds. *Genetic Explanations: Sense and Nonsense*. Cambridge, MA: Harvard University Press, 2012.
2. S. W. Ewen, A. Pusztai (October 1999). "Effect of diets containing genetically modified potatoes expressing Galanthus nivalis lectin on rat small intestine." *Lancet* 354 (9187): 1353–4. doi:10.1016/S0140-6736(98)05860-7. PMID 10533866.
3. G.-Eric Séralini, D. Cellier, and J. S. Vendomois. "New analysis of a rat feeding study with a genetically modified maize reveals signs of hepatorenal toxicity." *Arch. Environ. Contam. Toxicology* 52:596-602 (2007); J. Spirous de Vendomois, F. Roullier, D. Cellier, and G.-E Séralini, "A comparison of the effects of three GM corn varieties on mammalian health." *International Journal of Biological Sciences* 5(7):706-726 (2009).
4. M. Malatesta, F. Boraldi, G. Annovi, et al. "A long-term study on female mice fed on a genetically modified soybean: effects on liver ageing." *Histochem Cell Biol* 130: 967-77 (2008).
5. M. Malatesta, F. Boraldi, G. Annovi et al. "A long-term study on female mice fed on genetically modified soybean: effects on liver ageing." *Histochem Cell Biology* 130: 967-977 (2008).
6. J. S. de Vendomois, F. Roulier, D. Cellier, and G-Eric Séralini. "A comparison of the effects of three GM corn varieties on mammalian health." *International Journal of Biological Sciences* 5: 706-26 ((2009).
7. F.B. Brasil, L. L. Soares, T. S. Faria et al. "The impact of dietary organic and transgenic soy on the reproductive system of female adult rat." *The Anatomical Record* 292: 587-594 (2009).

8. R. Tudisco, P. Lumbardi, F. Bovera et al. "Genetically modified soya bean in rabbit feeding: detection of DNA fragments and evaluation of metabolic effects by enzymatic analysis." *Animal Science* 82: 193-199 (2006).
9. J. L. Domingo and J. G. Bordonaba. "A literature review on the safety assessment of genetically modified plants." *Environment International* 37: 734-742 (2011).
10. C. M. Benbrook. *Impacts of Genetically Engineered Crops on Pesticide Use in the United States: The First Thirteen Years.* The Organic Center, November 2009.
11. Ibid, p. 1.
12. John M. Pleasants and Karen S. Oberhauser. "Milkweed loss in agricultural fields because of herbicide use: effect on the monarch butterfly population." *Insect Conservation and Diversity*, doi: 10.1111/j.1752-4598.2012.00196.x (2012).
13. J. A. Nordlee, S. L Taylor, J. A. Townsend, et al. "Identification of a Brazil-nut allergen in transgenic soybeans." *NEJM* 334 (11): 688-692 (1996).
14. V. E. Prescott, P. M. Campbell, A. Moore et al. "Transgenic expression of Bean alpha-amylase inhibitor in peas results in altered structure and immunogenicity." *Journal of Agricultural and Food Chemistry* 53: 9023-9030 (2005).
15. Branum, A. M., Lukacs, S. L. "Food allergy among U.S. children: trends in prevalence and hospitalizations." *NCHS Data Brief*. 2008: 10:1-8.
16. Liu, A. H., Jaramillo, R, Sicherer, S. H., et al. "National prevalence and risk factors for food allergy and relationship to asthma: results from the National Health and Nutrition Examination Survey 2005-2006." *J Allergy Clin Immunol*. 2010; 126 (4): 798-806.e13.

INDEX

To subscribe to print and electronic versions of *GeneWatch*, please send a subscription request with your full name, mailing address and email address, along with a check* or credit card (MasterCard or Visa) information, to Council for Responsible Genetics, 5 Upland Road, Suite 3, Cambridge, MA 02140. Orders are also welcome online at www.councilforresponsiblegenetics.org or by phone at (617) 868-0870. Additional donations are welcome. Thank you!

*Annual print subscription rates are as follows:

- Individual: $35
- Nonprofit: $50
- Library: $70
- Corporation: $100